Genetic Manipulation in Plants for Mitigation of Climate Change

Pawan Kumar Jaiwal
Rana Pratap Singh
Om Parkash Dhankher
Editors

Genetic Manipulation in Plants for Mitigation of Climate Change

Springer

Editors
Pawan Kumar Jaiwal
Center for Biotechnology
M.D. University
Rohtak, Haryana, India

Om Parkash Dhankher
Stockbridge School of Agriculture
University of Massachusetts Amherst
Amherst, MA, USA

Rana Pratap Singh
Department of Environmental Science
Babasaheb Bhimrao Ambedkar University
Lucknow, UP, India

ISBN 978-81-322-2660-4 ISBN 978-81-322-2662-8 (eBook)
DOI 10.1007/978-81-322-2662-8

Library of Congress Control Number: 2015958661

Springer New Delhi Heidelberg New York Dordrecht London

Printed on acid-free paper

Springer (India) Pvt. Ltd. is part of Springer Science+Business Media (www.springer.com)

Preface

Climatic variability and global warming observed during the past few decades have become a frightening reality and a matter of deep concern worldwide especially for the developing countries. Extreme climatic conditions adversely affect quantitative and qualitative characteristics of plants. Food and nutritional security have emerged as a major challenge worldwide due to overexpansion of population and cultivated areas to less fertile fields as a result of extreme climatic changes. The extreme hot and cold seasons and subsequent drought and flood as well as the elevated greenhouse gases are generally considered to be the major factors resulting to such climatic variability globally. Continues industrialization aggravated by anthropogenic activities has added to the severity of climate change. Therefore, there is an urgent need to find out ways to mitigate the negative effects of climate change on biodiversity and crop productivity. The works related to the biotechnological strategies like breeding and genetic engineering to mitigate the problems of climate change are in the initial stage, and the efforts made in this direction may be limited to one specific trait, whereas the mitigation measures for these complex climatic factors need handling of multiple stresses in one go. The recent advances in molecular breeding (identification of tightly liked markers, QTLs/genes), transgenesis (introduction of exogenous genes or changing the expression of endogenous stress-responsive genes) and genomics approaches have made it easier to identify and isolate several key genes involved in abiotic stresses and their regulation. This book was planned with an objective to understand the detailed overviews and recent approaches in genetic manipulation studies in plants for mitigation of climate change.

Elevated O_3 in the troposphere due to industrialization and anthropogenic activities suppresses plant productivity, product quality and competitive ability of crop plants and forest trees. In order to ensure food security for the future, O_3 tolerance/resistance is to be incorporated in crop plants. The physiological, biochemical and molecular responses of plants to this toxic pollutant that provides rational targets for phenotypic screening and genetic manipulation of plants for tolerance to O_3 have been discussed by Li et al. (pages 1–14).

Heat stress affects many aspects of plant growth and development. Thermotolerance depends on the activation and coordination of various sensing, signalling and regulatory pathways. Roth et al. (pages 15–42) have discussed the effects of high temperature on plant growth and development, methods for thermotolerance screening, heat-sensing mechanism, metabolic changes and strategies to improve thermotolerance in plants.

Breeding crops for drought or flood tolerance/resistance has been a priority for a long time; however, the progress has been slow due to the physiological and genetic complexity of these traits. Fundamental insights into the stress sensing, downstream signal transduction and metabolic alterations that promote tolerance are key to increased crop production in drought and flood-prone environments. Rice, a staple food crop of more than half of the world population, is facing the problem of water shortage and drought. You and Xiong (pages 43–72) have provided an overview and perspective of the strategies, resources, progress and challenges of drought resistance in rice. Zhang and co-workers (pages 73–102) have summarized the recent progress about the major constraints affecting plant performance in waterlogged soils and discussed the mechanism employed by the plants to deal with the stress.

Polyamines are small polycationic compounds involved in numerous metabolic pathways in plants including stress protection. Macro et al. (pages 103–116) have discussed approaches undertaken to demonstrate the involvement of polyamines in the stress response and genetic manipulation of polyamine levels as an efficient tool for plant resistance to abiotic stress along with recent advances in the potential mechanism of action by which polyamines could contribute to stress protection.

Vinod (pages 117–142) has given an overview of the recent developments in breeding efforts towards nutrient deficiency-tolerant rice as a sustainable solution for future agriculture. Rice is the predominant stable crop providing more than 21 % of the daily calories of the word. However, in rice-growing regions nutrient starvation in the soil (either due to excessive farming or less availability/absorption of nutrients in marginal land) is putting rice crop/plants under stress. To avoid this, excessive input of fertilizers is used which is causing environmental degradation. Because there is enough variability available for nutrient response within rice gene pool, breeding may be a good alternative for this crop.

An elevated concentration of heavy metals adversely affects growth and productivity of crop plants. Climate change is related to heavy metal contamination due to its effects on the bioavailability of metals and its ability to alter the environmental fate and intensity of pollutants. In the future, it is expected that climate change may increase the heavy metals in the agricultural soils. So, developing crops resistant to heavy metals will be required for marginal and moderately contaminated soils. In this context, Tomar et al. (pages 143–168) have emphasized on the production of new cultivars with enhanced heavy metals and metalloids tolerance using transgenic approaches. Sudhakar and Suprassana (pages 169–186) have focused on the issue of climate change keeping in view the phytoremediation potential of plants and have highlighted the prospects of genetic manipulation for enhanced heavy metals tolerance.

Most of the abiotic stress tolerance mechanisms involve multiple metabolic pathways, and manipulating such complex traits through conventional breeding remains a big challenge. Hence, genetic transformation with regulatory genes especially transcription factors is a promoting alternative. Baudh and co-workers (pages 187–204) have assessed the potentials and limitations of biotechnological approaches for mitigation of various multigenic stresses like enhanced temperature and water scarcity which are affecting the crop productivity.

Overall physiology of C_3 plants is affected by increasing atmospheric CO_2 levels in the post industrialization era, and hence, dependent herbivorous insects are also affected. Thus, climate change affects plant-insect interactions. The basic regulatory connections between primary and secondary metabolism of plants need to be understood more to predict how insect populations respond to changes in host plant. Zavala and Gog (pages 205–222) have discussed mechanisms controlling insect herbivory related to global rise in atmospheric CO_2 levels and have suggested that advances in genetic manipulation techniques afford the prospects of affecting changes in ecosystem downstream from plant physiology.

Increased input costs, low benefits and marginalization of land as a result of extreme climates have all added to low income and hence poverty and malnutrition culminating into food insecurity in developing countries. GM crops have been proved beneficial to farmers and society especially in developing countries by increasing productivity, enhancing nutrient quality and reducing food and input costs. Sainger et al. (pages 223–241) have discussed the essence of GM crops for the developing world for increasing farmers' income, their role in poverty alleviation, nutrition and health and feasibility of GM crops in the time of climate change.

The chapters critically evaluate the current knowledge, state of the art and future prospects of genetic manipulation of plants to minimize the adverse effect of climate change on plant growth and productivity. The book is valuable for the scientist, academicians, researchers, students, planners and industrialists working in the area of biotechnology, plant agriculture, agronomy, horticulture, plant physiology, molecular biology, plant sciences and environmental sciences. Each chapter in this volume has been written by an expert group in plant stress biology. We are grateful to the contributors for their efforts in preparing insightful and authoritative accounts of various aspects of the knowledge in this area. We express our sincere thanks and gratitude to all these colleagues and warm appreciations and thanks to Springer for their keen interest in bringing out this title with quality work. We are also thankful to our family members and Ph.D. students for their understanding and patience during planning and preparations of this title.

Rohtak, India
October, 2015

Pawan Kumar Jaiwal
Rana Pratap Singh
Om Prakash Dhankher

Contents

Contributors

Rubén Alcázar Unitat de Fisiologia Vegetal, Facultat de Farmàcia, Universitat de Barcelona, Barcelona, Spain

Kuldeep Bauddh Centre for Environmental Sciences, Central University of Jharkhand, Brambe, Ranchi, India

Marta Bitrián Institut de Biologie Moléculaire des Plantes, Centre National de la Recherche Scientifique, Strasbourg, France

Pedro Carrasco Departament de Bioquímica i Biologia Molecular. Facultat de Ciències Biològiques, Universitat de València, Burjassot,València, Spain

Darshna Chaudhary Centre for Biotechnology, Maharshi Dayanand University, Rohtak, India

Om Parkash Dhankher Stockbridge School of Agriculture, University of Massachusetts Amherst, Amherst, MA, USA

Anirudha R. Dixit John F. Kennedy Space Center, Merritt Island, FL, USA

Sotirios Fragkostefanakis Department of Biosciences, Molecular Cell Biology of Plants, Goethe University, Frankfurt am Main, Germany

Linus Gog Department of Plant Biology, University of Illinois at Urbana-Champaign, Champaign, IL, USA

Xin Huang School of Land and Food, University of Tasmania, Hobart, Australia

Pawan Kumar Jaiwal Center for Biotechnology, Maharshi Dayanand University, Rohtak, India

Ranjana Jaiwal Department of Zoology, Maharshi Dayanand University, Rohtak, India

Anthony Koutoulis School of Biological Sciences, University of Tasmania, Hobart, Australia

Sanjeev Kumar Department of Environmental Science, Babasaheb Bhimrao Ambedkar University (A Central University), Lucknow, Uttar Pradesh, India

Yongfang Li College of Life Sciences, Henan Normal University, Xinxiang, Henan, China

246 Noble Research Center, Department of Biochemistry and Molecular Biology, Oklahoma State University, Stillwater, OK, USA

Ramamurthy Mahalingam 246 Noble Research Center, Department of Biochemistry and Molecular Biology, Oklahoma State University, Stillwater, OK, USA

USDA-ARS, Cereal Crops Research Unit, Madison, WI, USA

Francisco Marco Departament de BiologiaVegetal, Facultat de Farmàcia, Universitat de València, Burjassot, València, Spain

Meenakumari Muthuramalingam 246 Noble Research Center, Department of Biochemistry and Molecular Biology, Oklahoma State University, Stillwater, OK, USA

Puneet Paul Department of Biosciences, Molecular Cell Biology of Plants, Goethe University, Frankfurt am Main, Germany

Sascha Röth Department of Biosciences, Molecular Cell Biology of Plants, Goethe University, Frankfurt am Main, Germany

Manish Sainger Centre for Biotechnology, Maharshi Dayanand University, Rohtak, India

Poonam Ahlawat Sainger Department of Environmental Sciences, Maharshi Dayanand University, Rohtak, India

Lana Shabala School of Land and Food, University of Tasmania, Hobart, Australia

Sergey Shabala School of Land and Food, University of Tasmania, Hobart, Australia

Rana Pratap Singh Department of Environmental Science, Babasaheb Bhimrao Ambedkar University (A Central University), Lucknow, Uttar Pradesh, India

Sudhakar Srivastava Nuclear Agriculture and Biotechnology Division, Bhabha Atomic Research Centre, Mumbai, Maharashtra, India

Institute of Environment & Sustainable Development, Banaras Hindu University, Varanasi, India

P. Suprasanna Nuclear Agriculture and Biotechnology Division, Bhabha Atomic Research Centre, Mumbai, Maharashtra, India

Antonio F. Tiburcio Unitat de Fisiologia Vegetal, Facultat de Farmàcia, Universitat de Barcelona, Barcelona, Spain

Parul Rana Tomar Stockbridge School of Agriculture, University of Massachusetts Amherst, MA, USA

K.K. Vinod Division of Genetics, ICAR-Indian Agricultural Research Institute, Rice Breeding & Genetics Research Centre, Tamil Nadu, India

Lizhong Xiong National Key Laboratory of Crop Genetic Improvement and National Center of Plant Gene Research (Wuhan), Huazhong Agricultural University, Wuhan, China

Jun You National Key Laboratory of Crop Genetic Improvement and National Center of Plant Gene Research (Wuhan), Huazhong Agricultural University, Wuhan, China

Jorge A. Zavala Cátedra de Bioquímica/Instituto de Investigación en Biociencias Agrícolas y Ambientales, Facultad de Agronomía, University of Buenos Aires-Consejo Nacional de Investigaciones Científicas y Técnicas de Argentina, Buenos Aires, Argentina

Xuechen Zhang School of Land and Food, University of Tasmania, Hobart, Australia

Meixue Zhou School of Land and Food, University of Tasmania, Hobart, Australia

About the Editors

Dr. Pawan Kumar Jaiwal is presently working as a professor in the Centre for Biotechnology at the M.D. University, Rohtak, India, and has served at various administrative positions as the director, dean (presently) and chairman/member of statuary bodies at the University. He has 30 years of PG teaching and 35 years of research experience and has published about 85 original research papers, 27 book chapters, review articles and 14 books. He has guided over 68 M.Sc. and M. Phil. students for their dissertations and 20 Ph.D. students for their degree. He is a member of several academic bodies and is in the editorial board of four international research journals. Dr. Jaiwal has been awarded DST Young Scientist Project; Visiting Fellowship by INSA, New Delhi; Overseas Associateship by DBT, New Delhi; and Prof. H. S. Srivastava Gold Medal by the National Academy of Environmental Sciences, Lucknow, India. He has several international fellowships to his credit and has visited many countries for academic contributions including the laboratory of Prof. Ingo Potrykus at the Institute of Plant Sciences, ETH, Zurich, Switzerland. His current research interests are metabolic engineering for resistance to abiotic and biotic stresses, nutrient use efficiency and nutritional quality improvement in pulses especially *Vigna* species, oil crops (*Brassica juncea* and sesame) and cereals (wheat).

Dr. Rana Pratap Singh is presently working as a professor in the Department of Environmental Science, Babasaheb Bhimrao Ambedkar (A Central) University, Lucknow, India. He has contributed significantly in the understanding of ammonia assimilation and N-metabolism in plants. Besides, he has contributed some new knowledge on toxicity and remediation of soil and water ecosystems. He has 30 years of PG teaching, 35 years of research and 10 years of administrative experiences and has published about 100 original research papers, 14 review articles, 24 book chapters and 16 books. He has guided over 100 M.Sc. and M. Phil. students for their dissertations and 27 Ph.D. students for their degree. Professor Singh is editor in chief of an international journal *Physiology and Molecular Biology of Plants* (www.springer.com/journal/12298) and editor of *Climate Change and Environmental Sustainability* (www.indianjournals.com). He has 9 academic awards and international fellowships to his credit and has visited many countries for academic contributions.

Prof. Om Parkash Dhankher is a plant biotechnologist (associate professor) in the Stockbridge School of Agriculture, University of Massachusetts, Amherst (USA). He developed the first transgenic plant-based approach for arsenic phytoremediation. Prof. Dhankher has contributed more than 40 publications in high impact factor journals including *Nature Biotech* and was awarded several patents. His research was also featured in headlines on the National Geographic Channel, ABC, Reuter, etc. His research focus is multidisciplinary in nature ranging from crop improvements to phytoremediation to biofuels. He is editor of the *International Journal Plant Biology Research*, associate editor of *Frontier's Agricultural Biological Chemistry*, editorial board member of two international journals and a member of the executive committee of the American Society of Plant Biologists (ASPB). He has supervised seven Ph.D. students, three M.Sc. students, four visiting professors, five postdoctoral research associates and over a dozen undergraduate honour thesis students and has several ongoing collaborations with researchers in India, China, Italy, Egypt and the USA.

Plant Responses to Tropospheric Ozone

1

Yongfang Li, Meenakumari Muthuramalingam, and Ramamurthy Mahalingam

Abstract

Tropospheric ozone is the second most abundant air pollutant and an important component of the global climate change. Over five decades of research on the phytotoxicity of ozone in model plant systems, crop plants, and forest trees have provided some insight into the physiological, biochemical, and molecular responses to this toxic pollutant. Majority of the studies on ozone have been conducted using acute treatment regimes on model plant systems. Several omics platforms have been used to investigate the ozone responses in plants. Some efforts have been undertaken to understand the genetic basis of ozone resistance using *Arabidopsis* and rice. The omics and mapping studies have shown that resistance to ozone impinges upon multiple pathways including phytohormones, stress/defense, secondary metabolism, and redox signaling. Furthermore, ozone stress is more likely to co-occur with other global climate change factors including increasing CO_2 levels, high temperatures, and other abiotic stresses such as drought, UV light, and salinity. Crop germplasm screening for ozone resistance should consider using Free-Air Carbon Dioxide Enrichment technology in conjunction with state-of-the-art remote-sensing reflectance spectroscopy. Cisgenic and transgenic approaches to develop ozone-resistant crops will be more successful if combinations of ozone and other stresses are taken into consideration.

Y. Li
College of Life Sciences, Henan Normal University, Xinxiang, Henan 453007, China

246 Noble Research Center, Department of Biochemistry and Molecular Biology, Oklahoma State University, Stillwater, OK 74078, USA

M. Muthuramalingam
246 Noble Research Center, Department of Biochemistry and Molecular Biology, Oklahoma State University, Stillwater, OK 74078, USA

R. Mahalingam (✉)
246 Noble Research Center, Department of Biochemistry and Molecular Biology, Oklahoma State University, Stillwater, OK 74078, USA

USDA-ARS, Cereal Crops Research Unit, Madison, WI 537262, USA
e-mail: mali.mahalingam@ars.usda.gov

P.K. Jaiwal et al. (eds.), *Genetic Manipulation in Plants for Mitigation of Climate Change*, DOI 10.1007/978-81-322-2662-8_1

Keywords
Arabidopsis • Tropospheric ozone • Acute ozone • Chronic ozone • Climate change • FACE • Omics • Phytohormones • QTL mapping • Combined stresses

Abbreviations

2-DE	two-dimensional gel electrophoresis
2D-DiGE	two-dimensional difference gel electrophoresis
ABA	abscisic acid
APX	ascorbate peroxidase
AsA	ascorbate
ET	ethylene
FACE	Free-Air Carbon Dioxide Enrichment
GST	glutathione S-transferase
GSH	glutathione
JA	jasmonic acid
O_3	ozone
OTC	open-top chamber
PR	pathogenesis related
rcd1	*radical cell death 1*
SA	salicylic acid
SOD	superoxide dismutase
VOC	volatile organic compounds
NO	nitric oxide
NO_x	oxides of nitrogen
QTL	quantitative trait loci
ROS	reactive oxygen species
vtc1	*vitamin C deficient 1*

1.1 Introduction

Ozone (O_3) is the triatomic form of oxygen that forms in the stratosphere (10–30 miles above sea level) spontaneously by the light-driven reactions of atmospheric oxygen. The stratospheric ozone or the "good ozone" blocks the sun's harmful ultraviolet radiation from reaching the earth's surface and thus protects both the plant and animal life (http://www.epa.gov/airquality/gooduphigh/).

Tropospheric ozone also referred to as "bad ozone" is formed mostly by anthropogenic activities. Fossil fuel combustion and industrial pollution lead to release of volatile organic compounds (VOCs), methane, and oxides of nitrogen (NO_x) that react in the presence of sunlight to form ozone (Heath and Taylor 1997). In the northern hemisphere ozone concentrations have more than doubled since the preindustrial era (Fuhrer 2009; Vingarzan 2004). Tropospheric ozone is a major health threat to infants, asthmatics, and older populations, and it has been estimated that 0.7 million respiratory mortalities are due to this second largest air pollutant (Anenberg et al. 2010).

Ozone adversely affects vegetation and entire forest ecosystems (Miller 1983). Effects of this phytotoxic pollutant on crop yields have been studied over the last three decades (Emberson et al. 2009; Feng and Kobayashi 2009; Mills et al. 2007). In Europe, economic losses due to ozone on 23 crops in the year 2000 were estimated to be 6.7 billion euros (Holland et al. 2006). Global losses due to ozone in 2000 were estimated to be \$11–18 billion (Avnery et al. 2011). Estimated reduction of global yields due to ozone phytotoxicity varied considerably for crops ranging from 2.2 to 5.5 % for maize, 3.9 to 15 % for wheat, and 8.5 to 14 % for soybeans (Avnery et al. 2011). These studies demonstrate that tropospheric ozone is a major threat to crop production. In order to ensure food security for the future, it is necessary to incorporate ozone tolerance/resistance in crop plants.

1.2 Assessing Phenotypic Effects of Ozone on Plants

Three different exposure technologies have been used for the studies of ozone tolerance/sensitivity in plants. Most of the earlier studies were conducted in controlled growth or environmental chambers that were connected to ozone generators. The popular ozone generator models are simple hollow UV bulb/tube through which the oxygen gas flows through the inlet valve and is spontaneously converted into ozone that can then be shunted via the outlet valve fitted with tubing leading into the plant growth chamber. Majority of these studies were conducted with the model plant *Arabidopsis* or in plants of economic importance in the seedling stages. Accommodating plants like soybeans or wheat in sufficient numbers for a thorough analysis of ozone symptoms over an entire growing season in growth chambers is difficult. Nonetheless, growth chambers have been extremely useful for assessing the impact of short-term high-ozone dose studies, also referred to as the acute ozone treatment. These studies clearly demonstrated the impact of ozone as an oxidative stressor and in fact led to the use of ozone as a model system to understand oxidative signaling in plants (Rao et al. 2000a, b).

From a climate change perspective, it was recognized that crop plant responses to low ozone does over extended periods of time, also referred to, as chronic ozone treatment was more important. The open-top chambers (OTCs) are just transparent enclosures of glass or plastic with an open top. These OTCs were installed in field plots so that ozone responses could be monitored under realistic field conditions throughout the crop-growing season. In the late 1990s the advent of Free-Air Carbon Dioxide Enrichment (FACE) technology provided a pragmatic means for assessing the impact of changes in ozone and CO_2 under agronomic conditions (Long et al. 2005). FACE avoids any changes to the microenvironment imposed by chambers or OTCs, thereby providing the most reliable estimates of plant responses to climate change factors.

1.3 Physiological Impact of Ozone in Crop Plants

Crops show a wide variability in their phenotypic responses to ozone, and this includes both intra- and interspecific variation (Maggs and Ashmore 1998; Biswas et al. 2008; Brosche et al. 2010). Several studies have reported lists of species or genotypes that are resistant or sensitive to ozone (Heagle 1989; Mills et al. 2007). It is important to understand the differences between sensitive and resistant responses to ozone exposure. The visible injury symptoms due to ozone are mostly assessed by damage to the foliage. It appears as small chlorotic or necrotic lesions on leaves that can coalesce into larger patches of injured area, and such leaves usually senesce early. This reduces the effective biomass that in turn will take a toll on crop yields (Wilkinson et al. 2012). Apart from leaves, ozone exposure has been reported to have negative impact on reproductive development in plants (Black et al. 2000). In field crops such as wheat, rice, maize, soybean, and sorghum, ozone stress reduces several important grain traits such as size, weight, nutritional quality, and number (Mulholland et al. 1998; Biswas and Jiang 2011). In pod crops such as beans, ozone reduces pod number and size, and in tuber crops like potato, tuber size is reduced (Wilkinson et al. 2012).

From an agronomic point of view, the definition of crop sensitivity to ozone is misleading. For example, crops can be sensitive to ozone with reference to visible foliar damage at early stages of growth but may not have a net impact on the grain yield during harvest. In rice and wheat, plants with the least visible foliar symptoms showed maximum yield losses (Sawada and Kohno 2009; Picchi et al. 2010), and this was explained on the basis of stomatal closure response. Cultivars in which ozone causes stomatal closure prevent the influx of ozone and reduced the foliar injury. Thus, based on the damage to leaves, these cultivars are resistant to ozone. However, prolonged stomatal closure affects carbon fixation and in turn the amount of assimilates required for grain filling. Thus, with reference to yield these cultivars are ozone sensitive. Other mechanisms for the negative effect of ozone could be due to reduction of new growth (McKee and Long 2001), reduced root biomass (Grantz et al. 2006), reduced phloem translocation efficiency, or reduced carbon portioning to grains over synthesis of protective chemicals (Betzelberger et al. 2010).

1.4 Biochemical Changes in Response to Ozone Stress

1.4.1 Changes in Reactive Oxygen Species, Nitric Oxide, and Calcium

Ozone entry through the cuticle is negligible (Kerstiens and Lendzian 1989). The stomatal guard cells act as the first structures that regulate the entry of ozone. Thus, if stomatal aperture is wider, it may allow a larger influx of gases like ozone. Hence, stomatal conductance was considered a vital component of plant resistance or sensitivity response to ozone (Heath and Taylor 1997; Sandermann and Matyssek 2004). Stomatal closure was reported within 12 min of ozone exposure suggesting that this rapid response may be due to a direct effect of this pollutant (Moldau et al. 1990). However, the ozone concentration inside the cells is close to zero (Laisk et al. 1989). This suggested a rapid degradation of the ozone into various reactive oxygen species or ozone-derived ROS and is observed both in the ozone-tolerant and ozone-sensitive plant species (Kangasjarvi et al. 2005). The presence of free radicals in ozone-fumigated pea and bean leaves was shown using electron spin studies (Mehlhorn et al. 1990). In aqueous phase O_3^- degrades further to $O_2^{\bullet -}$, O_2, and OH (Grimes et al. 1983). Even though ozone-mediated hydroxyl radical formations are less prone to occur at physiological pH, some phenolic compounds in cell wall can induce formation of ozone-derived radicals. It has been reported that a ROS burst of plant origin several hours after the end of the ozone treatment is a hallmark feature of only the ozone-sensitive plants (Wohlgemuth et al. 2002). Plasma membrane-bound NADPH oxidase, peroxidase, oxalate oxidase, and polyamine oxidases are the major sites of ROS production in the apoplast (Bolwell 1999; Sebela et al. 2001; Scheel 2002). Other organelles such as chloroplasts, mitochondria, and peroxisomes are involved in metabolic activities with high rates of electron flow and are also considered to be major ROS producers.

The importance of nitric oxide (NO) in cellular signaling, detoxification, and metabolism has been illustrated by previous studies (Parani et al. 2004; Zeidler et al. 2004; Delledonne 2005). A role for NO in modifying gene expression was reported in ozone-exposed plants (Mahalingam et al. 2006; Ahlfors et al. 2009). Accumulation of NO along with H_2O_2 was shown in ozone-fumigated tobacco plants using NO-specific fluorescent dye (Ederli et al. 2006). Due to its lipophilic nature, NO has the ability to diffuse through the membranes. One of the important biological reactions of nitric oxide is S-nitrosylation, the posttranslational modification of thiol groups to form S-nitrosothiols (RSNO). In addition to reacting with H_2O_2 and forming $HO^{\bullet}$, superoxides can also react with $NO^{\cdot}$ to form peroxynitrite. Peroxynitrite anion is relatively stable and is known to oxidize proteins and non-protein thiol groups (Quijano et al. 1997; Landino et al. 2002). However, in plants peroxynitrite may not be toxic (Delledonne et al. 2001).

Perturbations to ROS and NO homeostasis can trigger changes in calcium influx and also lead to activation of MAP kinase cascade reported in response to many stresses accompanied by ROS accumulation such as wounding, pathogen infection, cold, and drought (Seo et al. 1995; Usami et al. 1995; Suzuki and Shinshi 1995; Jonak et al. 1996; Zhang and Klessig 1998). Studies have demonstrated the activation of MAP kinases in both *Arabidopsis* and *tobacco* plants under ozone stress within minutes, representing one of the early responses to ozone (Samuel et al. 2000; Ahlfors et al. 2004). Lowered expression of MPK4 in transgenic tobacco plants led to higher sensitivity to ozone, and on the other hand its overexpression rendered ozone tolerance to plants (Gomi et al. 2005). Some studies speculated ozone-induced calcium might activate MAPKs, which trigger the phosphorylation of NADPH oxidase, which is one of the known sources of ROS (Keller et al. 1998). Addition of ROS-scavenging chemicals, especially for singlet oxygen, inhibited O_3-induced cell death in tobacco cell suspension culture. Similarly the presence of calcium ion chelators inhibited the ozone-mediated cell death, demonstrating the importance of calcium and ROS homeostasis in response to ozone (Kadono et al. 2006).

1.4.2 Phytohormone Changes in Response to Ozone Stress

Various phytohormones like jasmonic acid (JA), salicylic acid (SA), abscisic acid (ABA), ethylene, and intracellular components such as calcium ions are reported to be major regulators of oxidative signaling pathway. These hormones play synergistic and antagonistic roles in ozone-induced defense response (Overmyer et al. 2003). For example, SA was induced rapidly in ozone-treated *Arabidopsis* and tobacco plants (Yalpani et al. 1994; Sharma and Davis 1997). In the same study similar kinetics was observed for the defense-related gene transcripts suggesting a key role for SA in activating O_3-mediated defense response. Several studies demonstrated the induction of genes involved in JA biosynthetic pathway in ozone-treated plants (Maccarrone et al. 1997; Mahalingam et al. 2005) that lead to the finding of JA accumulation in response to ozone (Koch et al. 1998, 2000). Furthermore, JA biosynthesis-defective mutants displayed increased sensitivity to ozone (Rao et al. 2000a, b). Exogenous treatment of wild-type or ozone-sensitive plants with JA diminished the spread of cell death and ozone damage. On the contrary, SA and ethylene alleviate the ROS production resulting in cell death upon ozone exposure (Orvar et al. 1997; Overmyer et al. 2000; Beers and McDowell 2001; Rao et al. 2002). For example, JA application before ozone treatment protected tobacco plants (Orvar et al. 1997); another study demonstrated O_3 sensitivity of poplar clone due to JA and SA insensitivity (Koch et al. 2000). Ozone-sensitive *Arabidopsis* species Cvi-0 produced large amounts of SA by ozone treatment but displayed low JA sensitivity suggesting negative regulation of SA by JA signaling pathway. Furthermore, ozone-induced ethylene accumulation was also shown to decrease following JA treatment. Hence JA signaling negatively regulates both SA accumulation and ethylene biosynthesis. Increase in ethylene production is considered as one of the early response marker for ozone sensitivity (Tingey et al. 1976; Mehlhorn and Wellburn 1987; Grantz and Vu 2012). Ozone-sensitive Bel-W3 tobacco plants showed increased level of ethylene compared to tolerant Bel-B line upon ozone exposure (Langebartels et al. 1991). Several studies have postulated the role of ethylene as a regulator of programmed cell death, and mutant studies support the role of ethylene in cell death by plant–pathogen interaction and in O_3 response (Bent et al. 1992; Greenberg and Ausubel 1993; Orzaez and Granell 1997; Young et al 1997). Ozone-sensitive *Arabidopsis* mutant *radical cell death 1* (*rcd1*) accumulated high levels of ethylene and SA (Overmyer et al. 2005). Thus, using exogenous hormone treatments, hormone biosynthetic and signaling mutants, it is now widely accepted that oxidative cell death cycle is a central component in the ozone-signaling pathway (Overmyer et al. 2003; Kangasjarvi et al. 2005).

1.4.3 Antioxidant Defense Response to Ozone

Plants have developed an innate mechanism to regulate ROS levels via various enzymatic and nonenzymatic antioxidants. Low-molecular antioxidants, such as ascorbate (AsA), glutathione (GSH), α-tocopherols, β-carotenoids, and polyamines, are classified as nonenzymatic defense system that are able to scavenge distinct ROS molecules and ensure the integrity of biological membranes (Kangasjarvi et al. 1994; Noctor and Foyer 1998; Chaudiere and Ferrari-Iliou 1999; Baroli and Niyogi 2000; Blokhina et al. 2003). Since ozone does not enter the cell, it has been argued that the apoplastic antioxidant capacity is a key factor for ozone resistance in plants. In fact, apoplastic ascorbate was shown to be the first line of defense against ozone using the *Arabidopsis* ascorbate-deficient, *vitamin C deficient 1* (*vtc1*), mutant (Conklin et al. 1996). AsA reacts not only with H_2O_2 but also with superoxides, hydroxyl radical, singlet oxygen, and lipid hydroperoxides (Buettner and Jurkiewicz 1996). It is quantitatively the most predominant antioxidant and acts as cofactor for the regeneration of some enzymatically important thiols (Mandl et al. 2009). Additionally it regulates the cellular H_2O_2 level together with glutathione called

Asada–Halliwell pathway or ascorbate–glutathione cycle. The importance of ascorbate in ozone tolerance is shown by its amount in leaves; it can represent over 10 % of all soluble carbohydrates (Noctor and Foyer 1998). Investigations on fumigated spinach apoplasts with ozone indicated that the flux of reduced ascorbate from cytosol to the apoplast is linked to oxidative stress while discontinued fumigation with ozone lowered the ascorbate export rate into the apoplast below detection level (Luwe et al. 1993). Expression level of cytosolic ascorbate peroxidase (APx) was shown in tobacco to be upregulated due to the oxidative stress, caused by O_3 (Willekens et al. 1994). Ozone stress also affects glutathione levels in plants (Gupta et al. 1991). A study showed higher GSH level and glutathione reductase activity that determined the O_3 sensitivity in bean plants (Guri 1983). Plants exposed to moderate rate of ozone exhibited increase in transcripts corresponding to GST, SOD, and peroxidase involved in antioxidant defense system (Sharma and Davis 1994). Several studies have been done in generating transgenic plants with low or higher levels of antioxidant enzymes (SOD, catalase, APx) which exhibited differential response to ozone or biotic elicitor mediated PCD (Mittler et al. 1999). Such transgenic studies support the importance of tightly regulated redox homeostasis needed for the optimal cell function and provide novel tools for engineering tolerance to ozone in crop plants.

Besides the enzymatic antioxidant defense system, polyamines (PA) may play a role in scavenging radicals. PA are aliphatic nitrogen derivatives which are involved in various regulatory processes such as enhancing growth, DNA replication, and cell division and differentiation (Evans and Malmberg 1989). They protect plants against environmental stresses due to its dual anion- and cation-binding properties by indirectly chelating ions that catalyze the peroxidation reaction (Bors et al. 1989). Plants tolerate ozone toxicity by increasing levels of PA (Langebartels et al. 1991). The same study postulated differential levels of PA in ozone-sensitive versus ozone-tolerant cultivars of tobacco. Few studies have delineated the connection of PA biosynthesis with ethylene emission in ozone-treated plants (Bouchereau et al. 1999). These studies demonstrate the intricate connections between ROS, phytohormones, and antioxidants, and the interplay between these labile molecules determines the cellular biochemical response to ozone stress.

1.5 Molecular Studies of Ozone Stress in Plants

Plants exposed to ozone stress use multiple mechanisms, including gene expression which results in profound changes in composition of transcripts, proteins, and metabolism, to restore or reestablish cellular homeostasis and maintain physiological and biochemical state necessary to survive and continue growing under stress condition (Baier et al. 2005; Kosova et al. 2011; Singh and Jwa 2013). The high-throughput omics technologies (transcriptomics, proteomics) have been widely used to identify the O_3-responsive components in model and non-model plants. Here, we review the recent progress in ozone-responsive transcriptome and proteome and some progress that has been made recently in quantitative trait loci (QTL) mapping.

1.5.1 Ozone-Responsive Transcriptome

Extensive transcriptome profiles responding to ozone stress have been conducted in model plant *Arabidopsis* (Miller et al. 1999; Matsuyama et al. 2002; Tamaoki et al. 2003; Ludwikow et al. 2004; Mahalingam et al. 2005). Using mini-microarray, *PR-1*, *AtGST-2*, and *EDS1* were found upregulated strongly by exposure to O_3 in *Arabidopsis* (Matsuyama et al. 2002). Ludwikow et al. (2004) found expression of 2385 genes changed more than twofold in response to ozone stress. Tamaoki et al. found 205 nonredundant expressed sequence tags (ESTs) which were regulated by ozone exposure. Of these, 157 were induced and 48 were suppressed by ozone; multiple signaling pathways act mutually antagonistically to induce alteration of gene expression. Many defense genes were induced by ethylene (ET) and jasmonic acid (JA) signal pathway but suppressed by salicylic acid (SA) signal pathway. It is well known that ozone stress induces accelerated foliar senescence in many plant species. Expression of eight senescence-related genes were induced by ozone exposure in *Arabidopsis*, while levels of photosynthetic gene transcripts declined (Miller et al. 1999). Microarray expression profiling of a 4-week-old *Arabidopsis* plant over a 12-h period commencing from the onset of a 6-h ozone exposure presented a temporal evolution of the oxidative stress response (Mahalingam et al. 2005). Eighty-one genes belong to an "early upregulated" cluster; their expression was induced immediately after the onset of ozone exposure and reached maximal amplitude within the first 1.5–3 h and declined thereafter; among these genes, 36 genes are integral membrane protein involved in transporting small molecules across membrane and in detoxification reaction, 15 signaling genes, and 10 transcription factors. Sixty genes were designed as "late upregulated"; expression of these genes increased slowly, reached peak level at 3–4.5 h or later, and then returned to preexposure level by 12 h; genes related with redox homeostasis (GST and APX) and pathogenesis-related (PR) proteins fall in this cluster; these "late upregulated" transcripts were also confirmed in another study (D'haese et al. 2006). Fifty-nine genes, designated as "downregulated" cluster, showed a transient decrease in transcript abundance upon the onset of ozone exposure and then returned to preexposure level gradually; most of these genes are chloroplast-targeted protein which are associated with photosynthesis (Mahalingam et al. 2005).

In another study, two-week-old rice seedlings were exposed to 0.2 ppm O_3 for 24 h. A total of 1535 nonredundant genes were found altered in leaves more than fivefold over the control, representing eight main functional categories (Cho et al. 2008). Among these transcripts, cellular processing and signaling categories were highly represented within 1 h of ozone treatment; transcription factor and signal transduction were the main subcategories. Genes involved in information storage and processing, cellular processing, signaling, and metabolism were mainly regulated at 12 and 24 h; their main subcategories were ribosomal protein, posttranslational modification, and signal transduction and secondary metabolite biosynthesis (Cho et al. 2008). Ozone-induced transcriptome signatures in rice panicles and grains reported majority of the genes were associated with hormonal signaling, proteolysis, transcription, cell wall, and defense signaling (Cho et al. 2013).

Different ozone-responsive transcription profiles were reported between ozone-tolerant cultivar and ozone-sensitive cultivar (Lee and Yun 2006; Puckette et al. 2008). Comparative analysis of ozone-tolerant JE154 and ozone-sensitive Jemalong cultivar of *Medicago truncatula* showed rapid changes within 1 h after treatment initiation in the transcriptome of JE154, suggesting that the initial signals responding to the oxidant were rapidly perceived in the tolerant line (Puckette et al. 2008). Among 180 up- or down-regulated genes, 67 % (120) were regulated differently in ozone-sensitive and ozone-tolerant peppers. Interestingly, most of the ozone-responsive genes were specifically upregulated in the ozone-sensitive cultivar. Many of these ozone-responsive genes were found associated with pathogen infection or responding to cold, drought, and salinity stresses (Lee and Yun 2006). These studies show the broad overlap between plant responses to ozone and other abiotic and biotic stresses.

1.5.2 Ozone-Responsive Proteome

Unlike transcripts, proteins are direct effectors of plant stress response. Proteins not only include enzymes but also include components of transcription and translation machinery and regulate plant stress response at both transcript and protein levels (Kosova et al. 2011). Previous studies have demonstrated that the changes in gene transcription level do not often correspond with the changes at protein level (Gygi et al. 1999; Bogeat-Triboulot et al. 2007). Therefore, studies of changes in proteome composition under stress are

Table 1.1 Proteins responsive to ozone stress

Pathway	Protein	Plants	References
Antioxidant protein	SOD, catalase, APX, GST	Rice	Agrawal et al. (2002) and Cho et al. (2008)
		Soybean	Torres et al. (2007) and Ahsan et al. (2010)
		Arabidopsis	Sharma and Davis (1997)
		Maize	Torres et al. (2007)
Carbon metabolism		Soybean	Ahsan et al. (2010)
		Poplar	Bohler et al. (2007)
Photosynthesis and carbon assimilation	RuBisCO, RuBisCO activase	Rice	Agrawal et al. (2002) and Cho et al. (2008)
		Soybean	Ahsan et al. (2010)
		Maize	Torres et al. (2007)
		Bean	Torres et al. (2007)
		Poplar	Bohler et al. (2007)
Pathogenesis-related (PR) protein	Chitinase	Rice	Agrawal et al. (2002) and Cho et al. (2008)
		Poplar	Bohler et al. (2007)
		Arabidopsis	Sharma and Davis (1997)
Protein folding	Chaperone protein	Soybean	Ahsan et al. (2010)
		Pepper	Bohler et al. (2007)
	Heat shock proteins	Maize, bean	Bohler et al. (2007) and Ahsan et al. (2010)
		Rice	Cho et al. (2008)
		Pepper	Bohler et al. (2007)

highly important since they are helpful to unravel the possible relationships between protein abundance and plant stress tolerance (Kosova et al. 2011). Employing two-dimensional electrophoresis (2-DE), amino acid sequencing, MALDI-TOF, 2D-DiGE, and immunoblot analysis, proteins responding to ozone stress have been investigated in many plant species. Here, we summarize the common ozone-responsive proteins (Table 1.1).

1.5.2.1 Accumulation of Antioxidant Proteins

As indicated earlier, to reduce the oxidative stress generated by O_3, plants employ strong antioxidant defense mechanisms, including antioxidative enzymes, secondary metabolites, carotenoids, ascorbate, and glutathione to attenuate O_3 injury (Baier et al. 2005). Superoxide dismutase (SOD), including Cu/Zn SOD in chloroplasts and MnSOD in mitochondria, converts $O_2^{\bullet-}$ to H_2O_2. Induction of SOD has been found in many plant species exposed to ozone stress (Agrawal et al. 2002, Ahsan et al. 2010; Singh and Jwa 2013). Ascorbate peroxidase (APX) and catalase subsequently reduce H_2O_2 to water. Several proteomic studies have found accumulation of APX in plants under ozone stress, and APX has been considered to be a marker protein for O_3 stress in plants (Agrawal et al. 2002, Bohler et al. 2007; Torres et al. 2007; Ahsan et al. 2010). Glutathione S-transferase (GST), an important enzyme involved in detoxifying toxic products of lipid peroxidation, is also induced by elevated O_3 exposure (Sharma and Davis 1997; Bohler et al. 2010; Ahsan et al. 2010).

1.5.2.2 Downregulation of Photosynthetic Proteins

Many studies have shown that carbon-fixation rate or net photosynthesis was greatly reduced under ozone exposure (Renaut et al. 2009; Ahsan et al. 2010; Singh and Jwa 2013; Khan et al. 2013). Proteins associated with primary carbon assimilation and Calvin cycle (i.e., RuBisCo large and small subunit, RuBisCO activase, sedoheptulose-1, 7-bisphosphatase, triose-phosphate isomerase) were predominantly downregulated. In addition, a number of proteins associated with photosystem I/II and electron transport were also downregulated (Agrawal et al. 2002; Cho et al. 2008; Renaut et al. 2009; Ahsan et al. 2010; Bohler et al. 2010). The reduction in photosynthetic apparatus proteins correlates with reduction in carbon fixation observed in the leaves exposed to ozone.

1.5.2.3 Increased Carbon Metabolism

Upon ozone stress, proteins associated with glucose catabolism were increased in poplar (Bohler et al. 2007). In soybean, decrease in starch and increase in sucrose concentration were detected, which is correlated with the upregulation of proteins associated with carbon catabolism/metabolism, such as sucrose synthase, invertase, fructokinase, and fructose-1,6-bisphosphate aldolase 1 (Ahsan et al. 2010). The upregulation of sugar catabolism might feed the tricarboxylic acid cycle for energy production when the photosynthesis is severely impaired under ozone stress.

1.5.2.4 Other Stress-Related Proteins

Pathogenesis-related proteins have been found associated with different types of abiotic stress, including ozone stress. Increased abundance of multiple isoforms of chitinase was found in poplar (Bohler et al. 2010). A significant overlap between proteins induced by ozone stress and pathogenesis-related proteins was also found in other plant species (Sharma and Davis 1997; Agrawal et al. 2002; Cho et al. 2008). Other stress-related proteins like chaperones (Ahsan et al. 2010; Bohler et al. 2010) and heat shock proteins (HSP70, HSP90, and HSP30) were also found upregulated under ozone stress (Torres et al. 2007; Brosche et al. 2010). Other proteins, like ATP synthase, calcium-binding protein, calreticulin, and proteins related to nitrogen metabolism, were altered by ozone exposure (Agrawal et al. 2002; Ahsan et al. 2010; Khan et al. 2013). Comparison of the proteome profile between soybean ozone-tolerant and ozone-sensitive cultivars showed that ATP synthase α-subunit and ATP synthase β-subunits were among the significantly changing proteins. Further validation confirmed that extracellular ATP signaling in ATP synthase-dependent manner plays a pivotal role in inducing ozone stress tolerance and preventing injuries in soybean cultivars (Khan et al. 2013).

1.5.3 QTL Mapping

Genotypic difference in resistance to ozone has been evaluated in *Arabidopsis* (Brosche et al. 2010), wheat (Rawlings and Cure 1985), tobacco (Schraudner et al. 1998), pepper (Lee and Yun 2006), and soybean (Mulchi et al. 1988; Khan et al. 2013). Several studies have shown that resistance to O_3 is inherited as a quantitative trait (Chernikova et al. 2000; Kim et al. 2004; Singh and Jwa 2013). Marker-assisted selection has been broadly used to increase the precision and efficiency of plant breeding. However, the available ozone-resistant quantitative trait loci (QTL) mapping is limited to *Arabidopsis* and rice (Kim et al. 2004; Brosche et al. 2010). Cvi-0 is the most O_3-sensitive accession among 93 natural *Arabidopsis* accessions (Brosche et al. 2010). Three QTLs for O_3 sensitivity, $O_3$1-18, $O_3$2-13, and $O_3$3-14, were identified by QTL mapping in a Col-0 × Cvi-0 inbred line population and located on chromosomes 1, 2, and 3, respectively. Interestingly, the major O_3 QTL ($O_3$2-13) maps to the same position as one water loss QTL (WL 2–13), revealing the role of stomata in regulating O_3 entry and damage (Brosche et al. 2010). Using a RIL population of 150 plants derived from a cross between the ozone-sensitive Ws ecotype and ozone-resistant Landsberg erecta, only one QTL was mapped to the top of chromosome 4 close to the brevipedicellus gene (Tabassum 2010). Using 164 recombinant inbred (R1) lines from a cross between ozone-resistant "Milyang 23" and ozone-sensitive "Gihobyeo," QTLs associated with the O_3 resistance in rice were mapped on chromosomes 1, 7, and 11. These QTLs were tightly linked to the markers RG109, C507, and RG1094 and were detected in each of three biological replications. The association between these markers and O_3 resistance was further evaluated in 26 rice cultivars and doubled-haploid (DH) populations (Kim et al. 2004). The rapid advances in sequencing and genome-wide association studies provide a novel opportunity to clone the QTLs for ozone resistance in the near future.

1.6 Conclusions and Perspectives

Tropospheric ozone is a contemporary problem brought about by rapid industrialization and associated anthropogenic activities. We speculate that naturally occurring germplasm diversity of various crop plants has to be systematically evaluated for ozone response phenotypes using the FACE technology in conjunction with novel phenotyping tools such as the near-surface remote-sensing reflectance spectroscopy to assess changes in foliar properties (Ainsworth et al. 2013). Ozone-resistant germplasm from such phenotypic screening should be integrated into breeding programs.

It is important to note that ozone does not occur in isolation in nature. There have been several reports on the interactive effects of ozone and CO_2 in plants (Gupta et al. 2005; Kontunen-Soppela et al. 2010; Gillespie et al. 2012). Since ozone is usually a summertime problem, it is likely to co-occur in combination with other abiotic stresses such as drought and/or high temperature stress. In *Medicago truncatula* a combination of ozone and drought stress together is perceived as a "new stress state" and is not a simple additive effect of ozone and drought (Iyer et al. 2013). Genetic manipulation of crops to mitigate global climate change should consider the combination of multiple stresses simultaneously. Further, an integration of various omics platforms (transcriptome, sRNome, proteome, and metabolome) will provide rational targets for genetic manipulation to incorporate tolerance to multiple climate change factors simultaneously.

References

Agrawal GK, Rakwal R, Yonekura M, Kubo A, Saji H (2002) Proteome analysis of differentially displayed proteins as a tool for investigating ozone stress in rice (*Oryza sativa* L.) seedlings. Proteomics 2:947–959

Ahlfors R, Macioszek V, Rudd J, Brosche M, Schlichting R et al (2004) Stress hormone-independent activation and nuclear translocation of mitogen-activated protein kinases in *Arabidopsis thaliana* during ozone exposure. Plant J 40:512–522

Ahlfors R, Brosche M, Kollist H, Kangasjarvi J (2009) Nitric oxide modulates ozone induced cell death, hormone biosynthesis and gene expression in *Arabidopsis thaliana*. Plant J 58:1–12

Ahsan N, Nanjo Y, Sawada H, Kohno Y, Komatsu S (2010) Ozone stress-induced proteomic changes in leaf total soluble and chloroplast proteins of soybean reveal that carbon allocation is involved in adaptation in the early developmental stage. Proteomics 10:2605–2619

Ainsworth EA, Serbin SP, Skoneczka JA, Townsend PA (2013) Using leaf optical properties to detect ozone effects on foliar biochemistry. Photosynth Res 119:65–76

Anenberg SC, Horowitz LW, Tong DQ, West JJ (2010) An estimate of the global burden of anthropogenic ozone and fine particulate matter on premature human mortality using atmospheric modeling. Environ Health Perspect 118:1189–1195

Avnery S, Mauzerall DL, Liu J, Horowitz LW (2011) Global crop yield reductions due to surface ozone exposure: 1. Year 2000 crop production losses and economic damage. Atmos Environ 45:2284–2296

Baier M, Kandlbinder A, Golldack D, Dietz KJ (2005) Oxidative stress and ozone: perception, signalling and response. Plant Cell Environ 28:1012–1020

Baroli I, Niyogi KK (2000) Molecular genetics of xanthophyll-dependent photoprotection in green algae and plants. Philos Trans R Soc Lond B Biol Sci 355:1385–1393

Beers EP, McDowell JM (2001) Regulation and execution of programmed cell death in response to pathogens, stress and developmental cues. Curr Opin Plant Biol 4:561–567

Bent AF, Innes RW, Ecker JR, Staskawicz BJ (1992) Disease development in ethylene-insensitive *Arabidopsis thaliana* infected with virulent and avirulent Pseudomonas and Xanthomonas pathogens. Mol Plant Microbe Interact 5:372–378

Betzelberger AM, Gillespie KM, Mcgrath JM, Koester RP, Nelson RL et al (2010) Effects of chronic elevated ozone concentration on antioxidant capacity, photosynthesis and seed yield of 10 soybean cultivars. Plant Cell Environ 33:1569–1581

Biswas DK, Jiang GM (2011) Differential drought-induced modulation of ozone tolerance in winter wheat species. J Exp Bot 62:4153–4162

Biswas DK, Xu H, Li YG, Liu MZ, Chen YH et al (2008) Assessing the genetic relatedness of higher ozone sensitivity of modern wheat to its wild and cultivated progenitors/relatives. J Exp Bot 59:951–963

Black VJ, Black CR, Roberts JA, Stewart CA (2000) Impact of ozone on the reproductive development of plants. New Phytol 147:421–447

Blokhina O, Virolainen E, Fagerstedt KV (2003) Antioxidants, oxidative damage and oxygen deprivation stress: a review. Ann Bot 91:179–194

Bogeat-Triboulot MB, Brosche M, Renaut J, Jouve L, Le Thiec D et al (2007) Gradual soil water depletion results in reversible changes of gene expression, protein profiles, ecophysiology, and growth performance in *Populus euphratica*, a poplar growing in arid regions. Plant Physiol 143:876–892

Bohler S, Bagard M, Oufir M, Planchon S, Hoffmann L et al (2007) A DIGE analysis of developing poplar leaves subjected to ozone reveals major changes in carbon metabolism. Proteomics 7:1584–1599

Bohler S, Sergeant K, Lefevre I, Jolivet Y, Hoffmann L et al (2010) Differential impact of chronic ozone exposure on expanding and fully expanded poplar leaves. Tree Physiol 30:1415–1432

Bolwell GP (1999) Role of active oxygen species and NO in plant defence responses. Curr Opin Plant Biol 2:287–294

Bors W, Langebartels C, Michel C, Sandermann H (1989) Polyamines as radical scavengers and protectants against ozone damage. Phytochemistry 28:1589–1595

Bouchereau A, Aziz A, Larher F, Martin-Tanguy J (1999) Polyamines and environmental challenges: recent development. Plant Sci 140:103–125

Brosche M, Merilo E, Mayer F, Pechter P, Puzorjova I et al (2010) Natural variation in ozone sensitivity among *Arabidopsis thaliana* accessions and its relation to stomatal conductance. Plant Cell Environ 33:914–925

Buettner GR, Jurkiewicz BA (1996) Catalytic metals, ascorbate and free radicals: combinations to avoid. Radiat Res 145:532–541

Chaudiere J, Ferrari-Iliou R (1999) Intracellular antioxidants: from chemical to biochemical mechanisms. Food Chem Toxicol 37:949–962

Chernikova T, Robinson JM, Lee EH, Mulchi CL (2000) Ozone tolerance and antioxidant enzyme activity in soybean cultivars. Photosynth Res 64:15–26

Cho K, Shibato J, Agrawal GK, Jung YH, Kubo A et al (2008) Integrated transcriptomics, proteomics, and metabolomics analyses to survey ozone responses in the leaves of rice seedling. J Proteome Res 7:2980–2998

Cho K, Shibato J, Kubo A, Kohno Y, Satoh K et al (2013) Genome-wide mapping of the ozone-responsive transcriptomes in rice panicle and seed tissues reveals novel insight into their regulatory events. Biotechnol Lett 35:647–656

Conklin PL, Williams EH, Last RL (1996) Environmental stress sensitivity of an ascorbic acid-deficient Arabidopsis mutant. Proc Natl Acad Sci U S A 93:9970–9974

D'haese D, Horemans N, De Coen W, Guisez Y (2006) Identification of late O-3-responsive genes in Arabidopsis thaliana by cDNA microarray analysis. Physiol Plant 128:70–79

Delledonne M (2005) NO news is good news for plants. Curr Opin Plant Biol 8:390–396

Delledonne M, Zeier J, Marocco A, Lamb C (2001) Signal interactions between nitric oxide and reactive oxygen intermediates in the plant hypersensitive disease resistance response. Proc Natl Acad Sci U S A 98:13454–13459

Ederli L, Morettini R, Borgogni A, Wasternack C, Miersch O et al (2006) Interaction between nitric oxide and ethylene in the induction of alternative oxidase in ozone-treated tobacco plants. Plant Physiol 142:595–608

Emberson LD, Buker P, Ashmore MR (2009) A comparison of North American and 24 Asian exposure-response data for ozone effects on crop yields. Atmos Environ 43:1945–1953

Evans PT, Malmberg RL (1989) Do polyamines have roles in plant development? Annu Rev Plant Physiol Plant Mol Biol 40:235–269

Feng Z, Kobayashi K (2009) Assessing the impacts of current and future concentrations of surface ozone on crop yield with meta-analysis. Atmos Environ 43:1510–1519

Fuhrer J (2009) Ozone risk for crops and pastures in present and future climates. Naturwissenschaften 96:173–194

Gillespie KM, Xu F, Richter KT, McGrath JM, Markelz RJ et al (2012) Greater antioxidant and respiratory metabolism in field-grown soybean exposed to elevated O_3 under both ambient and elevated CO_2. Plant Cell Environ 35:169–184

Gomi K, Ogawa D, Katou S, Kamada H, Nakajima N et al (2005) A mitogen-activated protein kinase NtMPK4 activated by SIPKK is required for jasmonic acid signaling and involved in ozone tolerance via stomatal movement in tobacco. Plant Cell Physiol 46:1902–1914

Grantz DA, Vu HB (2012) Root and shoot gas exchange respond additively to moderate ozone and methyl jasmonate without induction of ethylene: ethylene is induced at higher O_3 concentrations. J Exp Bot 63:4303–4313

Grantz DA, Gunn S, Vu HB (2006) O_3 impacts on plant development: a meta-analysis of root/shoot allocation and growth. Plant Cell Environ 29:1193–1209

Greenberg JT, Ausubel FM (1993) Arabidopsis mutants compromised for the control of cellular damage during pathogenesis and aging. Plant J 4:327–341

Grimes HD, Perkins KK, Boss WF (1983) Ozone degrades into hydroxyl radical under physiological conditions – a spin trapping study. Plant Physiol 72:1016–1020

Gupta AS, Alscher RG, McCune D (1991) Response of photosynthesis and cellular antioxidants to ozone in populus leaves. Plant Physiol 96:650–655

Gupta P, Duplessis S, White H, Karnosky DF, Martin F et al (2005) Gene expression patterns of trembling aspen trees following long-term exposure to interacting elevated CO_2 and tropospheric O_3. New Phytol 167:129–141

Guri A (1983) Variation in glutathione and ascorbic-acid content among selected cultivars of *Phaseolus-vulgaris* prior to and after exposure to ozone. Can J Plant Sci 63:733–737

Gygi SP, Rochon Y, Franza BR, Aebersold R (1999) Correlation between protein and mRNA abundance in yeast. Mol Cell Biol 19:1720–1730

Heagle AS (1989) Ozone and crop yield. Ann Rev Phytopathol 27:397–412

Heath RL, Taylor GE (1997) Physiological processes and plant responses to ozone exposure. In: Sanderman H, Wellburn AR, Heath RL (eds) Forest decline and ozone. Springer, Berlin/Heidelberg/New York, pp 317–368

Holland M, Kinghorn S, Emberson LD, Cinderby S, Ashmore MR, et al (2006) Development of a framework for probabilistic assessment of the economic losses caused by ozone damage to crops in Europe. In: Part of the UNCE international cooperative programme on vegetation. Contract Report EPG 1/3/205. CEH Project No. C02309NEW Contract REport EPG 1/3/205. CEH Project No. C02309NEW. Center for Ecology and Hydrology, Bangor, Gwynedd

Iyer NJ, Tang Y, Mahalingam R (2013) Physiological, biochemical and molecular responses to a combination of drought and ozone in *Medicago truncatula*. Plant Cell Environ 36:706–720

Jonak C, Kiegerl S, Ligterink W, Barker PJ, Huskisson NS et al (1996) Stress signaling in plants: a mitogen-activated protein kinase pathway is activated by cold and drought. Proc Natl Acad Sci U S A 93:11274–11279

Kadono T, Yamaguchi Y, Furuichi T, Hirono M, Garrec JP et al (2006) Ozone-induced cell death mediated with oxidative and calcium signaling pathways in tobacco bel-w3 and bel-B cell suspension cultures. Plant Signal Behav 1:312–322

Kangasjarvi J, Talvinen J, Utriainen M, Karjalainen R (1994) Plant defense systems induced by ozone. Plant Cell Environ 17:783–794

Kangasjarvi J, Jaspers P, Kollist H (2005) Signalling and cell death in ozone-exposed plants. Plant Cell Environ 28:1021–1036

Keller T, Damude HG, Werner D, Doerner P, Dixon RA et al (1998) A plant homolog of the neutrophil NADPH oxidase gp91phox subunit gene encodes a plasma membrane protein with Ca^{2+} binding motifs. Plant Cell 10:255–266

Kerstiens G, Lendzian KJ (1989) Interactions between ozone and plant cuticles. 1. Ozone deposition and permeability. New Phytol 112:13–19

Khan NA, Komatsu S, Sawada H, Nouri MZ, Kohno Y (2013) Analysis of proteins associated with ozone stress response in soybean cultivars. Protein Pept Lett 20:1144–1152

Kim KM, Kwon YS, Lee JJ, Eun MY, Sohn JK (2004) QTL mapping and molecular marker analysis for the resistance of rice to ozone. Mol Cells 17:151–155

Koch JR, Scherzer AJ, Eshita SM, Davis KR (1998) Ozone sensitivity in hybrid poplar is correlated with a lack of defense-gene activation. Plant Physiol 118:1243–1252

Koch JR, Creelman RA, Eshita SM, Seskar M, Mullet JE et al (2000) Ozone sensitivity in hybrid poplar correlates with insensitivity to both salicylic acid and jasmonic acid. The role of programmed cell death in lesion formation. Plant Physiol 123:487–496

Kontunen-Soppela S, Riikonen J, Ruhanen H, Brosche M, Somervuo P et al (2010) Differential gene expression in senescing leaves of two silver birch genotypes in response to elevated CO_2 and tropospheric ozone. Plant Cell Environ 33:1016–1028

Kosova K, Vitamvas P, Prasil IT, Renaut J (2011) Plant proteome changes under abiotic stress – contribution of proteomics studies to understanding plant stress response. J Proteomics 74:1301–1322

Laisk A, Kull O, Moldau H (1989) Ozone concentration in leaf intercellular air spaces is close to zero. Plant Physiol 90:1163–1167

Landino LM, Hasan R, McGaw A, Cooley S, Smith AW et al (2002) Peroxynitrite oxidation of tubulin sulfhydryls inhibits microtubule polymerization. Arch Biochem Biophys 398:213–220

Langebartels C, Kerner K, Leonardi S, Schraudner M, Trost M et al (1991) Biochemical plant responses to ozone: I. Differential induction of polyamine and ethylene biosynthesis in tobacco. Plant Physiol 95:882–889

Lee S, Yun SC (2006) The ozone stress transcriptome of pepper (*Capsicum annuum* L.). Mol Cells 21:197–205

Long SP, Ainsworth EA, Leakey AD, Morgan PB (2005) Global food insecurity. Treatment of major food crops with elevated carbon dioxide or ozone under large-scale fully open-air conditions suggests recent models may have overestimated future yields. Philos Trans R Soc Lond B Biol Sci 360:2011–2020

Ludwikow A, Gallois P, Sadowski J (2004) Ozone-induced oxidative stress in Arabidopsis: transcription profiling by microarray approach. Cell Mol Biol Lett 9:829–842

Luwe MWF, Takahama U, Heber U (1993) Role of ascorbate in detoxifying ozone in the apoplast of spinach (*Spinacia-oleracea* L) leaves. Plant Physiol 101:969–976

Maccarrone M, Veldink GA, Vliegenthart FG, Finazzi Agro A (1997) Ozone stress modulates amine oxidase and lipoxygenase expression in lentil (*Lens culinaris*) seedlings. FEBS Lett 408:241–244

Maggs R, Ashmore MR (1998) Growth and yield responses of Pakistan rice (*Oryza sativa* L.) cultivars to O_3 and NO_2. Environ Pollut 103:159–170

Mahalingam R, Shah N, Scrymgeour A, Fedoroff N (2005) Temporal evolution of the Arabidopsis oxidative stress response. Plant Mol Biol 57:709–730

Mahalingam R, Jambunathan N, Gunjan SK, Faustin E, Weng H et al (2006) Analysis of oxidative signaling induced by ozone in *Arabidopsis thaliana*. Plant Cell Environ 29:1357–1371

Mandl J, Szarka A, Banhegyi G (2009) Vitamin C: update on physiology and pharmacology. Br J Pharmacol 157:1097–1110

Matsuyama T, Tamaoki M, Nakajima N, Aono M, Kubo A et al (2002) cDNA microarray assessment for ozone-stressed *Arabidopsis thaliana*. Environ Pollut 117:191–194

McKee IF, Long SP (2001) Plant growth regulators control ozone damage to wheat yield. New Phytol 152:41–51

Mehlhorn H, Wellburn AR (1987) Stress ethylene formation determines plant-sensitivity to ozone. Nature 327:417–418

Mehlhorn H, Tabner BJ, Wellburn AR (1990) Electron-spin-resonance evidence for the formation of free-radicals in plants exposed to ozone. Physiol Plant 79:377–383

MIller PR (1983) Ozone effects in the San Bernardino National Forest. In: Davis DD, Miller AA, Dochinger L (eds) Air pollution and the productivity of the forest. Isaak Walton League of America, Arlington, pp 161–193

Miller JD, Arteca RN, Pell EJ (1999) Senescence-associated gene expression during ozone-induced leaf senescence in Arabidopsis. Plant Physiol 120:1015–1023

Mills G, Buse A, Gimeno B, Bermejo V, Holland M et al (2007) A synthesis of AOT40-based response functions and critical levels of ozone for agricultural and horticultural crops. Atmos Environ 41:2630–2643

Mittler R, Herr EH, Orvar BL, van Camp W, Willekens H et al (1999) Transgenic tobacco plants with reduced capability to detoxify reactive oxygen intermediates are hyperresponsive to pathogen infection. Proc Natl Acad Sci U S A 96:14165–14170

Moldau H, Sober J, Sober A (1990) Differential sensitivity of stomata and mesophyll to sudden exposure of bean shoots to ozone. Photosynthetica 24:446–458

Mulchi CL, Lee E, Tuthill K, Olinick EV (1988) Influence of ozone stress on growth-processes, yields and grain quality characteristics among soybean cultivars. Environ Pollut 53:151–169

Mulholland BJ, Craigon J, Black CR, Colls JJ, Atherton J et al (1998) Growth, light interception and yield responses of spring wheat (*Triticum aestivum* L.) grown under elevated CO_2 and O_3 in open-top chambers. Glob Chang Biol 4:121–130

Noctor G, Foyer CH (1998) Ascorbate and glutathione: keeping active oxygen under control. Annu Rev Plant Physiol Plant Mol Biol 49:249–279

Orvar BL, McPherson J, Ellis BE (1997) Pre-activating wounding response in tobacco prior to high-level ozone exposure prevents necrotic injury. Plant J 11:203–212

Orzaez D, Granell A (1997) DNA fragmentation is regulated by ethylene during carpel senescence in *Pisum sativum*. Plant J 11:137–144

Overmyer K, Tuominen H, Kettunen R, Betz C, Langebartels C et al (2000) Ozone-sensitive Arabidopsis rcd1 mutant reveals opposite roles for ethylene and jasmonate signaling pathways in regulating superoxide-dependent cell death. Plant Cell 12:1849–1862

Overmyer K, Brosche M, Kangasjarvi J (2003) Reactive oxygen species and hormonal control of cell death. Trends Plant Sci 8:335–342

Overmyer K, Brosche M, Pellinen R, Kuittinen T, Tuominen H et al (2005) Ozone-induced programmed cell death in the Arabidopsis radical-induced cell death1 mutant. Plant Physiol 137:1092–1104

Parani M, Rudrabhatla S, Myers R, Weirich H, Smith B et al (2004) Microarray analysis of nitric oxide responsive transcripts in Arabidopsis. Plant Biotechnol J 2:359–366

Picchi V, Iriti M, Quaroni S, Saracchi M, Viola P et al (2010) Climate variations and phenological stages modulate ozone damages in field-grown wheat. A three-year study with eight modern cultivars in Po Valley (Northern Italy). Agr Ecosyst Environ 135:310–317

Puckette MC, Tang Y, Mahalingam R (2008) Transcriptomic changes induced by acute ozone in resistant and sensitive *Medicago truncatula* accessions. BMC Plant Biol 8:46

Quijano C, Alvarez B, Gatti RM, Augusto O, Radi R (1997) Pathways of peroxynitrite oxidation of thiol groups. Biochemical J 322:167–173

Rao MV, Koch JR, Davis KR (2000a) Ozone: a tool for probing programmed cell death in plants. Plant Mol Biol 44:345–358

Rao MV, Lee H, Creelman RA, Mullet JE, Davis KR (2000b) Jasmonic acid signaling modulates ozone-induced hypersensitive cell death. Plant Cell 12:1633–1646

Rao MV, Lee HI, Davis KR (2002) Ozone-induced ethylene production is dependent on salicylic acid, and both salicylic acid and ethylene act in concert to regulate ozone-induced cell death. Plant J 32:447–456

Rawlings JO, Cure WW (1985) The Weibull function as a dose-response model to describe ozone effects on crop yields. Crop Sci 25:807–814

Renaut J, Bohler S, Hausman JF, Hoffmann L, Sergeant K et al (2009) The impact of atmospheric composition on plants: a case study of ozone and poplar. Mass Spectrom Rev 28:495–516

Samuel MA, Miles GP, Ellis BE (2000) Ozone treatment rapidly activates MAP kinase signaling in plants. Plant J 22:367–376

Sandermann H, Matyssek R (2004) Scaling up from molecular to ecological processes. In: Sandermann H (ed) Molecular ecotoxicology of plants. Springer, Berlin/Heidelberg, pp 207–226

Sawada H, Kohno Y (2009) Differential ozone sensitivity of rice cultivars as indicated by visible injury and grain yield. Plant Biol 11:70–75

Scheel D (2002) Oxidative burst and the role of reactive oxygen species. In: Inze D, Montagu MV (eds) Oxidative stress in plants. Taylor and Francis, London/New York, pp 137–153

Schraudner M, Moeder W, Wiese C, Van Camp W, Inze D et al (1998) Ozone-induced oxidative burst in the ozone biomonitor plant, tobacco Bel W3. Plant J 16:235–245

Sebela M, Radova A, Angelini R, Tavladoraki P, Frebort II et al (2001) FAD-containing polyamine oxidases: a timely challenge for researchers in biochemistry and physiology of plants. Plant Sci 160:197–207

Seo S, Okamoto N, Seto H, Ishizuka K, Sano H et al (1995) Tobacco map kinase – a possible mediator in wound signal-transduction pathways. Science 270:1988–1992

Sharma YK, Davis KR (1994) Ozone-induced expression of stress-related genes in *Arabidopsis thaliana*. Plant Physiol 105:1089–1096

Sharma YK, Davis KR (1997) The effects of ozone on antioxidant responses in plants. Free Radic Biol Med 23:480–488

Singh R, Jwa NS (2013) Understanding the responses of rice to environmental stress using proteomics. J Proteome Res 12:4652–4669

Suzuki K, Shinshi H (1995) Transient activation and tyrosine phosphorylation of a protein kinase in tobacco cells treated with a fungal elicitor. Plant Cell 7:639–647

Tabassum N (2010) Mapping the ozone tolerance trait in *Arabidopsis thaliana*. MS thesis, Oklahoma State University, Stillwater

Tamaoki M, Nakajima N, Kubo A, Aono M, Matsuyama T et al (2003) Transcriptome analysis of O_3-exposed Arabidopsis reveals that multiple signal pathways act mutually antagonistically to induce gene expression. Plant Mol Biol 53:443–456

Tingey DT, Standley C, Field RW (1976) Stress ethylene evolution: a measure of ozone effects on plants. Atmos Environ 10:969–974

Torres NL, Cho K, Shibato J, Hirano M, Kubo A et al (2007) Gel-based proteomics reveals potential novel protein markers of ozone stress in leaves of cultivated bean and maize species of Panama. Electrophoresis 28:4369–4381

Usami S, Banno H, Ito Y, Nishihama R, Machida Y (1995) Cutting activates a 46-kilodalton protein kinase in plants. Proc Natl Acad Sci U S A 92:8660–8664

Vingarzan R (2004) A review of surface ozone background levels and trends. Atmos Environ 38:3431–3442

Wilkinson S, Mills G, Illidge R, Davies WJ (2012) How is ozone pollution reducing our food supply? J Exp Bot 63:527–536

Willekens H, Vancamp W, Vanmontagu M, Inze D, Langebartels C et al (1994) Ozone, sulfur-dioxide, and ultraviolet-B have similar effects on mRNA accumulation of antioxidant genes in *Nicotiana-plumbaginifolia* L. Plant Physiol 106:1007–1014

Wohlgemuth H, Mittelstrass K, Kschieschan S, Bender J, Weigel HJ et al (2002) Activation of an oxidative burst is a general feature of sensitive plants exposed to the air pollutant ozone. Plant Cell Environ 25:717–726

Yalpani N, Enyedi AJ, Leon J, Raskin I (1994) Ultraviolet-light and ozone stimulate accumulation of salicylic-acid, pathogenesis-related proteins and virus-resistance in tobacco. Planta 193:372–376

Young TE, Gallie DR, DeMason DA (1997) Ethylene-mediated programmed cell death during maize endosperm development of wild-type and shrunken2 genotypes. Plant Physiol 115:737–751

Zeidler D, Zahringer U, Gerber I, Dubery I, Hartung T et al (2004) Innate immunity in *Arabidopsis thaliana*: Lipopolysaccharides activate nitric oxide synthase (NOS) and induce defense genes. Proc Natl Acad Sci U S A 101:15811–15816

Zhang S, Klessig DF (1998) Resistance gene N-mediated *de novo* synthesis and activation of a tobacco mitogen-activated protein kinase by tobacco mosaic virus infection. Proc Natl Acad Sci U S A 95:7433–7438

Plant Heat Stress Response and Thermotolerance

2

Sascha Röth, Puneet Paul,
and Sotirios Fragkostefanakis

Abstract

Climate change and global warming are considered to be major threats for agricultural production and food safety. Decreased yield of several important crops has already been related to frequently occurring extreme environmental conditions such as heat waves. Since most of the economically and dietary important crops are sensitive to high temperatures, the development of cultivars that can withstand adverse temperatures is a prerequisite for meeting the demands for increased food production. The processes of sensing and responding to heat are complex phenomena in plants which comprise the activation of numerous regulatory and signaling pathways that eventually lead to a fine metabolic adjustment to ensure cell survival. Currently, our knowledge of heat stress response is greatly advanced by the massive production of datasets derived from *-omics* studies which supplement the current models with new genes, proteins, and metabolites or even introduce whole new pathways. This information is essential for the improvement of plant thermotolerance either through breeding programs or approaches using genetic engineering. This chapter contains an assembly of several aspects regarding heat stress response and thermotolerance. The effects of high temperatures on major aspects of plant growth and development are described, and different methods for thermotolerance screening are presented. In addition, putative heat sensing mechanisms are discussed and the most important metabolic changes are elaborated. Last, a summary of efforts and strategies to improve thermotolerance by breeding or genetic engineering is given.

S. Röth • P. Paul • S. Fragkostefanakis (✉)
Department of Biosciences, Molecular Cell Biology of Plants, Goethe University, Max von Laue Str. 9, 60438 Frankfurt am Main, Germany
e-mail: fragkost@bio.uni-frankfurt.de

P.K. Jaiwal et al. (eds.), *Genetic Manipulation in Plants for Mitigation of Climate Change*,
DOI 10.1007/978-81-322-2662-8_2

Keywords
Heat stress • Thermotolerance • Abiotic stress • Sensing • Unfolded protein response

2.1 Introduction

Climate changes pose major social, economic, and ecological threats and are expected to affect every level of human activities in the near future. Crop production is already one major field influenced by the robust and coherent global patterns of rising temperature (Wheeler and Braun 2013). Speculations about a temperature increase of 2–5 °C by the end of the century or sooner (IPCC 2012) predict a reduction of the crop yield by 16 % (Cline 2007). Taking into account the recent projections of global warming, the worldwide food production needs to be increased by 70 % in order to feed the expected population of 9 billion by 2050 (Bita and Gerats 2013). Thereby, it becomes inevitable to comprehend and dissect the phenomenon of thermotolerance and develop crop plants that are resistant to even higher temperatures. Due to their sessile nature, plants are frequently exposed to abiotic stresses such as high temperatures, salinity, or drought, which often occur simultaneously. These stresses usually lead to a disruption of cellular homeostasis and in acute conditions may even lead to death (Mittler 2006).

Injuries by high temperature range from direct effects like denaturation and aggregation of proteins or increased membrane fluidity to indirect effects via inactivation of chloroplast or mitochondrial enzymes or even disruption of microtubule organization (Howarth 2005). Consequently, plants adjust their physiological and biochemical activities by altering their transcriptome, proteome, metabolome, and lipidome to counter heat stress injuries. The cellular adjustments aim to maintain protein homeostasis, protect structural components, and reduce the deleterious effects of some highly accumulating molecules such as reactive oxygen species (ROS). Therefore, plants respond to heat stress by promptly activating unfolded protein response (UPR) mechanisms which in turn induce several processes ranging from the recruitment of heat shock proteins (HSPs) to the induction of antioxidant defense systems and enzymes such as catalases and superoxide dismutases and the synthesis of osmoprotectants like proline and sugar alcohols (Liu and Huang 2000; Moreno and Orellana 2011). The inherent ability of plants to withstand temperatures above optimal is known as basal thermotolerance (Vierling 1991). Additionally, plants can also survive lethal temperatures if they are previously acclimated in sublethal conditions, known as acquired thermotolerance. In several studies, factors involved in acquired thermotolerance are not required for basal thermotolerance and vice versa, indicating that basic features of the two types of thermotolerance are similar but major differences also exist (for review, see Bokszczanin et al. 2013).

2.2 Morphological and Physiological Changes in Response to Heat Stress

Heat stress dramatically affects many aspects of plant growth and development. While developing seeds are delayed in germination and lose their vigor (Wahid et al. 2007), vegetative growth is halted resulting in reduced internode length and overall shoot dry mass (Ashraf and Hafeez 2004). To sustain a certain water level, plants close their stomata at the cost of reduced CO_2 levels which in turn affect the rate of photosynthesis (Bañon et al. 2004). The loss of photosynthetic activity during HS has been attributed to other factors as well, including structural reorganization of thylakoids (Karim et al. 1997). Grape plants exposed to high temperatures have more rounded chloroplasts, swollen lamellae, disrupted cristae, and empty mitochondria, suggesting that major anatomical changes affect survival by posing major impact on photosynthetic and respiratory activities (Zhang et al. 2005).

The sensitivity to different temperature regimes varies among different species, and discrepancies can be found within a species in different developmental stages or even within the same stage, among different organs. Research on heat stress response has been mainly focused on vegetative tissues such as leaves, although it has long been discussed that reproductive organs are more sensitive to temperature fluctuations (Giorno et al. 2013). Since yield and reproductive development strongly correlate in many crop plants, the sexual phase is considered to be among the most critical phases of a plant's life in regard to heat stress susceptibility and thus should be of major interest. The effects of elevated temperatures on flower development range from asynchrony between male and female development to altered structures and functions of the respective reproductive organs. For example, the receptivity of the stigma is shortened which in turn hampers the successful self-pollination (Kakani et al. 2005; Zinn et al. 2010). Furthermore, heat stress might lead to a reduced number and size of floral organs (Takeoka et al. 1991; Morrison and Stewart 2002), induce flower abortion, and cause defects in the structure and function of floral parts such as corolla, stigma, and stamens (Zinn et al. 2010).

In principle, the onset of flowering is controlled by environmental signals such as light and temperature (Blázquez et al. 2003). However, as a consequence of global warming, an earlier onset of flowering for mild and higher latitudes and an extension of the growing season have already been recorded (Menzel et al. 2006). Moreover, unfavorable conditions can also accelerate the transition from the vegetative to the reproductive stage, thereby disrupting fruit setting due to reduced availability of adequate resources (Zinn et al. 2010).

2.2.1 Sexual Reproduction under Heat Stress

Heat stress can damage both the male and female organs to different extents, although in general the male reproductive tissues are considered the weakest in regard to heat stress (Zinn et al. 2010). Female organs can be significantly affected even by small temperature changes. A 3 °C increase causes abnormalities in ovaries and the development of a shorter style in apricot (Rodrigo and Herrero 2002). Similarly, suppression of embryo sac development has been observed for peach (Kozai et al. 2004). In addition, the receptivity of the stigma might be significantly shortened by increased temperatures (Hedhly et al. 2003, 2005). Thus, despite the higher tolerance of the female organs, the effects of heat stress on female sporophyte and gametophyte need to be studied in detail.

Regarding the male reproductive organs, heat stress affects the quantity and morphology of pollen, anther dehiscence, architecture of pollen cell wall, and chemical composition and metabolism of pollen (Hedhly et al. 2009; Giorno et al. 2013). Moreover, elevated temperatures can result in an abnormal development of the external layers of anther walls such as epidermis, endothecium, stomium, and septum causing pollen sterility, although in some cases pollen remains viable (Sato et al. 2002). Since the tapetum is actively involved in microsporogenesis by providing nutrients to the developing microspores and by secreting the enzyme mixture for the release of microspores from tetrads (Scott et al. 2004), defects such as premature or delayed degradation, abnormal nuclear behavior, or its malfunction can cause male sterility (Ahmed et al. 1992). In agreement with these observations, heat stress during the meiotic phase of pollen development can cause the premature degradation of tapetum, which ultimately leads to male sterility (Sakata and Higashitani 2008).

2.2.2 Effect of Heat Stress on Photosynthesis

The photosynthetic apparatus is sensitive to high temperatures and the reduction of photosynthetic activity is considered to be a major factor of plant sensitivity to heat (Wise et al. 2004). Exposure to elevated temperatures causes structural alterations in chloroplast protein complexes and reduction in the activity of several enzymes related to photosynthetic machinery (Barnabas et al. 2008). Even under moderate temperatures, the permeability of the thylakoid membrane is altered and the activity of PSI is stimulated (Schrader et al. 2004). This results in the deactivation of Rubisco to maintain the thylakoid energy gradient and to extenuate the accumulation of photorespiratory intermediates (Schrader et al. 2004). The decline in photosynthesis also negatively affects the reproductive tissue development and successful fruit and seed setting, due to reduction in resource availability and energy reserves (Young et al. 2004).

Interestingly, the thermal stability of Rubisco is quite high and the rate of carboxylation increases with increasing temperatures (Salvucci and Crafts-Brandner 2004). Under such conditions, photorespiration is also increased in C3 plants, which eventually leads to a decrease in net photosynthetic activity (Ainsworth and Ort 2010). The reduction in the overall photosynthetic activity is mainly attributed to the weak activation of Rubisco and its poor discrimination rate between CO_2 and oxygen due to the increased solubility of the latter (Ainsworth and Ort 2010).

Ribulose-1,5-bisphosphate (RuBP) regeneration is the rate-limiting step for photosynthesis at higher temperatures and can be targeted to improve thermotolerance (Sage and Kubien 2007). One approach aims at the sedoheptulose-1,7-bisphosphatase (SBPase), a key enzyme in the regenerative branch of the Calvin cycle where carbon is committed to RuBP regeneration instead of starch synthesis (Raines et al. 2000). Feng et al. (2007) have shown that the overexpression of SBPase in rice plants increased CO_2 assimilation and enhanced tolerance to high temperatures in young seedlings by maintaining the activity of Rubisco.

The thermal inactivation of Rubisco is primarily attributed to the temperature-dependent inactivation of Rubisco activase (Portis 2003; Spreitzer and Salvucci 2002). Therefore, the activation state of the Rubisco and the CO_2 assimilation rate decrease in concert, once the leaf temperature exceeds the optimum for photosynthesis. Since under heat stress the rate of Rubisco deactivation exceeds the capacity of Rubisco activase to promote activation, one approach for enhanced thermotolerance could be a higher ratio of Rubisco activase to Rubisco (Ainsworth and Ort 2010). The overexpression of Rubisco activase from maize in the wild-type background of rice resulted in a slightly increased thermal stability of photosynthesis, although both enzymes from maize and rice have similar temperature optima (Yamori et al. 2012). This indicates that an increased ratio of Rubisco activase to Rubisco might be related to the increased thermotolerance.

The development of a Rubisco activase with higher thermostability is a major target for production of increased thermotolerance. In *Arabidopsis*, the exchange of the native Rubisco activase with a more heat-tolerant enzyme increased seed yield under moderate heat stress (Kurek et al. 2007; Kumar et al. 2009). However, it remains unclear whether this approach requires the complete exchange of endogenous enzyme by the heat-tolerant enzyme or whether the addition of the latter in the wild-type background is also beneficial (Parry et al. 2013). Interestingly, it has been shown that Rubisco activase associates with the chloroplast GroEL cpn60β, indicating that this chaperonin acts as a protector of the heat-sensitive Rubisco activase and provides an acclimation mechanism to high-temperature stress (Salvucci 2008).

2.3 The Molecular Basis of the Heat Stress Response

2.3.1 Heat Sensing and Signaling

Survival under elevated temperatures primarily depends on the function of "thermosensors" that can detect temperature changes and trigger downstream signaling routes (Ruelland and Zachowski 2010). Plants have evolved several temperature-sensitive machineries acting as thermostats which are mainly related to central cellular heat responses such as protein homeostasis and changes in membrane fluidity (Mittler et al. 2012). These "sensors" exhibit a differential sensitivity to temperature fluctuations but play significant roles for heat stress responses and thermotolerance (McClung and Davis 2010). An overview of different sensing and signaling mechanisms is given in the following section (Fig. 2.1).

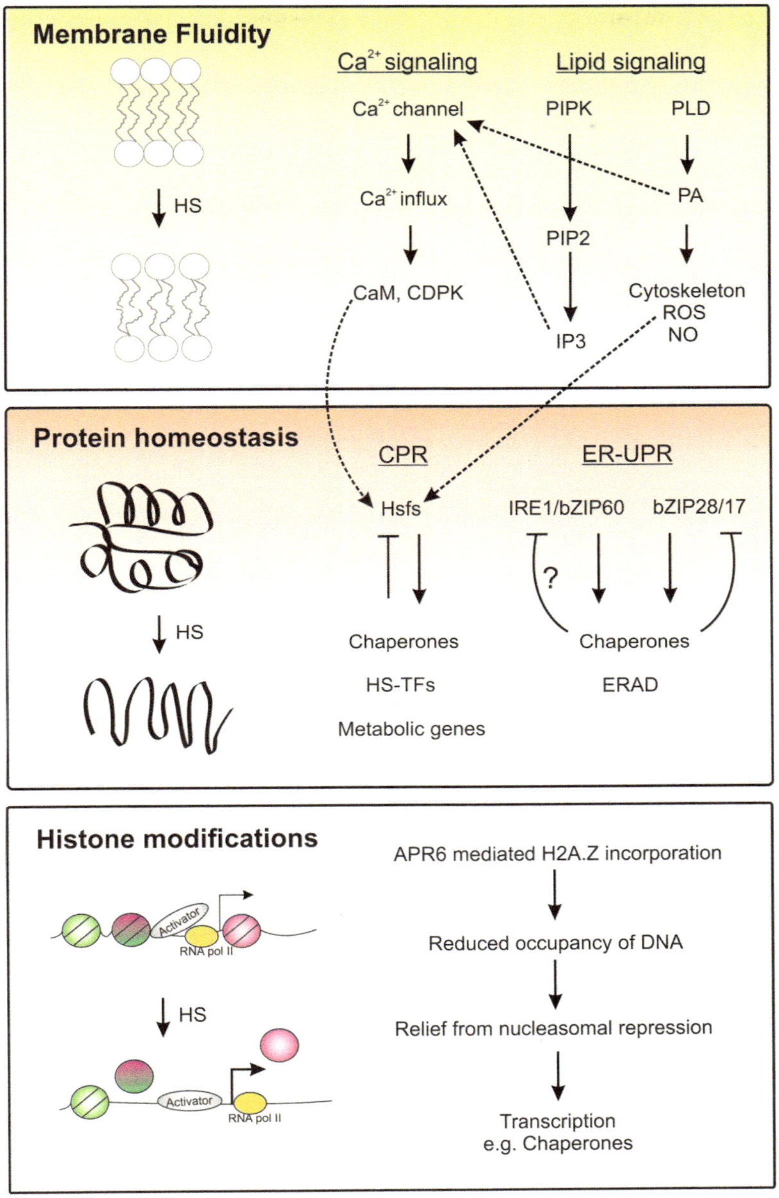

Fig. 2.1 Heat stress sensing mechanisms in plants. Changes in membrane fluidity activate calcium channels and trigger the transient influx of Ca^{2+}. Calcium binds to calmodulins (CaM) and activates kinases like CDPK. Changes in membrane fluidity affect lipid signaling as well. Heat stress-activated enzymes like PIPK and PLD trigger the synthesis of PIP2 and PA, respectively. In a subsequent step, PIP2 is converted to IP_3. PA is thought to affect cytoskeleton organization and mediate ROS and NO accumulation and signaling. PA and IP_3 can also influence the intracellular Ca^{2+} increase. Disturbance of protein homeostasis under high temperatures leads to the accumulation of misfolded proteins which triggers the cytosolic (CPR) and ER-UPR systems. CPR is regulated by Hsfs which control the transcription of molecular chaperones, as well as other heat stress-inducible TFs and genes encoding proteins involved in metabolic pathways. In a similar manner, nonconventional mRNA splicing of bZIP60 by IRE1 and proteolytic processing of bZIP28 and bZIP17 activate ER-UPR, leading to synthesis of ER-destined chaperones as well as the activation of ER-associated degradation (ERAD) mechanism, which is responsible for protein quality control and clearance in ER. Chaperones control the activity of Hsfs, bZIP28/17, and maybe IRE1 by direct protein-protein interactions. Changes of temperature can be sensed in the nucleus by alterations in the occupancy of H2A.Z histone (*pink*) which allows the transcriptional activation of heat stress-induced genes, like chaperones at higher temperatures, as proposed by Kumar and Wigge (2010)

2.3.1.1 Plasma Membrane as the Primary Sensor of Heat

The physicochemical properties and composition of cellular membranes are influenced by temperature fluctuations. During heat stress, the fluidity of all cellular membranes increases which in turn affects the activity of numerous membrane-associated proteins. The involvement of membrane fluidity in acclimation to higher or lower temperatures has been demonstrated in *Arabidopsis thaliana* and *Atriplex lentiformis* (Welti et al. 2002; Pearcy 1978). Moreover, Raison et al. (1982) have shown that the membrane fluidity of oleander (*Nerium oleander*) adapts to the temperatures in which the plants were grown. Membranes of plants grown at lower temperatures are more fluid at these temperatures than membranes of plants grown at higher temperatures. In turn, "high-temperature" plants have more rigid membranes at high temperatures than "low-temperature" plants. This exemplifies the reason why membrane signaling is an important part of stress sensing, as the fluidity of the membranes adapts to the ambient temperatures and senses small temperature fluctuations. While an effect of heat stress on the membrane and subsequent gene expression has been clearly seen, a membrane-associated or integral heat stress sensor leading to the downstream activation of target genes has to be identified yet (Los et al. 2013).

Studies with the moss *Physcomitrella patens* clearly demonstrate that the plasma membrane can sense even mild increase in temperature which eventually leads to a transient opening of a calcium channel to stimulate Ca^{2+} influx (Saidi et al. 2009). The signal of changing Ca^{2+} concentrations is perceived and further transduced by Ca^{2+}-binding proteins. In plants, the most prominent Ca^{2+}-binding proteins are calcineurin B-like proteins, calmodulin (CaM), calmodulin-like proteins (CML), and Ca^{2+}-dependent protein kinases (CDPKs). CDPKs act as a direct effector by phosphorylation of target proteins. CaM, CML, and CBLs on the other hand undergo structural changes when bound to Ca^{2+} which then leads to interaction with their specific targets (Snedden and Fromm 2001). In wheat, as a quick response to heat stress, calcium stimulates the expression of calmodulin (Liu et al. 2003). Upon interaction, calmodulin activates heat stress transcription factors (Hsfs), which can prime the expression of chaperones such as Hsp70 and Hsp26 (Liu et al. 2003; Li et al. 2004). In *Arabidopsis*, the calmodulin-binding protein kinase 3 and calmodulin-binding protein phosphatase 7 are involved in heat stress response as antagonistic linkers between calmodulin and Hsfs (Liu et al. 2007, 2008). Interestingly, preincubation of plants with Ca^{2+} chelators increases the sensitivity to heat stress, while pretreatment with Ca^{2+} enhances their thermotolerance (Gong et al. 1998).

Another membrane-associated signaling pathway involves the lipids phosphatidylinositol 4,5-bisphosphate (PIP2) and phosphatidic acid (PA). Mishkind et al. (2009) have demonstrated that both lipids are increased under heat stress conditions in tobacco cells. PIP2 is produced by activation of a phosphatidylinositol phosphate kinase, while PA levels increase due to reduced turnover and enhanced phospholipase D activity. The signal for activation of both enzymes is dependent on proper functioning of G-protein-based signaling. Interestingly, they could also show that PIP2 has different localizations during heat stress, ranging from the plasma membrane at the onset of stress to cytosolic, nuclear, and nucleolar localization in later stages, hinting at a regulatory role in gene transcription. This function of PIP2 is currently better understood in animal and yeast cells, where nuclear PI kinases have been identified that co-localize with PIP2 in so-called interchromatin granule clusters (IGCs) (Bunce et al. 2006). These clusters are processing sites of pre-mRNA, and it is proposed that PIP2 is either involved in direct regulation of the cytoskeleton or in the formation of the core of the IGC-localized spliceosome (Osborne et al. 2001). Alternatively, the processing of PIP2 by phospholipase C to DAG and IP_3 could act as a generation step for secondary messengers (Bunce et al. 2006). Nevertheless, several functions of PIP2 were also shown in plants, e.g., the regulation of potassium channels (Liu et al. 2005), implication in cytokinesis of plant cells, and organization of the cytoskeleton (van Leeuwen et al. 2007). Interestingly, certain lipid signaling molecules such as IP_3 can trigger calcium influx by stimulating channel opening, suggesting a cross talk between the two main plasma membrane sensing mechanisms (Mittler et al. 2012; Balogh et al. 2013).

2.3.1.2 Protein Homeostasis and Unfolded Protein Response

Heat Stress Transcription Factors and Cytosolic Unfolded Protein Response

Heat stress has a detrimental influence on protein stability and homeostasis, since high temperatures lead to denaturation or even irreversible aggregation of proteins. All organisms have evolved delicate signaling mechanisms which sense the amount of unfolded proteins and trigger the synthesis of molecular chaperones that maintain denatured proteins in a folding-competent state, protecting them from terminal aggregation. In plants, similar to other organisms, the synthesis of chaperones at the transcriptional level depends on the activity of heat stress-activated Hsfs. These Hsfs bind to conserved *cis-regulatory* elements called heat stress elements (HSEs) which can be found in the promoter region of many heat stress-inducible genes (Scharf et al. 2001). Remarkably, compared to other organisms, plants have a highly complex Hsf system with members ranging from 21 in *Arabidopsis* to 52 in soybean (Scharf et al. 2012). Interestingly, some Hsfs are constitutively expressed even under non-stress conditions, during which they remain in an inactive state through interaction with high molecular weight chaperones such as Hsp70s and Hsp90s (Hahn et al. 2011). This implies that protein homeostasis in the cytosol is finely controlled by a feedback mechanism involving Hsfs and Hsps through transcriptional regulation and protein-protein interactions.

Apart from the difference in number of Hsfs in different plant species, discrepancies in several aspects of major regulatory mechanisms have also been detected. Tomato HsfA1a has been defined as a master regulator and the presence of active HsfA1a is an absolute requirement for the induction of the expression of other Hsfs and Hsps. In turn, the induced Hsfs and Hsps are required for heat stress response and thermotolerance (Mishra et al. 2002). On the other hand, there is no drastic effect caused by the deletion of any of the four *Arabidopsis* A1 Hsfs. However, the triple knockout of HsfA1a/b/d shows a highly heat-sensitive phenotype, while the quadruple knockout including HsfA1e shows developmental retardation as well. This implies that in *Arabidopsis* the class A1 Hsfs have overlapping functions and the function of a master regulator cannot be attributed to a single Hsf (Liu et al. 2011; Yoshida et al. 2011).

In tomato, the interaction of HsfA1a with other Hsfs like HsfA2 or HsfB1 modulates its activity in a positive manner by formation of superactivator or histone acetyltransferase (HAC)-dependent co-activator complexes, respectively (Bharti et al. 2004; Chan-Schaminet et al. 2009). Apart from their implication in the heat stress response, Hsfs have been related to regulatory pathways of other abiotic stresses such as oxidative and salinity stresses (Liu et al. 2011; Nishizawa et al. 2006). Even more, their transcript profiles indicate the involvement of specific Hsfs in developmental processes such as pollen and seed development (Frank et al. 2009; Giorno et al. 2010; Kotak et al. 2007). This is of particular interest, considering that different organs or even developmental stages often exhibit different tolerance to heat stress and this might be related to the temporal and spatial expression of certain Hsfs.

The abundance, localization, and activity of Hsfs are subjected to several levels of control including transcriptional and posttranslational regulation. For example, the levels of some Hsfs are controlled by degradation via proteasome, while posttranscriptional modifications including phosphorylation, sumoylation, and acetylation govern their state of activity (Cohen-Peer et al. 2010; Hahn et al. 2011).

Recently alternative splicing of Hsf transcripts has been identified as a new regulatory mechanism during heat stress (Liu et al. 2013). Differential splicing in *Arabidopsis* was found for five Hsfs, including A2, A4c, A7b, B1, and B2a (Sugio et al. 2009; Liu et al. 2013), although this might be a process also valid for other Hsfs. The significance of these splicing events for the function of Hsfs is not clear yet. However, alternative splicing of *Arabidopsis* HsfA2 specifically at very high temperatures results in the generation of an isoform with a truncated DNA-binding domain and a C-terminally fused leucine-rich hydrophobic motif (Liu et al. 2013). Interestingly, the truncated Hsf is still able to mediate Hsf-like transcriptional activity, especially on the HsfA2 promoter, indicating a possible autoregulatory feedback loop. Another splicing isoform encoding for a truncated HsfA2 but lacking the hydrophobic C-terminus extension is subjected to nonsense-mediated mRNA decay (NMD; Sugio et al. 2009). Remarkably, rice OsHsfA2a has six transcript variants which show differential expression in different tissues under control conditions but also in response to different abiotic stresses, indicating distinctive roles for individual isoforms (Wang et al. 2013).

An additional Hsf regulating heat sensing mechanism was discovered in tobacco. The 5'UTR of the HsfB1 transcript includes two mini open reading frames (μORFs), the second of which represses the translation of the major ORF. Mutation of its start ATG led to translation of the transcript, as did heat stress treatment (Zhu et al. 2012). However, whether this upstream ORF is expressed as a peptide which suppresses HsfB1 translation or whether the mRNA sequence at this position forms specific secondary structures repressing the translation of the main ORF is not clear.

Endoplasmic Reticulum UPR

Heat stress affects the protein homeostasis in all cellular compartments. Therefore, similar to cytosolic protein response, mechanisms to prevent the accumulation of aggregated proteins at toxic levels either by protection of denaturated proteins or by clearance of protein aggregates have evolved in all organelles. Among these, the unfolded protein response of the endoplasmic reticulum is the best studied, considering the importance of the ER for protein synthesis and trafficking. In plants, exposure to elevated temperatures triggers two different pathways in the ER involving the activation of specific bZIP transcription factors (TFs) through nonconventional mRNA splicing and proteolytic cleavage which further leads to the transcriptional induction of ER-destined chaperones and the activation of endoplasmic reticulum (ER)-associated protein degradation (ERAD) mechanism for removal of terminally misfolded proteins (Howell 2013).

The inositol-requiring enzyme 1 (IRE1) is located in the ER membrane, with its catalytically active C-terminus facing the cytosol. IRE1 has a dual function as serine/threonine protein kinase and endonuclease (Sidrauski and Walter 1997). The kinase activity is important for the activation of the dimer's endonuclease activity. Upon heat stress, the mRNA of *Arabidopsis* bZIP60 is spliced, causing a frame shift in the coding sequence. The isoform has a substituted C-terminal portion which contains a nuclear localization sequence (Deng et al. 2011). By this, bZIP60 enters the nucleus and activates the transcription of ER-resident chaperones such as the Hsp70 homologue "binding protein 3" (BIP3) (Deng et al. 2011).

The second mechanism by which higher temperatures are perceived and processed by UPR utilizes the proteolytic processing of two ER-resident transmembrane TFs. In *Arabidopsis*, bZIP17 and bZIP28 are both involved in the regulation of ER and the heat stress response (Che et al. 2010). Upon heat stress, both TFs are relocated to the Golgi, where they are processed by serine proteases. The cytosol-facing domain, which comprises the DNA-binding domain and the transcriptional activator domain, is released and directed to the nucleus (Liu and Howell 2010). Interestingly, the maintenance of inactive bZIP17 and bZIP28 attached to ER membrane is regulated by interactions with ER chaperones in the lumen. Furthermore, the activation of bZIPs requires their release from these complexes which is stimulated by the accumulation of unfolded proteins (Srivastava et al. 2013). The activation of bZIPs by their release from chaperones highly resembles the cytosolic Hsf-Hsp system, highlighting the importance of the accumulation of unfolded proteins as an additional sensing mechanism of heat stress.

2.3.1.3 Nuclear Histone Modification as Sensor of Temperature Fluctuations

A heat sensing mechanism close to the transcriptional machinery is the modification of histones to achieve chromatin remodeling. Generally, chromosomal DNA is wrapped around heterooctameric histones (each twice: H2A, H2B, H3, and H4) and is thereby more or less inaccessible for transcription. After stimulation, covalent histone modifications are changed, e.g., by acetylation, phosphorylation, or demethylation, which leads to changes in the DNA packaging. In a subsequent step, chromatin-remodeling ATPases can move the histone octamer or substitute single components, while the compact nucleosomal structure is released and the DNA becomes accessible for components of the transcription machinery (Pfluger and Wagner 2007).

Several chromatin modification and chromatin-remodeling proteins involved in abiotic stress responses have been identified, although for heat stress the knowledge is restricted to two genes (Kumar and Wigge 2010; Luo et al. 2012). In a screening of *Arabidopsis* mutants impaired in heat sensing, the chromatin-remodeling ATPase APR6 was identified as a putative thermal sensor that is required for the incorporation of H2A.Z instead of H2A into histones (Mizuguchi et al. 2004; Kumar and Wigge 2010). The preferential insertion of H2A.Z results in more tightly packed nucleosomes, thereby prohibiting transcription. However, with rising temperatures, H2A.Z occupancy of the DNA decreases, pointing to H2A.Z as the thermal sensor in chromatin remodeling (Kumar and Wigge 2010). However, it is not clear yet if this mechanism applies only for temperatures considered as warming or also for high temperatures that are qualified as heat stress (Mittler et al. 2012).

Another chromatin-remodeling ATPase involved in heat stress regulation in plants has been identified in *Arabidopsis* as AtCHR12. Overexpression of this ATPase leads to a more pronounced growth arrest of plants during both heat and drought stress (Mlynarowa et al. 2007). Genes that were identified as targets for CHR12 are mainly signal transducers, TFs, and dormancy-/auxin-associated genes. The effect of growth retardation upon stress, which also occurs in wild type, is thought to be beneficial for plants, e.g., for redirection of resources (Achard et al. 2006).

2.3.2 Reactive Oxygen Species (ROS) Signaling

Reactive oxygen species, such as singlet oxygen ($^1O_2$), superoxide anion (O_2^-), hydrogen peroxide (H_2O_2), and hydroxyl radicals (HO), can act as signaling molecules in low amounts but are toxic at higher concentrations. Virtually any major component of the cell, i.e., lipids, proteins, and DNA, can be damaged by excess amounts of ROS (Gill and Tuteja 2010) which is why plants have evolved a delicate regulatory network for balancing ROS levels by production and scavenging (Mittler et al. 2004). This balancing of ROS levels enables their function as general stress signaling molecules, since most of them are generated due to an imbalance in metabolic pathways, a common side effect of various stresses. Plant ROS production takes place in nearly any compartment, most notably though in chloroplasts and peroxisomes (Asada and Takahashi 1987).

A role of ROS in the acquired thermotolerance was shown by enhancing cellular mechanisms that prevent oxidative damage under heat stress (Bergmüller et al. 2003; Larkindale and Huang 2004). Heat pretreatment in cool-season turfgrass was accompanied by an increased synthesis of ascorbate and glutathione, i.e., ROS scavengers, and an overall lower production of ROS (Xu et al. 2006). In the same experiment, cell membrane thermostability was higher in heat-acclimated samples compared to the non-acclimated samples.

In regard to heat stress, recent studies have identified Hsfs as potential targets for redox regulation by ROS in *Arabidopsis*. Apparently, the oxidation of Hsfs leads to a structural change which exposes the nuclear localization sequence and enables translocation into the nucleus. This was found for HsfA1d upon high-light treatment, as well as for HsfA8 upon H_2O_2 treatment (Miller and Mittler 2006; Jung et al. 2013; Dietz 2014). The existence of a cross talk between oxidative and heat stress signaling is supported in *Arabidopsis* plants overexpressing HsfA2, which showed increased levels of thermotolerance as well as tolerance to oxidative stress (Zhang et al. 2009). In contrast, the expression of ascorbate peroxidase APX2 was reduced in HsfA2 knock-out plants. A dominant negative mutation in *Arabidopsis* HsfA4a negatively affected the response of plants to oxidative stress and decreased the transcript levels of APX1 (Shim et al. 2009). Furthermore, in rice, a possible cross interaction between the ROS and nitric oxide signaling pathways was postulated that influenced the expression of at least one heat stress protein, Hsp26 (Uchida et al. 2002).

2.3.3 Nitric Oxide (NO) Signaling

Due to its chemical properties as a gas that is both water and lipid soluble, nitric oxide functions in a concentration-dependent manner as a major signaling molecule in various biotic and abiotic stresses (Wendehenne et al. 2001; Neill et al. 2003) or as a cytotoxic agent (Beligni and Lamattina 1999). The involvement of NO signaling in plant thermotolerance has been nicely shown in *Arabidopsis* heat stress-sensitive mutant (*hot5*), which encodes for an S-nitrosoglutathione reductase (GSNOR) (Lee et al. 2008). Studies in the unicellular green algae *Chlamydomonas reinhardtii* suggested that NO is important for signaling during heat stress, since photosynthesis is restricted under these conditions. This leads to a relay of excess NAD(P)H acting with a nitrate reductase to produce higher amounts of NO (Sakihama et al. 2002; Gould et al. 2003). In vitro experiments have shown that this reaction can also be carried out by plant nitrate reductase (Yamasaki and Sakihama 2000). Since several studies in different plants have identified elevated NO levels under both abiotic and biotic stresses (Gould et al. 2003), NO is thought to be a general stress signal molecule. However, it should be noted that opposed to the normal synthesis via the nitric oxide synthase, this pathway creates excessive NO as a by-product (Sakihama et al. 2002). Alternatively, in the presence of superoxide radicals, the reactive peroxynitrite can be generated as well, although it is thought to be less damaging than superoxide radicals and might be considered as an ROS-scavenging molecule (Radi et al. 1991; Wink et al. 1993).

In its role as signaling molecule, NO has actually different modes of action. Willmott et al. (1996) have shown that NO is an activator for calcium channels of animal cells, which has been proposed for plants as well, since excess NO leads to elevated concentrations of Ca^{2+} in the cytoplasm of tobacco (Gould et al. 2003). Exemplary other second messengers which are induced by NO are MAP kinases or cyclic guanosine monophosphate (cGMP). The induction of the latter probably occurs due to a rise in guanylate cyclase (GC) levels (Isner and Maathuis 2011), although no GC has been identified in plants so far (Gaupels et al. 2011).

In recent years, the involvement of NO in the general expression of several target genes has begun to be uncovered (Grün et al. 2006; Astier et al. 2011; Mengel et al. 2013). The mode of regulation was found to be S-nitrosylation of nuclear proteins, most notably TFs, e.g., MYB-TFs, NPR1, and TGA1, as well as histone deacetylases. Since several stress TFs are only upregulated in the respective stress, it remains unclear whether these might be modified by NO as well during the respective stress situation. Piterková et al. (2013) showed a more direct influence of both NO and H_2O_2 on the accumulation of the chaperone Hsp70 in tomato, both under control and heat stress conditions. Nevertheless, from the study, it remains unclear whether this regulation targets the expression on either the transcriptional or translational level or whether the turnover of Hsp70 is influenced by the reagents.

2.3.4 Hormone Signaling

Hormone biosynthesis and signaling are important components of heat stress response. Changes in hormone homeostasis including alterations in their stability, content, synthesis, and compartmentalization have been correlated with changes in heat stress tolerance (Maestri et al. 2002). The levels of ethylene, abscisic acid (ABA), and salicylic acid (SA) increase in response to high temperatures, while the levels of auxin, cytokinin, and gibberellins decline (Talanova et al. 2003; Larkindale and Huang 2004; Larkindale et al. 2005; Liu and Huang 2002). The involvement of the differential hormone levels in regulating the stress response has been shown in *Arabidopsis*. For example, the heat stress-induced oxidative damage is mitigated by exogenous application of the ethylene precursor 1-aminocyclopropane-1-carboxylic acid, SA, and ABA (Larkindale and Knight 2002). Furthermore, application of the synthetic cytokinin, zeatin riboside, reduces heat stress injury of creeping bent grass, indicating that cytokinin is also implicated in heat stress tolerance (Liu et al. 2002). In addition, the role of different hormones in heat stress response has been shown in studies with mutant lines. The ethylene-insensitive mutant *etr-1*, the ABA-insensitive mutant *abi-1*, and an SA-deficient transgenic line expressing *nahG* showed increased sensitivity to high temperatures (Larkindale and Knight 2002). ABA is related to cellular osmotic adjustment under stress conditions and is mediator of intracellular dehydration signaling pathways. ABA induction is especially important under conditions of both heat and drought stresses which are often simultaneously present in the field (Maestri et al. 2002).

Several lines of evidence suggest that hormone signaling has a major role in thermotolerance of reproductive organs. The transcription profile of many genes related to hormone biosynthesis and signaling is altered in heat-stressed pollen cells (Frank et al. 2009). In tomato, the ethylene-insensitive mutant Never ripe (Nr) is more sensitive to heat stress, while pollen quality increased when plants were pretreated with an ethylene releaser, suggesting that ethylene is an important hormone for pollen thermotolerance (Firon et al. 2012). Moreover, rice anthers exposed to high temperatures had decreased IAA and GA levels but increased levels of ABA (Tang et al. 2008). In another study, heat stress-induced male sterility in *Arabidopsis* and barley was reversed by application of auxin (Sakata et al. 2010). Similarly, the application of brassinosteroids on germination medium increased pollen germination under heat stress (Singh and Shono 2005). Despite their importance, the role of hormones on heat stress response and thermotolerance is not well understood and requires further investigation.

2.4 Metabolic Responses to Heat Stress

The abundance of several metabolites is affected by heat stress through modulation of their biosynthesis and stability. The vast information derived from metabolite-targeted studies, exploring the role of specific metabolites, are further enriched by nontargeted studies, generating valuable data that boosts our understanding of the heat stress response and reveals potential targets that can be used for crop improvement (Kaplan et al. 2004; Rizhsky et al. 2004; Arbona et al. 2013).

Plants respond to heat stress mainly by producing metabolites with osmoprotecting and/or antioxidant functions (Fig. 2.2). Compatible solutes such as amino acids, quaternary amines, and polyol/sugars accumulate under stress conditions, which serve primarily to protect the cell by different actions. These include the maintenance of the cell turgor, the increase of the cellular antioxidant capacity, or the stabilization of membranes or proteins as chemical chaperones (Hare et al. 1998; McNeil et al. 1999; Diamant et al. 2001). The accumulation of osmolytes acts not only as an adjustment system for the intracellular osmotic potential but also to protect protein structures, maintain higher energy status, and make the cell environment more reductive (Grover et al. 2013).

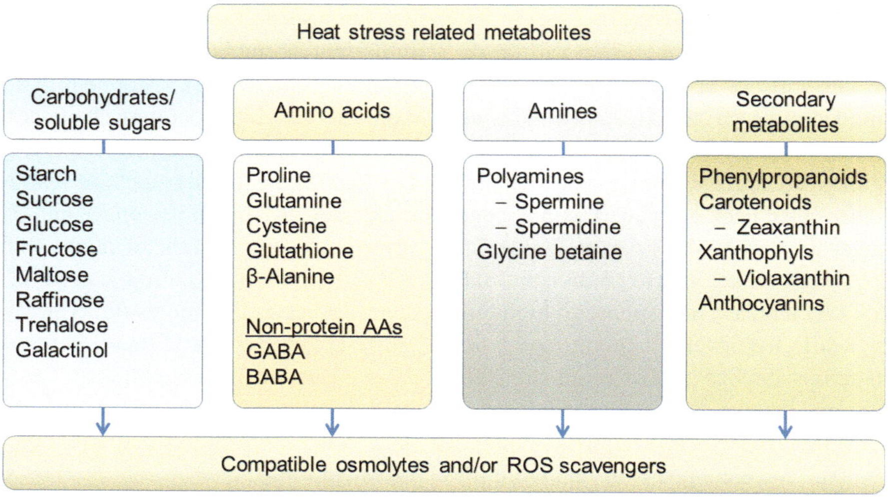

Fig. 2.2 Metabolites that have been associated with plant tolerance to heat stress, mainly due to their function as compatible osmolytes and/or ROS scavengers

2.4.1 Soluble Sugars

Soluble sugars do not only function as metabolic resources and structural constituents of cells but also as signals regulating various processes associated with plant growth and development, including the protection of cellular membranes (Valluru and Van den Ende 2008). High-temperature stress affects the photosynthetic activity in source tissues, subsequently reducing the supply of soluble sugars to sink tissues. Under conditions of low sugar availability, physiological and biochemical changes occur to sustain respiration and other metabolic processes.

The availability of carbohydrates is critical in specific tissues at certain developmental stages like seed germination, seedling growth, and pollen development (Pressman et al. 2002; Wahid et al. 2007). Under heat stress, the starch and soluble sugar contents are even further reduced which have been correlated in tomato with a reduced viability of heat-stressed tomato pollen grains (Firon et al. 2006). Therefore, thermotolerance is also related to a milder reduction in starch and sugar levels which can be then utilized for pollen germination (Pressmann et al. 2002; Firon et al. 2006), while similar observations have been made for sorghum and barley (Wallwork et al. 1998; Jain et al. 2007).

At high temperatures, the starch and sucrose synthesis is negatively regulated on several levels. Enzymes like the sucrose phosphate synthase, ADP-glucose pyrophosphorylase, and invertase are less active (Djianaguiramman et al. 2009; Rodriguez et al. 2005), while others like the acid invertase have already reduced mRNA levels in developing tomato anthers exposed to heat stress (Sato et al. 2006). Other sugars including maltose, nonstructural carbohydrates, starch, fructose, and raffinose are reduced in heat-stressed barley grains (Vasseur et al. 2011).

Cultivars or mutants with higher thermotolerance showed increased levels of different sugars. A thermotolerant cultivar of creeping bent grass retained higher glucose and sucrose content under stress conditions (Liu and Huang 2000). The nonreducing disaccharide, trehalose, was found to be accumulated in *Arabidopsis* plants after a 4-h heat stress treatment at 40 °C (Kaplan et al. 2004). In yeast, trehalose protected denatured proteins from aggregation by maintaining them in a partially folded state which could then be reactivated by the action of molecular chaperones (Singer and Lindquist 1998). When transgenic *Arabidopsis* and *Medicago* plants expressing the yeast homologues of trehalose-6-phosphate synthase (TPS) and trehalose-6-phosphate phosphatase (TPP) were heat stressed, both showed an increased thermotolerance (Miranda et al. 2007; Suárez et al. 2009). Moreover, Luo et al. (2008) showed that trehalose could act as a concentration-dependent ROS scavenger in heat-stressed wheat (*Triticum aestivum*). A similar function was attributed to raffinose and galactinol. *Arabidopsis* plants overexpressing galactinol synthases GolS1 and GolS2 had increased levels of both sugars and exhibited tolerance in response to oxidative stress, suggesting that these two oligosaccharides function both as osmoprotectants and scavengers of hydroxyl radicals (Nishizawa et al. 2008).

2.4.2 Amino Acids

Proline accumulates in response to different environmental cues including cold, salt, and drought and is considered as a major osmoprotectant. Regarding heat stress, proline biosynthesis is enhanced in some plants but this is not the general rule and it is rather restricted to specific crops and tissues. For example, proline levels increase in tomato (Rivero et al. 2004) and tobacco (Cvikrová et al. 2012) leaves exposed to heat stress but not in *Arabidopsis* (Rizhsky et al. 2004). In contrast, proline content drops in heat-stressed germinating wheat seeds (Song et al. 2005). In the same direction, overproduction of proline using transgenic approaches is beneficial for some plant species, such as soybean, where overexpression of *Arabidopsis* pyrroline-5-carboxylate reductase gene (*AtP5CR*) improved thermotolerance (de Ronde et al. 2000, 2004). However, *Arabidopsis* plants overexpressing Δ(1)-pyrroline-5-carboxylate synthetase 1 gene (*AtP5CS1*) exhibited decreased thermotolerance, most likely due to higher ROS production and inhibition of ABA and ethylene biosynthesis (Lv et al. 2011). Therefore, the protective role of proline is also species specific. Proline, supplied by the tapetum cells, is related to pollen viability under heat stress conditions (Sato et al. 2006). Defects in pollen development have been related with reduced proline content in heat-stressed rice and the legume *Vigna unguiculata* (Mutters et al. 1989; Tang et al. 2008).

Synthesis of other amino acids is also enhanced under stress conditions. *Arabidopsis* plants subjected to heat or drought stresses convert glutamic acid preferentially to glutamine instead of proline (Rizhsky et al. 2004). Under the same stress conditions, synthesis of cysteine, a precursor of the antioxidant glutathione, is favored, and the transcripts of several genes involved in glutathione biosynthesis are enhanced (Rizhsky et al. 2004).

β-Alanine is also expected to increase tolerance to thermal stress. Using a plastid transformation approach, Fouad and Altpeter (2009) produced transgenic tobacco plants expressing the *E. coli panD* gene. Subsequently the elevated decarboxylation of L-aspartate resulted in higher levels of β-alanine and carbon dioxide. The transgenic plants performed better under heat stress conditions, showing increased thermotolerance of photosynthesis and higher biomass production, suggesting that β-alanine serves as an important marker or candidate for genetic engineering for the development of thermotolerant crops.

2.4.3 Nonprotein Amino Acids

The osmolyte γ-aminobutyric acid (GABA) is a non-proteinogenic amino acid which accumulates in response to different stress conditions, including heat stress (Kinnersley and Turano 2000). Biosynthesis of GABA is stimulated by the activity of the calmodulin-dependent glutamate decarboxylase, which is related to the increased cellular levels of Ca^{2+}. GABA is synthesized from glutamate in the cytosol by glutamate decarboxylase and then transported to mitochondria. GABA is converted to succinate by the action of GABA transaminase and semialdehyde dehydrogenase to feed the TCA cycle (Fait et al. 2008). Apart from its action as an osmolyte, GABA also functions against oxidative damage, as shown by the inability of *Arabidopsis* T-DNA knockout mutants of the mitochondrial succinic-semialdehyde dehydrogenase (SSADH), the ultimate enzyme of the GABA biosynthesis pathway, to prevent accumulation of H_2O_2 under heat stress conditions (Bouche and Fromm 2004). Additionally, *Arabidopsis* plants expressing the xenobiotic non-proteinogenic β-aminobutyric acid (BABA) also conferred increased acquired thermotolerance in seedlings (Zimmerli et al. 2008). Therefore, non-proteinogenic amino acids such as GABA and BABA can be also considered as targets for crop improvement.

2.4.4 Amines

Polyamines, like putrescine, spermidine, and spermine, are also important constituents of abiotic stress responses in plants. Higher levels of free and conjugated polyamines as well as enhanced activity of key enzymes of polyamine biosynthetic pathways such as arginine decarboxylase and polyamine oxidase are correlated with a heat stress-tolerant rice cultivar (Roy and Ghosh 1996). The importance of polyamines for stress responses has been demonstrated in several studies by overexpression of key biosynthetic genes but also by exogenous application of specific polyamines. Polyamines can increase photosynthetic efficiency by maintaining the thermostability of thylakoid membranes under heat stress. For example, expression of yeast S-adenosyl-l-methionine decarboxylase (SAMDC) in tomato increased both spermidine and spermine content under heat stress, which led to enhanced antioxidant enzyme activity, protection of membrane lipid peroxidation, improved CO_2 assimilation, and increased tolerance to high temperature (Cheng et al. 2009). The activity of SAMDC during heat stress is important for pollen germination as well which was shown by the application of inhibitors of SAMDC which reduced the in vitro pollen germination rate even at ambient temperatures (Song et al. 2002). In contrast, addition of spermine and spermidine reversed the negative effect of the inhibitors even at high temperatures (Song et al. 1999). Similarly, overexpression of the key polyamine biosynthetic gene arginine decarboxylase (ADC) in eggplant increased its tolerance to several abiotic stresses (Prabhavathi and Rajam 2007), while overexpression of spermidine synthase increased tolerance to heat and oxidative stress in sweet potato (*Ipomoea batatas*) (Kasukabe et al. 2006). These findings support the role of polyamines in heat stress tolerance even further.

Apart from genetic approaches, the importance of polyamines in improving stress tolerance has been demonstrated by exogenous application of polyamines. Application of 4 mM spermidine increased heat tolerance in a sensitive tomato cultivar by improving chlorophyll fluorescence properties and hardening and enhancing the resistance of the pigment-protein complex structure to thermal damage, thus mitigating the decrease of PSII activity (Murkowski 2001). Application of polyamines can also serve as an agricultural practice to increase thermotolerance and extend the planting period for some crops, as for example in wheat, where foliar application of 2.5 mM putrescine allowed the delayed sowing, which otherwise had detrimental effects on plant growth due to heat stress conditions during the farming period (Mostafa et al. 2010).

Glycine betaine (GB) is an N-methyl-substituted derivative of glycine, belonging to the group of compatible osmolytes (Sakamoto and Murata 2002). The beneficial effects of GB have been shown by genetic approaches as well as through exogenous applications in plants that naturally do not accumulate GB under stress conditions. GB is thought to stabilize the quaternary structures of proteins and promote maintenance of highly ordered states of membranes under nonphysiological temperatures (Papageorgiou and Murata 1995). Accumulation of GB in chloroplasts of tobacco plants, transformed with the gene encoding for aldehyde dehydrogenase (BADH) from spinach, increased the resistance of PSII to moderate heat stress-induced photoinhibition, due to stabilization of Rubisco and reduced ROS generation (Yang et al. 2005). Similarly, *Arabidopsis* plants, expressing an *Arthrobacter globiformis* choline oxidase (*codA*) gene, accumulated GB and exhibited enhanced thermotolerance and reduced accumulation of heat shock proteins during seed imbibition and germination as well as during growth of young seedlings (Alia et al. 1998). Collectively, these results show that GB can be used as an effective biotechnological tool for improving thermotolerance in different crops.

2.4.5 Secondary Metabolites

In line with the changes in primary metabolism, heat stress alters the levels of various secondary metabolites as a means for survival. The levels of phenolic compounds such as flavonoids and phenylpropanoids are enhanced upon exposure to increased temperatures (Wahid et al. 2007). This is due to an induced activity of phenylalanine ammonia-lyase (PAL), the key enzyme of the phenylpropanoid pathway, and decreased activity of peroxidase and polyphenol oxidase in tomato and watermelon (*Citrullus lanatus*) plants (Rivero et al. 2001). The subsequent accumulation of soluble phenols suggests that enhancement of biosynthesis and reduction of oxidation of phenolic compounds is an important mechanism to heat stress acclimation.

As part of the antioxidant mechanisms activated during heat stress response, the accumulation of carotenoids, such as zeaxanthin, can reduce the peroxidative damage of membrane lipids by ROS (Wahid et al. 2007). Xanthophylls, including violaxanthin, on the other hand can increase the thermostability by interacting with membrane lipids, decrease membrane fluidity, and increase resistance to lipid peroxidation (Havaux 1998). *Arabidopsis* mutants deficient in zeaxanthin had decreased acquired and basal thermotolerance (Larkindale et al. 2005). Transgenic plants overexpressing β-carotene hydroxylase, the enzyme catalyzing the conversion of β-carotene to zeaxanthin, exhibited increased tolerance to a combination of heat and high-light stress (Davison et al. 2002).

The accumulation of flavonoids such as anthocyanins in vegetative tissues exposed to high temperature can reduce the transpirational losses by lowering leaf osmotic potential (Chalker-Scott 2002). However, the changes in anthocyanin content vary among different tissues. High temperature decreases anthocyanin content in reproductive organs, including buds and fruits, due to decreased synthesis and stability (Shaked-Sachray et al. 2002), but increases in vegetative tissues such as leaves (Wahid and Ghazanfar 2006). Other metabolites have been less studied but might also have an important role in plant thermotolerance. When the *Arabidopsis* glucosinolate mutant TU8 is exposed to high temperatures, it produces reduced levels of Hsp90 and is less thermotolerant, indicating a role of glucosinolates in heat stress response (Ludwig-Müller et al. 2000).

In conclusion, specific metabolites play important roles in heat stress response and thermotolerance both in vegetative and reproductive tissues. Metabolomics studies which allow the qualitative and quantitative determination of numerous metabolic profiles in different plants and organs under various stress conditions, in combination with forward and reverse genetics, will give more insights into the role of metabolic pathways in thermotolerance.

2.5 Approaches for Improving Plant Thermotolerance

Genetic improvement led to significant increases in yield of many agricultural crops over the last decades, but for many important crops such as rice and wheat, the ongoing climate change threatens to halt this increase (Long and Ort 2010). Considering that the capacity to expand agricultural land is low and that production per unit land area has to be doubled in order to meet the demands of rising population and dietary changes, global food safety requires the development of improved crops that will maintain the capacity for high yields and will be able to withstand and perform well under both physiological and unfavorable environmental conditions. However, improving crops is more challenging, as in nature it is often that heat stress is accompanied by other abiotic stresses such as drought.

Farmers have long been aware of desirable cultural practices to minimize adverse effects of environmental stresses on crop production. Typical examples of adjustments or modifications in farming practices are planting time, density of plantation, or soil and irrigation management (Wahid et al. 2007). Most commonly, farmers synchronize the stress-sensitive stage of plants with the most favorable time period of the season. In some crops, adjustment of the fertilization and irrigation program according to predicted environmental conditions can moderate the effects of heat stress to some extent. In wheat, continuous water supply can sustain the grain-filling rate, duration, and size under moderate stress conditions, while application of nitrogen, phosphorus, and potassium can improve growth (Dupont et al. 2006).

Preconditioning or exogenous application of chemical inducers of heat stress response such as osmoprotectants and hormones can mitigate the negative effects of high temperature on existing high-yield but temperature-sensitive cultivars. Preconditioning can trigger acquired thermotolerance and improve the osmotic adjustment and growth of tomato plants (Morales et al. 2003) or increase thermostability by decreasing damages of chloroplasts in turfgrass leaves (Xu et al. 2006). Alternatively, the survival rate and growth under heat stress conditions were promoted by preconditioning of cucumber (*Cucumis sativus*) seedlings with low and ambient doses of UVB light (Teklemariama and Blake 2003). Furthermore, pretreatment of barley seeds with glycine betaine improved stress tolerance of developed plants (Wahid and Shabbir 2005), while exogenously applied spermidine decreased the inhibition of PSII activity in heat-stressed tomato (Murkowski 2001). In another study, soaking of sugarcane (*Saccharum* sp.) nodal buds in a solution of proline and glycine betaine reduced the production of H_2O_2, enhanced the accumulation of soluble sugars, and protected the developing tissues from heat stress effects (Rasheed et al. 2011).

2.5.1 Screening for Thermotolerance

Apart from the good knowledge of heat stress physiology and biology, crop improvement requires the existence of the appropriate protocols that will allow the evaluation of tolerance or sensitivity of a large population of plants. Due to the complexity of heat stress response and the differences among various plant species, there is currently no unique protocol that can be used for all crops.

One set of protocols utilizes the measurement of the growth potential of different plant organs in response to heat stress. In *Arabidopsis*, survival rate as a measure to estimate tolerance between genotypes and different temperature regimes has been widely used in young seedlings grown on MS medium (Larkindale et al. 2005). Hypocotyl recovery or mass production is considered as a good indicator for heat tolerance in many plants regarding vegetative growth. Heat tolerance index (HTI) defined as the ratio of resumed coleoptile growth after heat stress compared to normal growth has been successfully used in sorghum (*Sorghum bicolor*), which allows the rapid and cost-effective screening of a large plant population within a relatively short period (Setimela et al. 2005).

Another rapid genotype screen for thermotolerance monitors the electrolyte leakage to estimate the membrane thermostability (Ibrahim and Quick 2001). In wheat, membrane stability and grain yield under heat stress are well correlated under stress conditions (Blum et al. 2001). Maintenance of the photosynthetic rate under stress conditions is an important factor of tolerance against heat stress, and therefore chlorophyll content has been also used as an indicator of thermotolerance as a high-throughput screening method (Reynolds et al. 1994; Shah and Paulsen 2003).

However, the evaluation of tolerance derived only from measurements of the photosynthetic rate or relative changes in growth can lead to false conclusions in cases that heat stress is expected to affect reproductive organs. Considering that male gametophyte development is the most sensitive developmental stage, pollen quality tests, including number of pollen grains produced as well as viability and germination rates, can offer important information on thermotolerance (Abdul-Baki and Stommel 1995). In crops that are destined for fruit production such as tomato, fruit setting is an indicator of thermotolerance since it is highly correlated with yield (Abdul-Baki 1991; Berry and Rafique-Uddin

1988). Similarly, tolerance is related to seed set and yield in cereals (Prasad et al. 2008). Rice harvest index, defined as the ratio of grain weight to total above-the-ground crop dry weight, has been successfully used in screening programs (Prasad et al. 2006). In wheat, maintenance of grain weight during grain-filling period under heat stress is a good indicator of thermotolerance, and some thermotolerant lines are characterized by high grain-filling rate and increased potential grain weight (Dias and Lidon 2009). In addition, grain yield is positively correlated with canopy temperature depression (CTD) and has been proposed as a trait for selection of heat-tolerant plants (Pradhan et al. 2012).

It is important to note that stress conditions applied for the characterization of mutants for biological purposes in most of the cases do not resemble natural stress conditions but challenge plant survival by the rapid application of very-high-temperature regimes, in contrast to the more gradual temperature increase occurring on a hot summer day. Ideally, for field crops, screening should be performed on field trials which represent the actual environmental conditions. Unfortunately, in field trials, environmental conditions cannot be controlled and the experimental setup is dependent on weather predictions. Moreover, the presence of other stress factors, including biotic stressors, might also influence the thermotolerance evaluation process. Such constraints can be overcome by indoor trials in greenhouses or growth chambers that allow the control of important environmental factors such as temperature, humidity, and light, which then can be properly adjusted to mimic natural field stress conditions. Most importantly, such trials can be repeated several times over the year, which allows more rapid collection of statistically sound data. Currently, understanding plant stress response and tolerance has been accelerated by the use of phenomics technologies, which offer simultaneous large data collection using high-throughput and high-resolution phenotyping tools which in many cases include many of the abovementioned screening methods (Furbank and Tester 2011).

2.5.2 Enhancing Heat Tolerance by Genetic Engineering

Heat stress is a complex phenomenon and cell survival depends on the maintenance of cellular homeostasis and the coordinated activation of numerous processes. Reverse and forward genetic approaches have revealed a plethora of genes as candidates for genetic manipulation and improvement of thermotolerance. These genes are mainly involved in protein homeostasis, photosynthesis, energy production, and antioxidant defense mechanism, while in several cases, overexpression of genes related to specific metabolic routes has been also proved to be beneficial.

The introduction or overexpression of specific genes offers many advantages over conventional breeding approaches. Plant transformation is a routine method for most of the crops, and in many cases, transgenic plants can be obtained within a relatively short period. Genetic engineering allows the introduction of genes from genetically distant sources, overcoming the problem of cross-barrier species, distant relatives, or even transformation of genes from non-plant organisms. For example, introduction of SAMDC from *S. cerevisiae* to tomato enhanced polyamine biosynthesis, antioxidant enzyme activity, and CO_2 assimilation, resulting in increased thermotolerance of the transgenic plants (Cheng et al. 2009). Similarly, expression of maize *EFTu1* in wheat reduced protein aggregation in heat-stressed leaves and injury to photosynthetic membranes, enhanced the rate of CO_2 fixation, and improved the tolerance to heat stress (Fu et al. 2008). In another approach, Ahn and Zimmermann (2006) created several transgenic potato lines expressing *Hsp17.7* from carrot in order to make the cold crop plant more heat resistant. The different potato lines showed increased thermotolerance, measured by higher membrane stability as well as improved tuberization at high temperatures.

In some cases, the introduction or overexpression of a single gene is not sufficient to trigger thermotolerance, which might require the coordinated induction of whole gene networks controlled by specific TFs. In tomato, HsfA1a is an absolute requirement for the induction of other Hsfs and many important heat stress-inducible genes and therefore is considered as master regulator (Mishra et al. 2002). Overexpression of *HsfA1a* conferred tolerance to heat stress in both vegetative organs and fruits. Furthermore, the overexpression of specific TFs can improve tolerance to different stresses. ZmDREB2A transcripts accumulate in response to cold, dehydration, salt, and heat stress in *Zea mays*, and *Arabidopsis* plants overexpressing ZmDREB2A are more tolerant to heat and drought stress (Qin et al. 2007).

A successful transgenic approach depends also on the selection of the appropriate promoter for the desired transcriptional regulation of the transgene. In the majority of studies, the overexpression of a gene is usually driven by the control of a constitutively active promoter such as cauliflower mosaic virus CaMV35S or UBI (Grover et al. 2013). However, constitutive expression of a gene might have pleiotropic and undesired effects. Boosting plant endogenous defense mechanisms by overexpression of regulatory elements can be accompanied by reduced growth and lower yield due to cross talk between overlapping stress response and developmental networks (Cabello et al. 2014). In such cases, the use of inducible promoters offers the possibility of controlling the timing of expression, tissue specificity, or even the strength of the transcript level.

Furthermore, the activity of some promoters varies significantly among different cell types. For example, the activity of the extensively used CaMV35S is very low in pollen cells (Mascarenhas and Hamilton 1992). The use of appropriately active pollen-specific promoters is suggested. Interestingly, the developmental regulation of some Hsfs and Hsps suggests that the promoter of such genes might be good candidates for combination of developmental stage-specific and heat-induced expression. For example, HsfA2 and Hsp17.4CII are both strongly induced by heat stress but are also highly expressed during the early stages of male gametophyte development and could be therefore used for such purposes (Giorno et al. 2010).

Despite the advantages of genetic engineering for crop improvement, the adoption of such a strategy for commercial purposes is still at its early phase, due to legislation which prohibits the cultivation of genetically modified (GM) crops or even marketing of their products in many countries worldwide. Even more, public concern has led to the low acceptance of GM products by the majority of consumers.

2.5.3 Breeding for Heat-Tolerant Crops

Tolerance to heat stress is an important and desired trait and several seed companies make efforts to develop thermotolerant cultivars. This is essential for crops cultivated exclusively in the field like rice, wheat, and maize but also important for field-grown vegetables, like tomato. Breeding for stress tolerance is still at its early phase due to the limited knowledge on heat stress response but also to the absence of appropriate screening tools that will allow the determination of tolerant phenotypes.

Breeders and breeding programs are nowadays supported by the burst of *-omics* technologies, which can provide valuable information and allow selection of promising candidates from a pool of transcripts, proteins, and metabolites (Langridge and Fleury 2011). These candidates can be identified and directly mapped onto the segregating population, giving information for the genetic loci of interest. The vast amount of genomic information derived from the development of high-throughput DNA sequencing technologies allows the large collections of markers, high-density genetic maps, and high-throughput genotyping strategies, facilitating the development of new, improved, and more rapid breeding methodologies (Pérez-de-Castro et al. 2012).

Identification and mapping of quantitative trait loci (QTL) for different agriculturally important and complex traits is a very popular approach in plant breeding research. In rice, a recombinant inbred line (RIL) population was screened for thermotolerance based on spikelet fertility at high temperatures (Jagadish et al.

2010). They identified eight loci mapped to different chromosomes that were associated with spikelet fertility, while QTLs from similar chromosomal positions have also been associated with tolerance to other stresses including drought, cold, and salinity, indicating that these loci are probably involved in common responsive genomic regions. Similarly, QTLs related to heat and drought tolerance have been identified in wheat (Pinto et al. 2010). In maize, Frova and Sari-Gorla (1994) found QTLs related with heat stress tolerance by screening pollen germination rate and pollen tube growth, while Yang et al. (2002) identified two QTLs that controlled grain filling under stress conditions.

Genotyping for heat stress tolerance has led to the selection of populations that are suitable for cultivation and fruit production in environments with higher temperatures (Hanson et al. 2002). The selection is based on several phenotyping factors including days to 50 % flowering, flowers per cluster, percentage of fruit set, number of fruit clusters per plant, fruits per plant, individual fruit weight, fruit yield per plant, fruit length and diameter, seeds per fruit, and percentage viable pollen grain which were measured in tomato (Ahmad et al. 2010). Therefore, a successful breeding program highly depends on the genetic material used but also on the screening methods applied due to the polygenic nature of heat stress response and thermotolerance.

2.6 Conclusions

Extreme seasonal heat is expected to have dramatic effects on agricultural productivity and food security (Battisti and Naylor 2009). Development of plants that will be able to maintain a high yield under increased temperature conditions requires the detailed understanding of the molecular and physiological mechanisms of heat stress response. Thermotolerance depends on the activation and coordination of various sensing, signaling, and regulatory pathways. This complexity suggests that more holistic approaches are required for the improvement of thermotolerance rather than focusing on single genes or routes. Approaches including *-omics* technologies offer a wealth of information and unravel new aspects of tolerance to heat stress. The use of such high-throughput platforms has advanced plant genotyping and selection for breeding. Alternatively, the development of heat-tolerant plants can be achieved by genetic engineering, but this approach has been proven useful mainly for deciphering heat stress response mechanisms and not for development of cultivars for food production. However, efforts that combine different research fields, systems, and approaches are required to develop plants with increased thermotolerance.

Acknowledgments This chapter is dedicated to Prof. Lutz Nover, a pioneer of research on plant heat stress response. We would also like to thank Enrico Schleiff and Klaus-Dieter Scharf for critical comments during writing the manuscript and providing support to P. Paul and S. Fragkostefanakis in frame of the Solanaceae Pollen Thermotolerance Initial Training Network (SPOT-ITN) consortium. S. Röth was supported by a stipendium of the Buchmann Stiftung at the Goethe University Frankfurt.

References

Abdulbaki AA (1991) Tolerance of tomato cultivars and selected germplasm to heat-stress. J Am Soc Hortic Sci 116:1113–1116

Abdulbaki AA, Stommel JR (1995) Pollen viability and fruit-set of tomato genotypes under optimum-temperature and high-temperature regimes. Hortscience 30:115–117

Achard P, Cheng H, De Grauwe L et al (2006) Integration of plant responses to environmentally activated phytohormonal signals. Science 311:91–94. doi:10.1126/science.1118642

Ahmad S, Rahman M, Quamruzzaman A, Uddin M (2010) Genetic analysis of quantitative characters of in heat tolerant tomato (*Solanum lycopersicum* L.). J Innov Dev Strateg 4:27–33

Ahmed FE, Hall AE, Demason DA (1992) Heat injury during floral development in cowpea (*Vigna unguiculata*, Fabaceae). Am J Bot 79:784–791. doi:10.2307/2444945

Ahn YJ, Zimmerman JL (2006) Introduction of the carrot HSP17.7 into potato (*Solanum tuberosum* L.) enhances cellular membrane stability and tuberization *in vitro*. Plant Cell Environ 29:95–104. doi:10.1111/j.1365-3040.2005.01403.x

Ainsworth EA, Ort DR (2010) How do we improve crop production in a warming world? Plant Physiol 154:526–530. doi:10.1104/pp. 110.161349

Alia HH, Chen THH, Murata N (1998) Transformation with a gene for choline oxidase enhances the cold tolerance of Arabidopsis during germination and early growth. Plant Cell Environ 21:232–239. doi:10.1046/j.1365-3040.1998.00264.x

Arbona V, Manzi M, de Ollas C, Gomez-Cadenas A (2013) Metabolomics as a tool to investigate abiotic stress tolerance in plants. Int J Mol Sci 14:4885–4911. doi:10.3390/Ijms14034885

Asada K, Takahashi M (1987) Production and scavenging of active oxygen in photosynthesis. In: Photoinhibition. Elsevier, Amsterdam, pp 227–287

Ashraf M, Hafeez M (2004) Thermotolerance of pearl millet and maize at early growth stages: growth and nutrient relations. Biol Plant 48:81–86. doi:10.1023/B:Biop.0000024279.44013.61

Astier J, Rasul S, Koen E et al (2011) S-nitrosylation: an emerging post-translational protein modification in plants. Plant Sci 181:527–533. doi:10.1016/j.plantsci.2011.02.011

Balogh G, Peter M, Glatz A, Gombos I, Torok Z, Horvath I, Harwood JL, Vigh L (2013) Key role of lipids in heat stress management. FEBS Lett 587:1970–1980. doi:10.1016/j.febslet.2013.05.016

Banon S, Fernandez JA, Franco JA, Torrecillas A, Alarcon JJ, Sanchez-Blanco MJ (2004) Effects of water stress and night temperature preconditioning on water relations and morphological and anatomical changes of Lotus creticus plants. Sci Hortic-Amst 101(3):333–342. doi:10.1016/j.scienta.2003.11.007

Barnabas B, Jager K, Feher A (2008) The effect of drought and heat stress on reproductive processes in cereals. Plant Cell Environ 31:11–38. doi:10.1111/j.1365-3040.2007.01727.x

Battisti DS, Naylor RL (2009) Historical warnings of future food insecurity with unprecedented seasonal heat. Science 323:240–244. doi:10.1126/science.1164363

Beligni MV, Lamattina L (1999) Nitric oxide protects against cellular damage produced by methylviologen herbicides in potato plants. Nitric Oxide 3:199–208. doi:10.1006/niox.1999.0222

Bergmuller E, Porfirova S, Dormann P (2003) Characterization of an Arabidopsis mutant deficient in gamma-tocopherol methyltransferase. Plant Mol Biol 52:1181–1190. doi:10.1023/B:Plan.0000004307.62398.91

Berry SZ, Uddin MR (1988) Effect of high-temperature on fruit-set in tomato cultivars and selected germplasm. Hortscience 23:606–608

Bharti K, Von Koskull-Döring P, Bharti S et al (2004) Tomato heat stress transcription factor HsfB1 represents a novel type of general transcription coactivator with a histone-like motif interacting with the plant CREB binding protein ortholog HAC1. Plant Cell 16:1521–1535. doi:10.1105/tpc.019927

Bita CE, Gerats T (2013) Plant tolerance to high temperature in a changing environment: scientific fundamentals and production of heat stress-tolerant crops. Front Plant Sci 4:273. doi:10.3389/fpls.2013.00273

Blazquez MA, Ahn JH, Weigel D (2003) A thermosensory pathway controlling flowering time in *Arabidopsis thaliana*. Nat Genet 33:168–171. doi:10.1038/Ng1085

Blum A, Klueva N, Nguyen HT (2001) Wheat cellular thermotolerance is related to yield under heat stress. Euphytica 117:117–123. doi:10.1023/A:1004083305905

Bokszczanin KL, Solanaceae Pollen Thermotolerance Initial Training Network C, Fragkostefanakis S (2013) Perspectives on deciphering mechanisms underlying plant heat stress response and thermotolerance. Front Plant Sci 4:315. doi:10.3389/fpls.2013.00315

Bouche N, Fromm H (2004) GABA in plants: just a metabolite? Trends Plant Sci 9:110–115. doi:10.1016/j.tplants.2004.01.006

Bunce MW, Bergendahl K, Anderson RA (2006) Nuclear PI(4,5)P(2): a new place for an old signal. Biochim Biophys Acta 1761:560–569. doi:10.1016/j.bbalip.2006.03.002

Cabello JV, Lodeyro AF, Zurbriggen MD (2014) Novel perspectives for the engineering of abiotic stress tolerance in plants. Curr Opin Biotechnol 26:62–70. doi:http://dx.doi.org/10.1016/j.copbio.2013.09.011

Chalker-Scott L (2002) Do anthocyanins function as osmoregulators in leaf tissues? Adv Bot Res 37:103–127. doi:10.1016/S0065-2296(02)37046-0

Chan-Schaminet KY, Baniwal SK, Bublak D et al (2009) Specific interaction between tomato HsfA1 and HsfA2 creates hetero-oligomeric superactivator complexes for synergistic activation of heat stress gene expression. J Biol Chem 284:20848–20857. doi:10.1074/jbc.M109.007336

Che P, Bussell JD, Zhou W et al (2010) Signaling from the endoplasmic reticulum activates brassinosteroid signaling and promotes acclimation to stress in Arabidopsis. Sci Signal 3:ra69. doi:10.1126/scisignal.2001140

Cheng L, Zou YJ, Ding SL, Zhang JJ, Yu XL, Cao JS, Lu G (2009) Polyamine accumulation in transgenic tomato enhances the tolerance to high temperature stress. J Integr Plant Biol 51:489–499. doi:10.1111/j.1744-7909.2009.00816.x

Cline WR (2007) Global warming and agriculture: end-of-century estimates by country. Peterson Institute, Washington

Cohen-Peer R, Schuster S, Meiri D, Breiman A, Avni A (2010) Sumoylation of Arabidopsis heat shock factor A2 (HsfA2) modifies its activity during acquired thermotholerance. Plant Mol Biol 74:33–45. doi:10.1007/s11103-010-9652-1

Cvikrova M, Gemperlova L, Dobra J, Martincova O, Prasil IT, Gubis J, Vankova R (2012) Effect of heat stress on polyannine metabolism in proline-over-producing tobacco plants. Plant Sci 182:49–58. doi:10.1016/j.plantsci.2011.01.016

Davison PA, Hunter CN, Horton P (2002) Overexpression of beta-carotene hydroxylase enhances stress tolerance in Arabidopsis. Nature 418:203–206. doi:10.1038/Nature00861 de Ronde JA, Spreeth MH, Cress WA (2000) Effect of antisense L-Delta(1)-pyrroline-5-carboxylate reductase transgenic soybean plants subjected to osmotic and drought stress. Plant Growth Regul 32:13–26. doi:10.1023/A:1006338911617

De Ronde JA, Cress WA, Kruger GHJ, Strasser RJ, Van Staden J (2004) Photosynthetic response of transgenic soybean plants, containing an Arabidopsis *P5CR* gene, during heat and drought stress. J Plant Physiol 161:1211–1224. doi:10.1016/j.jplph.2004.01.014

Deng Y, Humbert S, Liu J-X et al (2011) Heat induces the splicing by IRE1 of a mRNA encoding a transcription factor involved in the unfolded protein response in Arabidopsis. Proc Natl Acad Sci U S A 108:7247–7252. doi:10.1073/pnas.1102117108

Diamant S, Eliahu N, Rosenthal D, Goloubinoff P (2001) Chemical chaperones regulate molecular chaperones *in vitro* and in cells under combined salt and heat stresses. J Biol Chem 276(43):39586–39591. doi:10.1074/jbc.M103081200

Dias AS, Lidon FC (2009) Evaluation of grain filling rate and duration in bread and durum wheat, under heat stress after anthesis. J Agron Crop Sci 195:137–147. doi:10.1111/j.1439-037X.2008.00347.x

Dietz K-J (2014) Redox regulation of transcription factors in plant stress acclimation and development. Antioxid Redox Signal 21:1–40. doi: 10.1089/ars.2013.5672

Djanaguiraman M, Annie Sheeba J, Durga Devi D, Bangarusamy U (2009) Cotton leaf senescence can be delayed by nitrophenolate spray through enhanced antioxidant defence system. J Agron Crop Sci 195:213–224

Dupont FM, Hurkman WJ, Vensel WH, Tanaka C, Kothari KM, Chung OK, Altenbach SB (2006) Protein accumulation and composition in wheat grains: effects of mineral nutrients and high temperature. Eur J Agron 25:96–107. doi:10.1016/j.eja.2006.04.003

Fait A, Fromm H, Walter D, Galili G, Fernie AR (2008) Highway or byway: the metabolic role of the GABA shunt in plants. Trends Plant Sci 13:14–19. doi:10.1016/j.tplants.2007.10.005

Feng LL, Han YJ, Liu G, An BG, Yang J, Yang GH, Li YS, Zhu YG (2007) Overexpression of sedoheptulose-1,7-bisphosphatase enhances photosynthesis and growth under salt stress in transgenic rice plants. Funct Plant Biol 34:822–834. doi:10.1071/Fp07074

Firon N, Shaked R, Peet MM, Pharr DM, Zamski E, Rosenfeld K, Althan L, Pressman E (2006) Pollen grains of heat tolerant tomato cultivars retain higher carbohydrate concentration under heat stress conditions. Sci Hortic-Amst 109:212–217. doi:10.1016/j.scienta.2006.03.007

Firon N, Pressman E, Meir S, Khoury R, Altahan L (2012) Ethylene is involved in maintaining tomato (*Solanum lycopersicum*) pollen quality under heat-stress conditions. AoB plants 2012:pls024. doi:10.1093/aobpla/pls024

Fouad WM, Altpeter F (2009) Transplastomic expression of bacterial l-aspartate-alpha-decarboxylase enhances photosynthesis and biomass production in response to high temperature stress. Transgenic Res 18:707–718. doi:10.1007/s11248-009-9258-z

Frank G, Pressman E, Ophir R, Althan L, Shaked R, Freedman M, Shen S, Firon N (2009) Transcriptional profiling of maturing tomato (*Solanum lycopersicum* L.) microspores reveals the involvement of heat shock proteins, ROS scavengers, hormones, and sugars in the heat stress response. J Exp Bot 60:3891–3908. doi:10.1093/Jxb/Erp234

Frova C, Sarigorla M (1994) Quantitative Trait Loci (QTLs) for pollen thermotolerance detected in maize. Mol Gen Genet 245:424–430. doi:10.1007/Bf00302254

Fu JM, Momcilovic I, Clemente TE, Nersesian N, Trick HN, Ristic Z (2008) Heterologous expression of a plastid EF-Tu reduces protein thermal aggregation and enhances CO(2) fixation in wheat (*Triticum aestivum*) following heat stress. Plant Mol Biol 68:277–288. doi:10.1007/s11103-008-9369-6

Furbank RT, Tester M (2011) Phenomics-technologies to relieve the phenotyping bottleneck. Trends Plant Sci 16:635–644. doi:10.1016/j.tplants.2011.09.005

Gaupels F, Kuruthukulangarakoola GT, Durner J (2011). Upstream and downstream signals of nitric oxide in pathogen defence. Curr Opin Plant Biol 14: 707-714. doi:10.1016/j.pbi.2011.07.005

Gill SS, Tuteja N (2010) Reactive oxygen species and antioxidant machinery in abiotic stress tolerance in crop plants. Plant Physiol Biochem 48:909–930. doi:10.1016/j.plaphy.2010.08.016

Giorno F, Wolters-Arts M, Grillo S, Scharf KD, Vriezen WH, Mariani C (2010) Developmental and heat stress-regulated expression of HsfA2 and small heat shock proteins in tomato anthers. J Exp Bot 61:453–462. doi:10.1093/Jxb/Erp316

Giorno F, Wolters-Arts M, Mariani C, Rieu I (2013) Ensuring reproduction at high temperatures: the heat stress response during anther and pollen development. Plants 2:489–506

Gong M, van der Luit AH, Knight MR, Trewavas AJ (1998) Heat-shock-induced changes in intracellular Ca^{2+} level in tobacco seedlings in relation to thermotolerance. Plant Physiol 116:429–437. doi:10.1104/Pp.116.1.429

Gould KS, Lamotte O, Klinguer A et al (2003) Nitric oxide production in tobacco leaf cells: a generalized stress response? Plant Cell Environ 26:1851–1862. doi:10.1046/j.1365-3040.2003.01101.x

Grover A, Mittal D, Negi M, Lavania D (2013) Generating high temperature tolerant transgenic plants: achievements and challenges. Plant Sci 205:38–47. doi:10.1016/j.plantsci.2013.01.005

Grün S, Lindermayr C, Sell S, Durner J (2006) Nitric oxide and gene regulation in plants. J Exp Bot 57:507–516. doi:10.1093/jxb/erj053

Hahn A, Bublak D, Schleiff E, Scharf K-D (2011) Crosstalk between Hsp90 and Hsp70 chaperones and heat stress transcription factors in tomato. Plant Cell 23:741–755. doi:10.1105/tpc.110.076018

Hanson PM, Chen J, Kuo G (2002) Gene action and heritability of high-temperature fruit set in tomato line CL5915. Hortscience 37:172–175

Hare PD, Cress WA, Van Staden J (1998) Dissecting the roles of osmolyte accumulation during stress. Plant Cell Environ 21:535–553. doi:10.1046/j.1365-3040.1998.00309.x

Havaux M (1998) Carotenoids as membrane stabilizers in chloroplasts. Trends Plant Sci 3:147–151

Hedhly A, Hormaza JI, Herrero M (2003) The effect of temperature on stigmatic receptivity in sweet cherry (*Prunus avium* L.). Plant Cell Environ 26:1673–1680. doi:10.1046/j.1365-3040.2003.01085.x

Hedhly A, Hormaza JI, Herrero M (2005) The effect of temperature on pollen germination, pollen tube growth, and stigmatic receptivity in peach. Plant Biol 7:476–483. doi:10.1055/s-2005-865850

Hedhly A, Hormaza JI, Herrero M (2009) Global warming and sexual plant reproduction. Trends Plant Sci 14:30–36. doi:10.1016/j.tplants.2008.11.001

Howarth CJ (2005) Genetic improvements of tolerance to high temperature. In: Ashraf M, Harris PJC (eds) Abiotic stresses: plant resistance through breeding and molecular approaches. Howarth Press Inc, New York

Howell SH (2013) Endoplasmic reticulum stress responses in plants. Annu Rev Plant Biol 64:477–499. doi:10.1146/annurev-arplant-050312-120053

Ibrahim AMH, Quick JS (2001) Heritability of heat tolerance in winter and spring wheat. Crop Sci 41:1401–1405

Isner J-C, Maathuis FJM (2011) Measurement of cellular cGMP in plant cells and tissues using the endogenous fluorescent reporter FlincG. Plant J 65:329–334. doi:10.1111/j.1365-313X.2010.04418.x

Jagadish SVK, Cairns J, Lafitte R, Wheeler TR, Price AH, Craufurd PQ (2010) Genetic analysis of heat tolerance at anthesis in rice. Crop Sci 50:1633–1641. doi:10.2135/cropsci2009.09.0516

Jain M, Prasad PVV, Boote KJ, Hartwell AL, Chourey PS (2007) Effects of season-long high temperature growth conditions on sugar-to-starch metabolism in developing microspores of grain sorghum (*Sorghum bicolor* L. Moench). Planta 227:67–79. doi:10.1007/s00425-007-0595-y

Jung H, Crisp PA, Estavillo GM et al (2013) Subset of heat-shock transcription factors required for the early response of Arabidopsis to excess light. Proc Natl Acad Sci U S A 110:14474–14479. doi:10.1073/pnas.1311632110

Kakani VG, Reddy KR, Koti S, Wallace TP, Prasad PV, Reddy VR, Zhao D (2005) Differences in *in vitro* pollen germination and pollen tube growth of cotton cultivars in response to high temperature. Ann Bot 96:59–67. doi:10.1093/aob/mci149

Kaplan F, Kopka J, Haskell DW, Zhao W, Schiller KC, Gatzke N, Sung DY, Guy CL (2004) Exploring the temperature-stress metabolome of Arabidopsis. Plant Physiol 136:4159–4168. doi:10.1104/pp. 104.052142

Karim MA, Fracheboud Y, Stamp P (1997) Heat tolerance of maize with reference of some physiological characteristics. Ann Bangladesh Agric 7:27–33

Kasukabe Y, He L, Watakabe Y, Otani M, Shimada T, Tachibana S (2006) Improvement of environmental stress tolerance of sweet potato by introduction of genes for spermidine synthase. Plant Biotechnol 23:75–83

Kinnersley AM, Turano FJ (2000) Gamma aminobutyric acid (GABA) and plant responses to stress. Crit Rev Plant Sci 19:479–509. doi:10.1080/07352680091139277

Kotak S, Larkindale J, Lee U, von Koskull-Doring P, Vierling E, Scharf KD (2007) Complexity of the heat stress response in plants. Curr Opin Plant Biol 10:310–316. doi:10.1016/j.pbi.2007.04.011

Kozai N, Beppu K, Mochioka R, Boonprakob U, Subhadrabandhu S, Kataoka I (2004) Adverse effects of high temperature on the development of reproductive organs in 'Hakuho' peach trees. J Hortic Sci Biotech 79:533–537

Kumar SV, Wigge PA (2010) H2A.Z-containing nucleosomes mediate the thermosensory response in Arabidopsis. Cell 140:136–147. doi:10.1016/j.cell.2009.11.006

Kumar A, Li C, Portis AR Jr (2009) Arabidopsis thaliana expressing a thermostable chimeric Rubisco activase exhibits enhanced growth and higher rates of photosynthesis at moderately high temperatures. Photosynth Res 100:143–153. doi:10.1007/s11120-009-9438-y

Kurek I, Chang TK, Bertain SM, Madrigal A, Liu L, Lassner MW, Zhu G (2007) Enhanced thermostability of Arabidopsis Rubisco activase improves photosynthesis and growth rates under moderate heat stress. Plant Cell 19:3230–3241. doi:10.1105/tpc.107.054171

Langridge P, Fleury D (2011) Making the most of 'omics' for crop breeding. Trends Biotechnol 29:33–40. doi:10.1016/j.tibtech.2010.09.006

Larkindale J, Huang BR (2004) Changes of lipid composition and saturation level in leaves and roots for heat-stressed and heat-acclimated creeping bentgrass (*Agrostis stolonifera*). Environ Exp Bot 51:57–67. doi:10.1016/S0098-8472(03)00060-1

Larkindale J, Knight MR (2002) Protection against heat stress-induced oxidative damage in arabidopsis involves calcium, abscisic acid, ethylene, and salicylic acid. Plant Physiol 128:682–695. doi:10.1104/Pp.010320

Larkindale J, Hall JD, Knight MR, Vierling E (2005) Heat stress phenotypes of Arabidopsis mutants implicate multiple signaling pathways in the acquisition of thermotolerance. Plant Physiol 138:882–897. doi:10.1104/pp. 105.062257

Lee U, Wie C, Fernandez BO, Feelisch M, Vierling E (2008) Modulation of nitrosative stress by S-nitrosoglutathione reductase is critical for thermotolerance and plant growth in Arabidopsis. Plant Cell 20:786–802. doi:10.1105/tpc.107.052647

Li B, Liu H-T, Sun D-Y, Zhou R-G (2004) Ca($^{2+}$) and calmodulin modulate DNA-binding activity of maize heat shock transcription factor in vitro. Plant Cell Physiol 45:627–634

Liu JX, Howell SH (2010) Endoplasmic reticulum protein quality control and its relationship to environmental stress responses in plants. Plant Cell 22:2930–2942. doi:10.1105/tpc.110.078154

Liu XZ, Huang BR (2000) Heat stress injury in relation to membrane lipid peroxidation in creeping bentgrass. Crop Sci 40:503–510

Liu XH, Huang BR (2002) Cytokinin effects on creeping bentgrass response to heat stress: II. Leaf senescence and antioxidant metabolism. Crop Sci 42:466–472

Liu XH, Huang BR, Banowetz G (2002) Cytokinin effects on creeping bentgrass responses to heat stress: I. Shoot and root growth. Crop Sci 42:457–465

Liu H-T, Li B, Shang Z-L et al (2003) Calmodulin is involved in heat shock signal transduction in wheat. Plant Physiol 132:1186–1195. doi:10.1104/pp.102.018564

Liu K, Li L, Luan S (2005) An essential function of phosphatidylinositol phosphates in activation of plant shaker-type K+ channels. Plant J 42:433–443. doi:10.1111/j.1365-313X.2005.02384.x

Liu H-T, Li G-L, Chang H et al (2007) Calmodulin-binding protein phosphatase PP7 is involved in thermotolerance in Arabidopsis. Plant Cell Environ 30:156–164. doi:10.1111/j.1365-3040.2006.01613.x

Liu H-T, Gao F, Li G-L et al (2008) The calmodulin-binding protein kinase 3 is part of heat-shock signal transduction in *Arabidopsis thaliana*. Plant J 55:760–773. doi:10.1111/j.1365-313X.2008.03544.x

Liu HC, Liao HT, Charng YY (2011) The role of class A1 heat shock factors (HSFA1s) in response to heat and other stresses in Arabidopsis. Plant Cell Environ 34(5):738–751. doi:10.1111/j.1365-3040.2011.02278.x

Liu J, Sun N, Liu M et al (2013) An autoregulatory loop controlling Arabidopsis HsfA2 expression: role of heat shock-induced alternative splicing. Plant Physiol 162:512–521. doi:10.1104/pp. 112.205864

Long SP, Ort DR (2010) More than taking the heat: crops and global change. Curr Opin Plant Biol 13:241–248. doi:10.1016/j.pbi.2010.04.008

Los DA, Mironov KS, Allakhverdiev SI (2013) Regulatory role of membrane fluidity in gene expression and physiological functions. Photosynth Res 116:489–509. doi:10.1007/s11120-013-9823-4

Ludwig-Muller J, Krishna P, Forreiter C (2000) A glucosinolate mutant of Arabidopsis is thermosensitive and defective in cytosolic Hsp90 expression after heat stress. Plant Physiol 123:949–958. doi:10.1104/Pp.123.3.949

Luo Y, Li WM, Wang W (2008) Trehalose: protector of antioxidant enzymes or reactive oxygen species scavenger under heat stress? Environ Exp Bot 63:378–384. doi:10.1016/j.envexpbot.2007.11.016

Luo M, Liu X, Singh P et al (2012) Chromatin modifications and remodeling in plant abiotic stress responses. Biochim Biophys Acta 1819:129–136. doi:10.1016/j.bbagrm.2011.06.008

Lv WT, Lin B, Zhang M, Hua XJ (2011) Proline accumulation is inhibitory to Arabidopsis seedlings during heat stress. Amino Acids 41:S69–S69

Maestri E, Klueva N, Perrotta C, Gulli M, Nguyen HT, Marmiroli N (2002) Molecular genetics of heat tolerance and heat shock proteins in cereals. Plant Mol Biol 48:667–681. doi:10.1023/A:1014826730024

Mascarenhas JP, Hamilton DA (1992) Artifacts in the localization of GUS activity in anthers of petunia transformed with a CaMV 35S-GUS construct. Plant J 2:405–408. doi:10.1111/j.1365-313X.1992.00405.x

McClung CR, Davis SJ (2010) Ambient thermometers in plants: from physiological outputs towards mechanisms of thermal sensing. Curr Biol 20:R1086–R1092. doi:10.1016/j.cub.2010.10.035

McNeil SD, Nuccio ML, Hanson AD (1999) Betaines and related osmoprotectants. Targets for metabolic engineering of stress resistance. Plant Physiol 120:945–949. doi:10.1104/Pp.120.4.945

Mengel A, Chaki M, Shekariesfahlan A, Lindermayr C (2013) Effect of nitric oxide on gene transcription-S-nitrosylation of nuclear proteins. Front Plant Sci 4:293. doi:10.3389/fpls.2013.00293

Menzel A, Sparks TH, Estrella N, Koch E, Aasa A, Ahas R, Alm-Kubler K, Bissolli P, Braslavska O, Briede A, Chmielewski FM, Crepinsek Z, Curnel Y, Dahl A, Defila C, Donnelly A, Filella Y, Jatcza K, Mage F, Mestre A, Nordli O, Penuelas J, Pirinen P, Remisova V, Scheifinger H, Striz M, Susnik A, Van Vliet AJH, Wielgolaski FE, Zach S, Zust A (2006) European phenological response to climate change matches the warming pattern. Glob Chang Biol 12:1969–1976. doi:10.1111/j.1365-2486.2006.01193.x

Miller G, Mittler R (2006) Could heat shock transcription factors function as hydrogen peroxide sensors in plants? Ann Bot 98:279–288. doi:10.1093/aob/mcl107

Miranda JA, Avonce N, Suarez R, Thevelein JM, Van Dijck P, Iturriaga G (2007) A bifunctional TPS-TPP enzyme from yeast confers tolerance to multiple and extreme abiotic-stress conditions in transgenic Arabidopsis. Planta 226:1411–1421. doi:10.1007/s00425-007-0579-y

Mishkind M, Vermeer JEM, Darwish E, Munnik T (2009) Heat stress activates phospholipase D and triggers PIP accumulation at the plasma membrane and nucleus. Plant J 60:10–21. doi:10.1111/j.1365-313X.2009.03933.x

Mishra SK, Tripp J, Winkelhaus S et al (2002) In the complex family of heat stress transcription factors, HsfA1 has a unique role as master regulator of thermotolerance in tomato. Genes Dev 16:1555–1567. doi:10.1101/gad.228802

Mittler R (2006) Abiotic stress, the field environment and stress combination. Trends Plant Sci 11:15–19. doi:10.1016/j.tplants.2005.11.002

Mittler R, Vanderauwera S, Gollery M, Van Breusegem F (2004) Reactive oxygen gene network of plants.

Trends Plant Sci 9:490–498. doi:10.1016/j.tplants.2004.08.009

Mittler R, Finka A, Goloubinoff P (2012) How do plants feel the heat? Trends Biochem Sci 37:118–125. doi:10.1016/j.tibs.2011.11.007

Mizuguchi G, Shen X, Landry J et al (2004) ATP-driven exchange of histone H2AZ variant catalyzed by SWR1 chromatin remodeling complex. Science 303:343–348. doi:10.1126/science.1090701

Mlynárová L, Nap J-P, Bisseling T (2007) The SWI/SNF chromatin-remodeling gene AtCHR12 mediates temporary growth arrest in *Arabidopsis thaliana* upon perceiving environmental stress. Plant J 51:874–885. doi:10.1111/j.1365-313X.2007.03185.x

Morales D, Rodriguez P, Dell'Amico J, Nicolas E, Torrecillas A, Sanchez-Blanco MJ (2003) High-temperature preconditioning and thermal shock imposition affects water relations, gas exchange and root hydraulic conductivity in tomato. Biol Plant 47:203–208

Moreno AA, Orellana A (2011) The physiological role of the unfolded protein response in plants. Biol Res 44:75–80

Morrison MJ, Stewart DW (2002) Heat stress during flowering in summer Brassica. Crop Sci 42:797–803

Mostafa H, Hassanein R, Khalil S, El-Khawas S, El-Bassiouny H, El-Monem A (2010) Effect of arginine or putrescine on growth, yield and yield components of late sowing wheat. J Appl Sci Res 6:177–183

Murkowski A (2001) Heat stress and spermidine: effect on chlorophyll fluorescence in tomato plants. Biol Plant 44:53–57. doi:10.1023/A:1017966203859

Murray V, Ebi KL (2012) IPCC special report on managing the risks of extreme events and disasters to advance climate change adaptation (SREX). J Epidemiol Community Health 66:759–760. doi:10.1136/jech-2012-201045

Mutters RG, Ferreira LGR, Hall AE (1989) Proline content of the anthers and pollen of heat-tolerant and heat-sensitive cowpea subjected to different temperatures. Crop Sci 29(6):1497–1500

Neill SJ, Desikan R, Hancock JT (2003) Nitric oxide signalling in plants. New Phytol 159:11–35. doi:10.1046/j.1469-8137.2003.00804.x

Nishizawa A, Yabuta Y, Yoshida E, Maruta T, Yoshimura K, Shigeoka S (2006) Arabidopsis heat shock transcription factor A2 as a key regulator in response to several types of environmental stress. Plant J 48:535–547. doi:10.1111/j.1365-313X.2006.02889.x

Nishizawa A, Yabuta Y, Shigeoka S (2008) Galactinol and raffinose constitute a novel function to protect plants from oxidative damage. Plant Physiol 147:1251–1263. doi:10.1104/pp. 108.122465

Osborne SL, Thomas CL, Gschmeissner S, Schiavo G (2001) Nuclear PtdIns (4,5) P2 assembles in a mitotically regulated particle involved in pre-mRNA splicing. J Cell Sci 114:2501–2511

Papageorgiou GC, Murata N (1995) The unusually strong stabilizing effects of glycine betaine on the structure and function of the oxygen-evolving photosystem-II complex. Photosynth Res 44:243–252. doi:10.1007/Bf00048597

Parry MAJ, Andralojc PJ, Scales JC, Salvucci ME, Carmo-Silva AE, Alonso H, Whitney SM (2013) Rubisco activity and regulation as targets for crop improvement. J Exp Bot 64:717–730. doi:10.1093/Jxb/Ers336

Pearcy RW (1978) Effect of growth temperature on the fatty acid composition of the leaf lipids in *Atriplex lentiformis* (Torr.) Wats. Plant Physiol 61:484–486

Perez-de-Castro AM, Vilanova S, Canizares J, Pascual L, Blanca JM, Diez MJ, Prohens J, Pico B (2012) Application of genomic tools in plant breeding. Curr Genomics 13:179–195

Pfluger J, Wagner D (2007) Histone modifications and dynamic regulation of genome accessibility in plants. Curr Opin Plant Biol 10:645–652. doi:10.1016/j.pbi.2007.07.013

Pinto RS, Reynolds MP, Mathews KL, McIntyre CL, Olivares-Villegas JJ, Chapman SC (2010). Heat and drought adaptive QTL in a wheat population designed to minimize confounding agronomic effects. Theor Appl Genet 121:1001-1021. doi:10.1007/s00122-010-1351-4

Piterková J, Luhová L, Mieslerová B et al (2013) Nitric oxide and reactive oxygen species regulate the accumulation of heat shock proteins in tomato leaves in response to heat shock and pathogen infection. Plant Sci 207:57–65. doi:10.1016/j.plantsci.2013.02.010

Portis AR Jr (2003) Rubisco activase – Rubisco's catalytic chaperone. Photosynth Res 75:11–27. doi:10.1023/A:1022458108678

Prabhavathi VR, Rajam MV (2007) Polyamine accumulation in transgenic eggplant enhances tolerance to multiple abiotic stresses and fungal resistance. Plant Biotechnol 24:273–282

Pradhan GP, Prasad PVV, Fritz AK, Kirkham MB, Gill BS (2012) High temperature tolerance in aegilops species and its potential transfer to wheat. Crop Sci 52:292–304. doi:10.2135/cropsci2011.04.0186

Prasad PVV, Boote KJ, Allen LH, Sheehy JE, Thomas JMG (2006) Species, ecotype and cultivar differences in spikelet fertility and harvest index of rice in response to high temperature stress. Field Crop Res 95:398–411. doi:10.1016/j.fcr.2005.04.008

Prasad PVV, Pisipati SR, Mutava RN, Tuinstra MR (2008) Sensitivity of grain sorghum to high temperature stress during reproductive development. Crop Sci 48:1911–1917. doi:10.2135/cropsci2008.01.0036

Pressman E, Peet MM, Pharr DM (2002) The effect of heat stress on tomato pollen characteristics is associated with changes in carbohydrate concentration in the developing anthers. Ann Bot-Lond 90:631–636. doi:10.1093/Aob/Mcf240

Qin F, Kakimoto M, Sakuma Y, Maruyama K, Osakabe Y, Tran LSP, Shinozaki K, Yamaguchi-Shinozaki K (2007) Regulation and functional analysis of ZmDREB2A in response to drought and heat stresses in Zea mays L. Plant J 50:54–69. doi:10.1111/j.1365-313X.2007.03034.x

Radi R, Beckman JS, Bush KM, Freeman BA (1991) Peroxynitrite-induced membrane lipid peroxidation: The cytotoxic potential of superoxide and nitric oxide. Arch Biochem Biophys 288:481–487

Raines CA, Harrison EP, Olcer H, Lloyd JC (2000) Investigating the role of the thiol-regulated enzyme sedoheptulose-1,7-bisphosphatase in the control of photosynthesis. Physiol Plant 110:303–308. doi:10.1034/j.1399-3054.2000.1100303.x

Raison JK, Pike CS, Berry JA (1982) Growth temperature-induced alterations in the thermotropic properties of Nerium oleander membrane lipids. Plant Physiol 70:215–218

Rasheed R, Wahid A, Farooq M, Hussain I, Basra SMA (2011) Role of proline and glycinebetaine pretreatments in improving heat tolerance of sprouting sugarcane (*Saccharum* sp.) buds. Plant Growth Regul 65:35–45. doi:10.1007/s10725-011-9572-3

Reynolds MP, Balota M, Delgado MIB, Amani I, Fischer RA (1994) Physiological and morphological traits associated with spring wheat yield under hot, irrigated conditions. Funct Plant Biol 21:717-730. doi:10.1071/PP9940717

Rivero RM, Ruiz JM, Garcia PC, Lopez-Lefebre LR, Sanchez E, Romero L (2001) Resistance to cold and heat stress: accumulation of phenolic compounds in tomato and watermelon plants. Plant Sci 160:315–321. doi:10.1016/S0168-9452(00)00395-2

Rivero RM, Ruiz JM, Romero LM (2004) Importance of N source on heat stress tolerance due to the accumulation of proline and quaternary ammonium compounds in tomato plants. Plant Biol 6:702–707. doi:10.1055/s-2004-821293

Rizhsky L, Liang HJ, Shuman J, Shulaev V, Davletova S, Mittler R (2004) When defense pathways collide. The response of Arabidopsis to a combination of drought and heat stress. Plant Physiol 134:1683–1696. doi:10.1104/pp. 103.033431

Rodrigo J, Herrero M (2002) Effects of pre-blossom temperatures on flower development and fruit set in apricot. Sci Hortic-Amst 92:125–135. doi:10.1016/S0304-4238(01)00289-8

Rodríguez M, Canales E, Borrás-Hidalgo O (2005) Molecular aspects of abiotic stress in plants. Biotecnol Apl 22:1–10

Roy M, Ghosh B (1996) Polyamines, both common and uncommon, under heat stress in rice (Oryza sativa) callus. Physiol Plant 98:196–200

Ruelland E, Zachowski A (2010) How plants sense temperature. Environ Exp Bot 69:225–232. doi:10.1016/j.envexpbot.2010.05.011

Sage RF, Kubien DS (2007) The temperature response of C_3 and C_4 photosynthesis. Plant Cell Environ 30:1086–1106. doi:10.1111/j.1365-3040.2007.01682.x

Saidi Y, Finka A, Muriset M, Bromberg Z, Weiss YG, Maathuis FJM, Goloubinoff P (2009) The heat shock response in moss plants is regulated by specific calcium-permeable channels in the plasma membrane. Plant Cell 21:2829–2843. doi:10.1105/tpc.108.065318

Sakamoto A, Murata N (2002) The role of glycine betaine in the protection of plants from stress: clues from transgenic plants. Plant Cell Environ 25:163–171. doi:10.1046/j.0016-8025.2001.00790.x

Sakata T, Higashitani A (2008) Male sterility accompanied with abnormal anther development in plants–genes and environmental stresses with special reference to high temperature injury. Intl J Plant Dev Biol 2:42–51

Sakata T, Oshino T, Miura S, Tomabechi M, Tsunaga Y, Higashitani N, Miyazawa Y, Takahashi H, Watanabe M, Higashitani A (2010) Auxins reverse plant male sterility caused by high temperatures. Proc Natl Acad Sci U S A 107:8569–8574. doi:10.1073/pnas.1000869107

Sakihama Y, Nakamura S, Yamasaki H (2002) Nitric oxide production mediated by nitrate reductase in the green alga *Chlamydomonas reinhardtii*: an alternative NO production pathway in photosynthetic organisms. Plant Cell Physiol 43:290–297

Salvucci ME (2008) Association of Rubisco activase with chaperonin-60 beta: a possible mechanism for protecting photosynthesis during heat stress. J Exp Bot 59:1923–1933. doi:10.1093/Jxb/Erm343

Salvucci ME, Crafts-Brandner SJ (2004) Mechanism for deactivation of Rubisco under moderate heat stress. Physiol Plant 122:513–519. doi:10.1111/j.1399-3054.2004.00419.x

Sato S, Peet MM, Thomas JF (2002) Determining critical pre- and post-anthesis periods and physiological processes in *Lycopersicon esculentum* Mill. exposed to moderately elevated temperatures. J Exp Bot 53:1187–1195. doi:10.1093/jexbot/53.371.1187

Sato S, Kamiyama M, Iwata T, Makita N, Furukawa H, Ikeda H (2006) Moderate increase of mean daily temperature adversely affects fruit set of *Lycopersicon esculentum* by disrupting specific physiological processes in male reproductive development. Ann Bot-Lond 97:731–738. doi:10.1093/Aob/Mcl037

Scharf KD, Siddique M, Vierling E (2001) The expanding family of *Arabidopsis thaliana* small heat stress proteins and a new family of proteins containing alpha-crystallin domains (Acd proteins). Cell Stress Chaperones 6:225–237. doi:10.1379/1466-1268(2001)006<0225:Tefoat>2.0.Co;2

Scharf K-D, Berberich T, Ebersberger I, Nover L (2012) The plant heat stress transcription factor (Hsf) family: structure, function and evolution. Biochim Biophys Acta 1819:104–119. doi:10.1016/j.bbagrm.2011.10.002

Schrader SM, Wise RR, Wacholtz WF, Ort DR, Sharkey TD (2004) Thylakoid membrane responses to moderately high leaf temperature in Pima cotton. Plant Cell Environ 27:725–735. doi:10.1111/j.1365-3040.2004.01172.x

Scott RJ, Spielman M, Dickinson HG (2004) Stamen structure and function. Plant Cell 16:S46–S60. doi:10.1105/Tpc.017012

Setimela PS, Andrews DJ, Partridge J, Eskridge KM (2005) Screening sorghum seedlings for heat tolerance using a laboratory method. Eur J Agron 23:103–107. doi:10.1016/j.eja.2004.09.008

Shah NH, Paulsen GM (2003) Interaction of drought and high temperature on photosynthesis and grain-filling of wheat. Plant Soil 257:219–226. doi:10.1023/A:1026237816578

Shaked-Sachray L, Weiss D, Reuveni M, Nissim-Levi A, Oren-Shamir M (2002) Increased anthocyanin accumulation in aster flowers at elevated temperatures due to magnesium treatment. Physiol Plant 114:559–565. doi:10.1034/j.1399-3054.2002.1140408.x

Shim D, Hwang JU, Lee J, Lee S, Choi Y, An G, Martinoia E, Lee Y (2009) Orthologs of the class A4 heat shock transcription factor HsfA4a confer cadmium tolerance in wheat and rice. Plant Cell 21:4031–4043. doi:10.1105/tpc.109.066902

Sidrauski C, Walter P (1997) The transmembrane kinase Ire1p is a site-specific endonuclease that initiates mRNA splicing in the unfolded protein response. Cell 90:1031–1039

Singer MA, Lindquist S (1998) Multiple effects of trehalose on protein folding *in vitro* and *in vivo*. Mol Cell 1:639–648. doi:10.1016/S1097-2765(00)80064-7

Singh I, Shono M (2005) Physiological and molecular effects of 24-epibrassinolide, a brassinosteroid on thermotolerance of tomato. Plant Growth Regul 47:111–119. doi:10.1007/s10725-005-3252-0

Snedden WA, Fromm H (2001) Calmodulin as a versatile calcium signal transducer in plants. New Phytol 151:35–66. doi:10.1046/j.1469-8137.2001.00154.x

Song JJ, Nada K, Tachibana S (1999) Ameliorative effect of polyamines on the high temperature inhibition of in vitro pollen germination in tomato (*Lycopersicon esculentum* Mill.). Sci Hortic-Amst 80:203–212. doi:10.1016/S0304-4238(98)00254-4

Song JJ, Nada K, Tachibana S (2002) Suppression of S-adenosylmethionine decarboxylase activity is a major cause for high-temperature inhibition of pollen germination and tube growth in tomato (*Lycopersicon esculentum* Mill.). Plant Cell Physiol 43:619–627. doi:10.1093/Pcp/Pcf078

Song SQ, Lei YB, Tian XR (2005) Proline metabolism and cross-tolerance to salinity and heat stress in germinating wheat seeds. Russ J Plant Physiol 52:793–800. doi:10.1007/s11183-005-0117-3

Spreitzer RJ, Salvucci ME (2002) Rubisco: structure, regulatory interactions, and possibilities for a better enzyme. Annu Rev Plant Biol 53:449–475. doi:10.1146/annurev.arplant.53.100301.135233

Srivastava R, Deng Y, Shah S, Rao AG, Howell SH (2013) BINDING PROTEIN is a master regulator of the endoplasmic reticulum stress sensor/transducer bzip28 in arabidopsis. Plant Cell 25:1416–1429. doi:10.1105/tpc.113.110684

Suarez R, Calderon C, Iturriaga G (2009) Enhanced tolerance to multiple abiotic stresses in transgenic alfalfa accumulating trehalose. Crop Sci 49:1791–1799. doi:10.2135/cropsci2008.09.0573

Sugio A, Dreos R, Aparicio F, Maule AJ (2009) The cytosolic protein response as a subcomponent of the wider heat shock response in Arabidopsis. Plant Cell 21:642–654. doi:10.1105/tpc.108.062596

Takeoka Y, Hiroi K, Kitano H, Wada T (1991) Pistil hyperplasia in rice spikelets as affected by heat-stress. Sex Plant Reprod 4:39–43

Talanova VV, Alkimova TV, Titov AF (2003) Effect of whole plant and local heating on the ABA content in cucumber seedling leaves and roots and on their heat tolerance. Russ J Plant Physiol 50:90–94. doi:10.1023/A:1021996703940

Tang RS, Zheng JC, Jin ZQ, Zhang D, Huang H, Chen LG (2008) Possible correlation between high temperature-induced floret sterility and endogenous levels of IAA, GAs and ABA in rice (*Oryza sativa* L.). Plant Growth Regul 54:37–43. doi:10.1007/s10725-007-9225-8

Teklemariam T, Blake TJ (2003) Effects of UVB preconditioning on heat tolerance of cucumber (*Cucumis sativus* L.). Environ Exp Bot 50:169–182. doi:10.1016/S0098-8472(03)00024-8

Uchida A, Jagendorf AT, Hibino T et al (2002) Effects of hydrogen peroxide and nitric oxide on both salt and heat stress tolerance in rice. Plant Sci 163:515–523. doi:10.1016/S0168-9452(02)00159-0

Valluru R, Van den Ende W (2008) Plant fructans in stress environments: emerging concepts and future prospects. J Exp Bot 59:2905–2916. doi:10.1093/Jxb/Ern164

Van Leeuwen W, Vermeer JEM, Gadella TWJ, Munnik T (2007) Visualization of phosphatidylinositol 4,5-bisphosphate in the plasma membrane of suspension-cultured tobacco BY-2 cells and whole Arabidopsis seedlings. Plant J 52:1014–1026. doi:10.1111/j.1365-313X.2007.03292.x

Vasseur F, Pantin F, Vile D (2011) Changes in light intensity reveal a major role for carbon balance in Arabidopsis responses to high temperature. Plant Cell Environ 34:1563–1576. doi:10.1111/j.1365-3040.2011.02353.x

Vierling E (1991) The Roles of heat-shock proteins in plants. Annu Rev Plant Physiol 42:579–620. doi:10.1146/annurev.arplant.42.1.579

Wahid A, Ghazanfar A (2006) Possible involvement of some secondary metabolites in salt tolerance of sugarcane. J Plant Physiol 163:723–730. doi:10.1016/j.jplph.2005.07.007

Wahid A, Shabbir A (2005) Induction of heat stress tolerance in barley seedlings by pre-sowing seed treatment with glycinebetaine. Plant Growth Regul 46:133–141. doi:10.1007/s10725-005-8379-5

Wahid A, Gelani S, Ashraf M, Foolad MR (2007) Heat tolerance in plants: an overview. Environ Exp Bot 61:199–223. doi:10.1016/j.envexpbot.2007.05.011

Wallwork MAB, Jenner CF, Logue SJ, Sedgley M (1998) Effect of high temperature during grain-filling on the structure of developing and malted barley grains. Ann Bot-Lond 82:587–599. doi:10.1006/anbo.1998.0721

Wang HT, Bian MD, Yang ZM, Lin CT, Shi WL (2013) Preliminary functional analysis of the isoforms of oshsfa2a (*oryza sativa* l.) generated by alternative splicing. Plant Mol Biol Rep 31:38–46. doi:10.1007/s11105-012-0471-1

Welti R, Li W, Li M et al (2002) Profiling membrane lipids in plant stress responses. Role of phospholipase D alpha in freezing-induced lipid changes in Arabidopsis. J Biol Chem 277:31994–32002. doi:10.1074/jbc.M205375200

Wendehenne D, Pugin A, Klessig DF, Durner J (2001) Nitric oxide: comparative synthesis and signaling in animal and plant cells. Trends Plant Sci 6:177–183

Wheeler T, von Braun J (2013) Climate change impacts on global food security. Science 341:508–513. doi:10.1126/science.1239402

Willmott N, Sethi JK, Walseth TF et al (1996) Nitric oxide-induced mobilization of intracellular calcium via the cyclic ADP-ribose signaling pathway. J Biol Chem 271:3699–3705. doi:10.1074/jbc.271.7.3699

Wink DA, Hanbauer I, Krishna MC et al (1993) Nitric oxide protects against cellular damage and cytotoxicity from reactive oxygen species. Proc Natl Acad Sci U S A 90:9813–9817

Wise RR, Olson AJ, Schrader SM, Sharkey TD (2004) Electron transport is the functional limitation of photosynthesis in field-grown Pima cotton plants at high temperature. Plant Cell Environ 27:717–724. doi:10.1111/j.1365-3040.2004.01171.x

Xu S, Li JL, Zhang XQ, Wei H, Cui LJ (2006) Effects of heat acclimation pretreatment on changes of membrane lipid peroxidation, antioxidant metabolites, and ultrastructure of chloroplasts in two cool-season turfgrass species under heat stress. Environ Exp Bot 56:274–285. doi:10.1016/j.envexpbot.2005.03.002

Yamasaki H, Sakihama Y (2000) Simultaneous production of nitric oxide and peroxynitrite by plant nitrate reductase: in vitro evidence for the NR-dependent formation of active nitrogen species. FEBS Lett 468:89–92

Yamori W, Masumoto C, Fukayama H, Makino A (2012) Rubisco activase is a key regulator of non-steady-state photosynthesis at any leaf temperature, and to a lesser extent, of steady-state photosynthesis at high temperature. Plant J 71:871–880. doi:10.1111/j.1365-313X.2012.05041.x

Yang J, Sears RG, Gill BS, Paulsen GM (2002) Quantitative and molecular characterization of heat tolerance in hexaploid wheat. Euphytica 126:275–282. doi:10.1023/A:1016350509689

Yang XH, Liang Z, Lu CM (2005) Genetic engineering of the biosynthesis of glycinebetaine enhances photosynthesis against high temperature stress in transgenic tobacco plants. Plant Physiol 138:2299–2309. doi:10.1104/pp. 105.063164

Yoshida T, Ohama N, Nakajima J et al (2011) Arabidopsis HsfA1 transcription factors function as the main positive regulators in heat shock-responsive gene expression. Mol Genet Genomics 286:321–332. doi:10.1007/s00438-011-0647-7

Young LW, Wilen RW, Bonham-Smith PC (2004) High temperature stress of *Brassica napus* during flowering reduces micro- and mega-gametophyte fertility, induces fruit abortion, and disrupts seed production. J Exp Bot 55:485–495. doi:10.1093/Jxb/Erh038

Zhang JH, Huang WD, Liu YP, Pan QH (2005) Effects of temperature acclimation pretreatment on the ultrastructure of mesophyll cells in young grape plants (*Vitis vinifera* L. cv. Jingxiu) under cross-temperature stresses. J Integr Plant Biol 47:959–970. doi:10.1111/j.1744-7909.2005.00109.x

Zhang LR, Li YS, Xing D, Gao CJ (2009) Characterization of mitochondrial dynamics and subcellular localization of ROS reveal that HsfA2 alleviates oxidative damage caused by heat stress in Arabidopsis. J Exp Bot 60:2073–2091. doi:10.1093/Jxb/Erp078

Zhu X, Thalor SK, Takahashi Y et al (2012) An inhibitory effect of the sequence-conserved upstream open-reading frame on the translation of the main open-reading frame of HsfB1 transcripts in Arabidopsis. Plant Cell Environ 35:2014–2030. doi:10.1111/j.1365-3040.2012.02533.x

Zimmerli L, Hou BH, Tsai CH, Jakab G, Mauch-Mani B, Somerville S (2008) The xenobiotic beta-aminobutyric acid enhances Arabidopsis thermotolerance. Plant J 53:144–156. doi:10.1111/j.1365-313X.2007.03343.x

Zinn KE, Tunc-Ozdemir M, Harper JF (2010) Temperature stress and plant sexual reproduction: uncovering the weakest links. J Exp Bot 61:1959–1968. doi:10.1093/jxb/erq053

3 Plant Breeding for Flood Tolerance: Advances and Limitations

Xuechen Zhang, Xin Huang, Meixue Zhou,
Lana Shabala, Anthony Koutoulis,
and Sergey Shabala

Abstract

Over 17 million km^2 of land surface is affected by flooding every year, resulting in severe damage to plants and associated losses in agricultural production around the globe. While importance of plant breeding for waterlogging stress tolerance has been long on the agenda, the progress in the field is handicapped by the physiological and genetic complexity of this trait. In this chapter, we summarise the recent knowledge about the major constraints affecting plant performance in waterlogged soils and discuss the mechanisms employed by plants to deal with the stress. The topics covered include oxygen availability in flooded soils; whole-plant responses to oxygen deprivation; biochemical alterations in hypoxic roots; mechanisms of aerenchyma formation; the role of ethylene signalling and programmed cell death in hypoxic roots; oxygen transport from shoot to root; formation of ROL barrier and control of oxygen loss; changes in soil redox potential under flooding; Mn and Fe toxicity in waterlogged plants; secondary metabolite toxicity and plant adaptation to organic phytotoxins; MAS approach to plant breeding for flooding stress tolerance; and emerging areas such as elucidating the role of membrane transporters in flooding tolerance, developing high-throughput technology platforms for fine QTL mapping and understanding ROS signalling in flooding stress tolerance.

Keywords

Waterlogging • Aerenchyma • Elemental toxicity • Ethylene • Membrane transport • Programmed cell death • ROS • Breeding • QTL

X. Zhang • X. Huang • M. Zhou • L. Shabala
S. Shabala (✉)
School of Land and Food, University of Tasmania, Hobart, Australia
e-mail: Sergey.Shabala@utas.edu.au

A. Koutoulis
School of Biological Sciences, University of Tasmania, Hobart, Australia

3.1 Introduction

Over 17 million km^2 of land surface is affected by flooding every year, double the size of the USA (Voesenek and Sasidharan 2013). This results in an estimated annual damage exceeding 60

P.K. Jaiwal et al. (eds.), *Genetic Manipulation in Plants for Mitigation of Climate Change*,
DOI 10.1007/978-81-322-2662-8_3

billion euro (www.dartmouth.edu/~floods/Archives/2005sum.htm). Crop losses due to excess water are second only to drought, and yield reductions as high as 80 % have been recorded in waterlogged soils (MacEwan et al. 1992; Shaw et al. 2013). Waterlogging is common in duplex, or texture contrast, soils (Setter and Waters 2003). Such soils are widespread in the world, covering ~20 % of the landscape in Australia, Eastern Europe and the Russian Federation (Shaw et al. 2013) and 16 % of soils in the USA (Setter and Waters 2003). Because of this, the production of crop plants that can combine high grain yield with an increased flooding tolerance phenotype has been an important objective for decades (Agarwal and Grover 2006). However, despite some significant efforts, the progress in plant breeding for waterlogging stress tolerance is much slower than one would hope for. Several factors contribute to this problem.

The first one is the physiological complexity of waterlogging stress. Because gas diffusion is 10,000-fold slower in solution than in air (Armstrong 1979), the depletion of O_2 is a major feature of flooded sites, which creates hypoxia or anoxia around plant tissues. This leads to acute energy crises and a very substantial alterations in cell metabolism (and associated yield penalties). There are several possible options plants can use to adjust to this energy crisis:

3.1.1 Maintaining Adequate Oxygen

Maintaining adequate oxygen supply by a series of anatomical and morphological alterations in the root (Perata and Voesenek 2007). One of these alterations is the formation of aerenchyma. Species with higher root porosity are more tolerant to soil flooding, and in many wetland plants, aerenchyma is well developed even in drained conditions (and can be further enhanced in waterlogged conditions), while dry land species often do not form aerenchyma at all (Colmer 2003). In mature zone of rice (tolerant species) roots, aerenchyma comprised about 45 % of the root volume (Colmer 2003), in stark contrast to only 3 % of aerenchyma in the seminal roots of (intolerant) wheat species (McDonald et al. 2001). Another possibility is to minimise radial oxygen losses (ROL) by the formation of a tight barrier in the root peripheral cell layers exterior to the aerenchyma (Visser et al. 2000; McDonald et al. 2002). Formed as a result of suberisation and/or lignification of the cell walls, such ROL will facilitate a longitudinal diffusion of O_2 towards the root apex (Nishiuchi et al. 2012), the most metabolically active part of the root. In layman's terms, this is a "business as usual" option that would be the most preferred but that is rather difficult to achieve in full. Nevertheless, in anatomically adapted rice roots, anoxia-induced decrease in the rate of ATP synthesis was only 25 %, in marked contrast to the tenfold decrease in maize (highly sensitive to flooding) root tips (Ratcliffe 1997). More detailed analysis of anatomical and physiological traits contributing to adequate oxygen supply in flood-affected plants is given in Sect. 3.3.1.

3.1.2 Economisation of ATP Consumption

This includes shutting down energy-demanding processes such as protein synthesis (Branco-Price et al. 2008) and redirecting available ATP resources towards the production of molecular chaperones (e.g. heat shock proteins; Banti et al. 2010). This option can be classified as a "survival" strategy, as plant growth will be severely affected, with major implications for yield. Therefore, it is highly unlikely that this option may be sustainable for adaptation to prolonged soil flooding.

3.1.3 Developing a Capacity to Generate ATP without Oxygen

This includes energy-efficient sucrose catabolism through sucrose synthase, the preferential use of PPi-dependent enzymes and constrained catabolism of storage compounds such as starch, lipids and proteins (Licausi 2011; Bailey-Serres et al. 2012).

The above complexity of adaptive strategies makes it highly unlikely that one specific "key gene" responsible for waterlogging tolerance could be found and then introduced in high-yielding varieties, by either genetic or classical breeding methods. More likely, a thorough pyramiding of suitable traits should be envisaged. This question is discussed in more detail in Sect. 3.5.

In addition to reduced oxygen availability and associated metabolic shifts, plants grown in flooded soils are exposed to the range of elemental (Mn, Fe, etc.) and phytotoxicities (Shabala 2011). Surprisingly, up to now, the focus of plant breeders was predominantly on detrimental effects of anoxia, targeting traits dealing with oxygen uptake and redistribution within plant tissues, while tolerance to these toxicities was essentially neglected. This issue is discussed in more detail in Sects. 3.3 and 3.4.

3.2 Oxygen Availability and Plant Metabolism under Hypoxia and Anoxia

3.2.1 Oxygen Availability in Flooded Soils

As mentioned above, oxygen diffusion in water is 10^4 times slower than the oxygen diffusion in the air, resulting in hypoxia (oxygen deficiency) or anoxia (complete lack of oxygen) conditions around the roots (Bailey-Serres and Voesenek 2008). As a result, ATP production is greatly declined in hypoxic conditions (Colmer and Voesenek 2009), resulting in decreased energy availability and reduced water use efficiency in plants (Barrett-Lennard 2003). Decreased energy availability also disturbs the metabolic balance in plant tissues (Voesenek and Sasidharan 2013). Oxygen is the final electron acceptor in the mitochondrial electron transport. The decreased oxygen eliminates ATP production in mitochondria due to the reduced NADH oxidation (Bailey-Serres and Voesenek 2008). Activities of many glycolytic and fermentative enzymes increased in adventitious roots to generate ATP and regenerate NAD^+ (Bailey-Serres et al. 2012). However, oxygen production through glycolysis or fermentation is significantly lower than the energy from mitochondrial respiration. Roots are directly injured by the deprivation of oxygen in waterlogged soils, with the rapid death of seminal roots reported (Malik et al. 2001). Reduced nitrogen uptake and transportation, resulted from energy deficiency, lead to the death and senescence of old leaves (Trought and Drew 1980). Many other nutritional disorders also occur as a result of reduced ability of plant roots to maintain negative membrane potential and "fuel" high-affinity transport systems relying on a cotransport with H^+ (Pang et al. 2007b; Pang and Shabala 2010; Shabala et al. 2014). Consequently, the reduced leaf area, leaf water potential, dry matter, shoot growth, root growth and chlorophyll contents are common symptoms of waterlogging injury (Yan et al. 1996; Zhou et al. 1997; Malik et al. 2001; Pang et al. 2004; Kumutha et al. 2009).

3.2.2 Whole-Plant Responses to Oxygen Deprivation

Stomatal closure usually occurs within a few hours after exposure to waterlogged soils, with root signalling driving stomatal closure (Jackson 2002), most likely as a result of increased accumulation of abscisic acid in foliage (Jackson and Hall 1987). Stomata closure leads to the decreased CO_2 assimilation and reduced transpiration (Zhang and Zhang 1994; Yordanova and Popova 2001; Yordanova et al. 2005). Further decline in photosynthesis may be due to decreased chlorophyll content (Malik et al. 2001) and high levels of ROS accumulation in leaves (Arbona et al. 2008). Leaf chlorosis has been widely used as the indicator of waterlogging stress and waterlogging tolerance (Drew and Sisworo 1977; Xu and Mackill 1996; Yu and Rengel 1999; Zhou et al. 2007). Early senescence of older leaves is another typical symptom of waterlogging stress (Malik et al. 2002). Yield loss by waterlogging is mainly dependent on the different plant growth stage. Waterlogging at early seedling stages has a long-term influence on plant growth (Malik et al. 2002), and results showed that yield loss was the most when waterlogging occurred at early vegetative growth, compared with waterlogging at late vegetative growth or flowering (Watson et al. 1976).

3.2.3 Biochemical Alterations in Hypoxic Roots

Under waterlogging conditions, roots of plants and waterlogged soils synthesise a larger amount of ethylene quickly (Jackson 1985). Ethylene production has been detected under hypoxia conditions in different plants roots (Jackson et al. 1984; He et al. 1992; Steffens et al. 2010). Not only oxygen but also ethylene diffusion in waterlogged soils is slower than aerated soils, resulting in the ethylene accumulation in roots (Sairam et al. 2008). Ethylene is produced with the enzymes 1-aminocyclopropane-1-carboxylate (ACC) synthase and oxidase in plants. Many plant-adaptive physiological and biochemical traits are induced by the production and accumulation of ethylene, including aerenchyma formation (Sairam et al. 2008).

Waterlogging also induces overproduction of reactive oxygen species (ROS) including superoxide radical ($O_2^{\bullet}$), hydroxyl radical ($OH^{\bullet}$), hydroperoxyl radical ($HO_2^{\bullet}$) and hydrogen peroxide (H_2O_2). These ROS can be harmful to cellular metabolism and may lead to oxidative stress (Karuppanapandian et al. 2011; Overmyer et al. 2003). ROS in plants can also be used as signalling molecules inducing tolerance mechanisms (Wong et al. 2004; Gechev et al. 2006). Antioxidative enzymes and non-enzymatic compounds that scavenge ROS are effective at conferring waterlogging tolerance (Apel and Hirt 2004; He et al. 2012).

During waterlogging, oxygen deprivation is the main factor inhibiting plant growth, reducing the availability of oxygen for plants. Oxygen is the final electron acceptor in the mitochondrial electron transport. With oxygen deprivation in waterlogged soils, ATP production mainly depends on anaerobic metabolism, glycolysis or fermentation with the enhanced activities of alcohol dehydrogenase and pyruvate decarboxylase (Bailey-Serres et al. 2012; Yamauchi et al. 2014). However, oxygen production through glycolysis or fermentation is significantly lower than the energy from mitochondrial respiration. Insufficient energy inhibits the delivery of nutrients, and plants show deficiency in N, P, K, Ca and Mg in roots, shoots and leaves (Pang et al. 2007b). Anaerobic metabolism also creates toxic secondary metabolites, such as fatty acids and phenolics. These toxic secondary metabolites affect root nutrient uptake and membrane transport activities (Pang et al. 2007a; discussed also in Sect. 3.4). The relatively waterlogging-tolerant varieties were less affected by organic phytotoxins suggesting importance of this trait for breeding (Pang et al. 2007a).

3.3 Plant Adaptation to Hypoxia

In general, all the diversity of plant responses can be attributed to one of two strategies: escaping or quiescence. *Escaping strategy* involves the rapid stem elongation with the internode elongation under waterlogging conditions. Stem elongation allows plants to gain access to the atmosphere, enhancing respiration and photosynthesis (Hattori et al. 2010). However, fast elongation also probably cause plant death because of the fast energy consumption (Bailey-Serres and Voesenek 2008). *Quiescence strategy* restricts plant growth by reducing energy consumption under waterlogging conditions. After waterlogging, plants start regrowth rapidly (Colmer and Voesenek 2009). *Sub1* is the major locus for submergence tolerance on chromosome 9 in rice, contributing to 69 % of phenotypic variance (Xu and Mackill 1996). The locus shows the quiescence strategy by conserving energy (Xu et al. 2006). Several physiological features improve oxygen conditions in roots of waterlogging-tolerant plants. Among the most important are adventitious roots development (Mcdonald et al. 2001; Garthwaite et al. 2003), aerenchyma formation (Setter and Waters 2003; Evans 2004), reduced radial oxygen loss (ROL) (Garthwaite et al. 2003; Colmer et al. 2005), and greater activity of glycolytic pathway (Sairam et al. 2008).

3.3.1 Aerenchyma Formation

Root aerenchyma is a special tissue with gas spaces, forming an internal system to improve the diffusion and concentration of oxygen (Barrett-Lennard et al. 1988; Colmer 2003;

Visser and Bögemann 2003; Sairam et al. 2008). The increased concentration of oxygen leads to higher root respiration rates, increasing energy (ATP) availability in roots (Kawase and Whitmoyer 1980; Drew et al. 1985; Suralta and Yamauchi 2008) and allowing plants to avoid adverse effects caused by waterlogging (Burgos et al. 2001; Shiono et al. 2008). Aerenchyma formation is the most effective waterlogging tolerance mechanism for plants exposed to prolonged waterlogging treatments (Setter and Waters 2003).

Under waterlogged conditions, seminal roots are likely to be destroyed and adventitious roots are developed (Watson et al. 1976; Malik et al. 2001); if seminal roots can develop aerenchyma, their death can be prevented (Thomson et al. 1990; Colmer and Greenway 2010). Root porosity, via intercellular gas spaces, further increases with aerenchyma formation (Colmer 2003). However, aerenchyma formation would reduce the mechanical strength of roots; this problem can be addressed with a multiseriate ring of cells which is able to help plants maintain the mechanical strength of roots (Striker et al. 2007).

There are two types of aerenchyma: schizogenous aerenchyma and lysigenous aerenchyma. Schizogenous aerenchyma is generally the constitutive aerenchyma, formed by the separation of cells and growth patterns that form the gas spaces. Lysigenous aerenchyma is formed under stresses, and the main factor is the programmed cell death of cells in the cortex (Wegner 2010). Further aerenchyma types were classified, depending on the shape of aerenchyma (Jung et al. 2008), into three (Justin and Armstrong 1987) or six (Seago et al. 2005) major types. Aerenchyma can also be formed by the radial elongation of cortical cells in sponge gourd, resembling expansigeny aerenchyma, and this greatly improves waterlogging tolerance in plants (Shimamura et al. 2007).

3.3.1.1 Lysigenous Aerenchyma Formation

Lysigenous aerenchyma is normally formed under hypoxia conditions among many species, but only a few plants have constitutive schizogenous aerenchyma (Barrett-Lennard 2003). The mechanism of lysigenous aerenchyma formation is widely explored. Rice cultivars and wild relatives of maize are able to form constitutive lysigenous aerenchyma even under aerobic conditions (Colmer 2002; Mano et al. 2006b; Joshi and Kumar 2012; Nishiuchi et al. 2012; Yamauchi et al. 2013). Waterlogging stress resulted in lysigenous aerenchyma formation in many species including barley (Benjamin and Greenway 1979), maize (Mano et al. 2006b), rice (John 1977) and wheat (Malik et al. 2001). The genes and proteins associated with lysigenous aerenchyma formation have been studied (Emdadul Haque et al. 2011; Rajhi et al. 2011), and possible signalling pathways identified.

The classical view is that lysigenous aerenchyma formation includes five steps (Evans 2004): (1) hypoxia perception and the beginning of ethylene biosynthesis; (2) ethylene perception by cells of the mid-cortex; (3) loss of ions, initiation of cell death, plasma membrane invagination and formation of vesicles; (4) chromatin condensation, enhanced wall hydrolytic enzymes and organelles surrounded by membrane; and (5) cell lysis, cell wall degradation and fluid absorption by the surrounding cells and then formation of dry gas spaces. Recently, an ethylene-independent pathway of aerenchyma formation was suggested (Shabala 2011; Shabala et al. 2014), mediated by the loss of cytosolic K^+ and resultant increase in activity of proteolytic enzymes in root cortical cells.

3.3.1.2 Ethylene Signalling in Aerenchyma Formation via PCD

Numerous lines of evidence suggest that ethylene accumulation induced by hypoxia is essential to trigger the formation of lysigenous aerenchyma in waterlogged roots (He et al. 1996a; Colmer et al. 2006; Geisler-Lee et al. 2010). When the root was treated with the inhibitors of ethylene biosynthesis, aerenchyma was not developed and programmed cell death was also inhibited (He et al. 1996b; Yin et al. 2013).

According to accepted models, ethylene signalling leads to the expression of RBOH (the NADPH oxidase respiratory burst homolog), which induces the production of ROS (Torres and Dangl 2005; Kumutha et al. 2009). RBOH expression is the factor triggering the production and accumulation of H_2O_2 (one of the ROS forms) in the apoplast. H_2O_2 is then easily diffused into the cytosol (Rajhi et al. 2011) and triggers programmed cell death (Nishiuchi et al. 2012), leading to aerenchyma formation. The latter process may differ between species (Webb and Jackson 1986). In maize roots, the loss of tonoplast integrity leads to cellular collapse. In rice, non-apoptotic programmed cell death was detected during aerenchyma formation (Webb and Jackson 1986; Joshi et al. 2010).

The expression of genes controlling cell wall decomposition also increases when plants are subjected to waterlogging hypoxia conditions (Colmer and Voesenek 2009; Nishiuchi et al. 2012). The genes associated with scavenging of ROS and cell wall degradation were candidate genes contributing to the root lysigenous aerenchyma formation (Rajhi et al. 2011). In *Arabidopsis*, H_2O_2 and ethylene signalling also induce lysigenous aerenchyma formation, and three controlled genes were identified: *LESIONSIMULATINGDISEASE1* (*LSD1*), *ENHANCEDDISEASESUSCEPTIBILITY1* (*EDS1*), and *PHYTOALEXINDEFICIENT4* (*PAD4*) (Muhlenbock et al. 2007). Specifically, a hypoxia-induced gene was found to encode a xyloglucan *endo*-transglycosylase homolog, a putative cell wall-loosening enzyme associated with aerenchyma formation under waterlogging conditions (Peschke and Sachs 1994; Saab and Sachs 1996). Cell walls are further degraded by xylanolytic, pectolytic, and cellulosolytic enzymes (Jackson and Armstrong 1999). The fluid is then absorbed by surrounding cells, and the dry gas space is formed.

The process of lysigenous aerenchyma formation has been mainly studied in rice and maize (Bailey-Serres and Voesenek 2008; Evans 2004), while limited research has been performed in wheat and barley (McDonald et al. 2001; Garthwaite et al. 2003; Malik et al. 2003; Jiang et al. 2010). In terms of aerenchyma formation in wheat, membrane invagination and vesicles between the plasma membrane and the cell wall could be detected. The following process included chromatin condensation, membrane invagination and vesicles between the plasma membrane (Jiang et al. 2010). This process is similar to aerenchyma formation in maize (Evans 2004). Acid phosphatases were essential in the programmed cell death in wheat (Jiang et al. 2010), which was not detected in maize.

3.3.1.3 Ca^{2+} Signalling in the Lysigenous Aerenchyma Formation

Lysigenous aerenchyma formation induced by ethylene is also associated with other signalling pathways including Ca^{2+}, heterotrimeric G protein and phospholipase C (Nishiuchi et al. 2012). Microarray analysis and laser microdissection showed that Ca^{2+} signalling-related genes were upregulated under waterlogging conditions in both rice and maize (Rajhi et al. 2011).

Under waterlogging conditions, the formation of aerenchyma is greatly influenced by the availability of free cytoplasmic calcium (He et al. 1996b; Fagerstedt 2010). The inhibitors of free Ca^{2+} release in roots were able to block aerenchyma formation (He et al. 1996b). The increase of H_2O_2 and free Ca^{2+} in hypoxic stress in roots leads to ethanolic fermentation and alcohol dehydrogenase gene expression, which is another important strategy for plants to be tolerant under waterlogging stress (Baxter-Burrell et al. 2002).

Exogenous applications of ethylene and H_2O_2 increase Ca^{2+} influx in different plants under normal conditions. Cytoplasmic Ca^{2+} activities were studied in tobacco, by exposing tobacco protoplasts to ACC and ethylene-related compound, ethephon. Results showed that ethylene induced Ca^{2+} permeable channels activities and then increased Ca^{2+} influx and elevation (Zhao et al. 2007). Ca^{2+} signalling is able to trigger NADPH oxidase, which will increase the production of H_2O_2. H_2O_2 further activates hyperpolarisation-activated Ca^{2+} channels, resulting in the Ca^{2+} influx, which is also the beginning of this cycle (Laohavisit and Davies 2007). In *Arabidopsis*, exogenous H_2O_2 application leads to the increase of Ca^{2+} influx in roots (Demidchik et al. 2007).

3.3.2 Variation in Aerenchyma Formation between and Within Species

Several researchers reported that there is large genetic variation in waterlogging tolerance and aerenchyma formation among different genotypes in wheat and barley (Setter et al. 1999; Garthwaite et al. 2003; Zhou et al. 2007). This genetic variation can be utilised in the waterlogging tolerance breeding programmes in wheat and barley. In maize, genetic variation of aerenchyma formation has been utilised in improving waterlogging tolerance (Mano and Omori 2013a, b).

Rice is able to form constitutive aerenchyma. The constitutive root porosity in rice can be 20–30 %, increasing to more than 40 % in waterlogged soils (Colmer 2002; Steffens et al. 2010). In wheat, seminal, nodal and adventitious roots were all capable of forming aerenchyma under waterlogged conditions (Barrett-Lennard et al. 1988; Thomson et al. 1990; Setter et al. 1999). The wild relative *H. marinum* and some wild relatives of maize are able to form adventitious root aerenchyma under well-aerated conditions. This ability is expected to be related to waterlogging-tolerant traits since plants with developed aerenchyma adapt to waterlogged soils quickly (Mano and Omori 2008; Malik et al. 2009).

Cultivated barley (*Hordeum vulgare*) and wheat (*Triticum aestivum*) are more waterlogging susceptible, compared with their wild relatives *H. marinum*, and do not form constitutive aerenchyma under normoxic conditions. However, aerenchyma formation is induced in these species under waterlogging conditions. Root porosities of seminal, nodal and adventitious roots of cultivated barley and wheat all increased with restricted oxygen conditions, although to the lesser extent compared with their wild relatives, rice, or wetland species (Benjamin and Greenway 1979; Thomson et al. 1990; McDonald et al. 2001; Garthwaite et al. 2003).

The faster development of aerenchyma is also important for waterlogging tolerance in plants. After 1 week of hypoxia conditions or waterlogging treatments, waterlogging-tolerant *Melilotus siculus* species (Teakle et al. 2011) and waterlogging-tolerant soybean varieties were able to form aerenchyma (Thomas et al. 2005; Shimamura et al. 2010). The porosity of these waterlogging-tolerant varieties reached more than 20 % after 7 days of waterlogging treatment. By contrast, in the relatively less tolerant wheat and canola, aerenchyma was also induced soon after waterlogging treatment, but the percentage of aerenchyma formation was only less than 10 % (Voesenek et al. 1999; Yamauchi et al. 2014).

The above results support the notion that the waterlogging-tolerant cultivars generally have significantly higher root porosity than the susceptible cultivars, with faster aerenchyma formation under waterlogging conditions (Gibberd et al. 1999; Voesenek et al. 1999; Gibberd et al. 2001). Therefore, screening wheat and barley cultivars for adventitious root porosity might provide a potential selection criterion for tolerance under water waterlogging conditions.

3.3.3 Oxygen Transport from Shoot to Root

In waterlogged plants, oxygen supply in roots depends mainly on the internal transport of oxygen from shoots (Kotula et al. 2014). Morphological features, such as root thickness, transporting length and small numbers of laterals, also influence the internal oxygen transport (Jackson and Armstrong 1999). The oxygen transport from shoot to root is also affected by temperature since the lower temperature decreases the respiration rates, causing the lower demand of oxygen in roots (Colmer 2003). Longitudinal diffusion of oxygen through aerenchyma from shoots towards the roots may be diffused to the rhizosphere or consumed for respiration (Yamauchi et al. 2013). Convection of gas is another possible way to increase oxygen transport in waterlogged plants. But compared with oxygen diffusion from shoot to root, convection had less influence on root aeration (Beckett et al. 1988).

3.3.4 Formation of ROL Barrier and Control of Oxygen Loss

Longitudinal diffusion of oxygen through aerenchyma towards the roots may be diffused to the rhizosphere (Colmer 2003; Yamauchi et al. 2013). Such diffusion is termed radial oxygen loss (ROL). The barrier to ROL in waterlogged soil improves the diffusion of oxygen towards roots through aerenchyma, resulting in the greater accumulation of oxygen in roots (Colmer 2003; Nishiuchi et al. 2012), and thus is regarded as an important adaptive trait in waterlogged soils (Armstrong 1979). In addition to preventing the oxygen loss from roots to the soils, ROL also impedes movement of toxins into the roots (Malik et al. 2001; Colmer 2003; Kotula et al. 2009b).

In rice roots, the suberised exodermis and lignified sclerenchyma cells act as the main ROL barrier under stagnant conditions (Kotula and Steudle 2009; Kotula et al. 2009a). The suberin and lignin contents in the outer parts of roots in deoxygenated solution were greatly higher than the contents of plants in aerated conditions (Kotula et al. 2009a; Ranathunge et al. 2011). Suberin deposits were accumulated earlier than lignin, indicating the possibility that suberin is more important than lignin in forming the barrier to ROL (Shiono et al. 2011). The oxygen permeability under aerated conditions was higher than the oxygen permeability under non-aerated conditions, indicating the barrier to ROL was effective in lowering the oxygen permeability (Kotula and Steudle 2009).

The factors inducing the formation of the ROL barrier are not discovered yet. Ethylene is not able to enhance the ROL barrier even though ethylene is the main factor contributing to aerenchyma formation (Colmer et al. 2006). In addition, the formation of the ROL barrier is affected by root length, while aerenchyma formation is not associated with root length (Shiono et al. 2011). These findings are indicative of the fact that the formation of aerenchyma and ROL barrier may occur via independent signalling pathways. Cell degradation during aerenchyma formation might also lead to the formation of ROL barrier (Garthwaite et al. 2008). Carboxylic acids and caproic acid had a great impact on forming the barrier to ROL in rice, while the barrier to ROL is mainly formed because toxins caused injury rather than the function of specific signals (Armstrong and Armstrong 2001, 2005). Silicon also induced the increase of suberisation and lignifications and thus was suggested as a component of the ROL barrier formation in rice (Fleck et al. 2011). Overall, compared with the detailed aerenchyma formation process, the process of ROL barrier formation remains less understood.

It was suggested that in wetland species, the ROL barrier could influence water and nutrient uptake (Armstrong 1979). Other results, however, showed that water uptake was not affected by the formation of the barrier to ROL. In both *Hordeum marinum* and rice, ROL barrier did not influence the hydraulic conductivity of root systems (Garthwaite et al. 2006; Ranathunge et al. 2011). When rice was subjected to hypoxia conditions, rice roots had much higher bulk water permeability than diffusional water permeability (Kotula et al. 2009a; Ranathunge et al. 2004). There seems to be a good balance between water uptake and oxygen loss in waterlogging-tolerant cultivars (Kotula et al. 2009a; Nishiuchi et al. 2012; Ranathunge et al. 2011). Hence, it seems to be possible to reduce the ROL while retaining water uptake in plants.

3.3.5 Gas Film Formation

Leaf gas film, termed plant plastrons (Raven 2008), is another trait that enables oxygen availability and enhances waterlogging tolerance. In submerged plants, leaf gas films facilitate the gas exchange between the submerged leaves and water due to enlarged water-gas interface (Colmer and Pedersen 2008). Formed leaf gas films improve the whole plant aeration via the oxygen transport from water to plants in darkness and enhance photosynthesis rate in light through the carbon dioxide transport (Pedersen et al. 2009). Relatively waterlogging-tolerant plants maintain higher sugar contents through an increased photosynthesis rate and whole plant aeration conditions (Pedersen et al. 2009). Enhanced growth of submerged rice was also observed with the leaf gas film formation (Colmer and Pedersen 2008).

3.4 Secondary Metabolite Toxicities in Flooded Soils

3.4.1 Changes in Soil Redox Potential under Flooding

As waterlogging occurs in the soil, progressive decrease in redox potential (Eh) occurs. The latter is generally regarded the most appropriate indicator of the chemical changes occurring during soil waterlogging (Pezeshki and DeLaune 1998). Not only it is a suitable indicator of O_2 level (Eh around +350 mV under anaerobic conditions), but Eh per se also considerably affects the availability and concentration of many nutrients in the soil (Pezeshki 2001). Following flooding, the solute O_2 in soil is rapidly consumed in aerobic microbial respiration, and other inorganic electron acceptors are used in microbial respiration in the well-known sequence NO_3^-, Mn^{4+}, Fe^{3+} and SO_4^{2-} (Kirk et al. 2003). As soon as oxygen is depleted in soils, nitrate is utilised by soil microorganisms as a primary alternative electron acceptor in respiration. As a result, nitrate is reduced to nitrite (NO_2^-) or various nitrous oxides (e.g. N_2O, NO) and molecular nitrogen (N_2) in the process of denitrification, which occurs at round +300 mV redox potential (corrected to pH 7) (Gambrell et al. 1991). Evidence also showed that nitrite can substitute oxygen at the terminal cytochrome oxidase step allowing electron transport to proceed in anaerobic metabolism (Izaguirre-Mayoral and Sinclair 2005).

Mn and Fe are essential trace elements for all higher organisms and used in a variety of metabolic processes, including enzyme activation (Marschner 1995; Alscher et al. 2002). The oxidised forms Mn(III) and Mn(IV) are not bioavailable to plants and cannot be utilised (Rengel 2000). Fe is mostly found to be presenting in its insoluble ferric form that is unable to be utilised under various physiological conditions although it is abundant in the earth's crust (Guerinot and Yi 1994). While redox potential drops to approximately +200 mV, Mn oxides are the next electron acceptors. In this case, the solubility of Mn is significantly enhanced under flooding condition, exceeding plant requirement for this micronutrient and often exceeding toxicity level (Khabaz-Saberi et al. 2006). As excess Mn and Fe can be toxic to plants, accumulation and segregation of both metals must be kept under rigorous homeostatic control. This includes control over both uptake and sequestration of Mn and Fe.

3.4.2 Elemental Toxicities in Flooded Soils

3.4.2.1 Mn and Fe Uptake by Roots

The molecular mechanisms for Mn^{2+} transport into the root from the soil solution are still poorly understood (Rengel 2000; Pittman 2005), even though several transporters have been implicated to be responsible for Mn accumulation (Pedas et al. 2008; Cailliatte et al. 2010; Sasaki et al. 2012). Manganese is probably absorbed and transported into the cell only in its reduced form of Mn(II) or divalent cation Mn^{2+}. The concentration of Mn^{2+} in the soil is strongly dependent on pH of the soil solution. Mn toxicity can be a crucial constraint to plant growth when the soils are too acidic ($pH<5.5$); on the contrary, Mn deficiency will occur in alkaline soils (Marschner 1995). Mn^{2+}-ethylenediaminetetraacetic acid (MnEDTA) complex is suggested to be taken up into root cells in barley (Laurie et al. 1995). Recently, IRT1 is revealed to have a capacity to transport a range of metals including Mn^{2+}, and it was suggested as the primary pathway for Mn^{2+} transport during Fe deficiency in barley (Pedas et al. 2008). Additionally, two other different plasma membrane transporters from the NRAMP family have been reported to affiliate with Mn uptake into root cells. NRAMP1, which is expressed in the root elongation zone of root tips, contributes high-affinity Mn uptake in *Arabidopsis* (Cailliatte et al. 2010). In comparison with IRT1, the expression of NRAMP1 is more widespread along the root, especially enhanced in Mn deficiency. However, OsNRAMP1 was not responsible for Mn uptake in rice roots (Ishimaru et al. 2012b), where this role was attributed to NRAMP5. The latter transporter is only expressed in the plasma membrane at the distal site of exodermis and endodermis; this differs from NRAMP1 which can be detected in all cell layers of the root. NRAMP5 knock-out lines of rice showed a significant reduction of Mn concentration in the shoot in plants cultivated under flooding conditions (Sasaki et al. 2012).

The molecular mechanisms of Fe transport in plants are known better than those of Mn, and most research on Fe in plants concentrate on deficiency rather than excess. Two distinct strategies are used by plants to acquire Fe from soils. Reduction (strategy I) is utilised by most types of plants (eudicots and non-grass monocots). Ferric reductase oxidase (FRO2, the key strategy I gene) encodes ferric chalet reductase in root, which is used to reduce chelate Fe(III) to soluble Fe^{2+} form (Robinson et al. 1999). Plants belonging to this strategy can transport iron across plasma membrane only in the reduced form (Fe^{2+}). AtIRT1 is expressed as the Fe^{2+} transporter in the root epidermis. In *Arabidopsis*, the loss-of-function IRT1-1 mutant line is responsible for Fe deficiency in plants, resulting in plant chlorosis and proving that IRT1 can be the primary transporter involved in Fe uptake from the soil (Henriques et al. 2002; Varotto et al. 2002; Vert et al. 2002). Expression of IRT1 is under the control of the FIT bHLH transcription factor, which is required for iron deficiency response (Colangelo and Guerinot 2004). Similar to IRT1, IRT2, which is an intracellular membrane protein, is upregulated under Fe deficiency and is capable of transporting Fe^{2+} (Vert et al. 2001, 2002). However, IRT2 is not immediately involved in the absorption of Fe^{2+}, which appears to sequester excess Fe that is accumulated through IRT1 activity, implying that IRT2 is not able to replace the role of IRT1 in iron accumulation from the soil (Varotto et al. 2002; Vert et al. 2009).

In contrast, strategy II plants (grasses that lack the reduction mechanism) usually accumulate chelated Fe(III) (Zaharieva and Römheld 2000). These plants can release Fe(III)-solubilising compounds termed phytosiderophores such as mugineic acid (Mino et al. 1983) and deoxyribonucleic acid into the rhizosphere. TOM1, identified as phytosiderophores transporter, is involved in the efflux of deoxyribonucleic acid in rice and barley (Nozoye et al. 2011). Fe(III)-phytosiderophore complexes (Fe(III)-PS) are taken into the root cells by Yellow Stripe1 or Yellow Stripe-Like1 (YS1/YSL1) transporters identified in maize (Curie et al. 2001) and barley (Murata et al. 2006). HvYSL1 transporter specifically transports Fe(III)-PS. However, ZmYSL1 transporter appears to possess extended transport capacities that can also transport of Co(II)-, Cu(II)-, Ni(II)-, Mn(II)- and Zn(II)-PS (Schaaf et al. 2004). In barley, Fe(III)-PS uptake can possibly be achieved without being disturbed by the uptake of Co(II)-, Cu(II)-, or Zn(II)-PS complexes. Strategy II plants can also acquire Fe^{2+} from the soil by IRT transporters. The functional Fe(II) transport system performing in rice has similar characteristics as observed in strategy I plants. OsIRT1 and OsIRT2 encode plasma membrane transporters which can be detected in roots under low-Fe conditions (Ishimaru et al. 2006). OsNRAMP5 is expressed as a bivalent cation transporter located in the root epidermis, exodermis and outer cortex cells and is considered to be a major contributor to the transportation of Fe, Mn and Cd (Ishimaru et al. 2012b).

3.4.2.2 Mn and Fe Transport to the Shoot

Following the initial uptake of Mn and Fe into root cells, they must be transported to the aboveground portion of the plant through phloem and xylem. Due to the poor solubility and high reactivity of Fe, translocation of Fe inside the plant should be affiliated with proper chelators and certain redox states between ferrous and ferric forms (Marschner 1995; Hell and Stephan 2003). Citrate has long been accepted as the main form responsible for long-distance transport in the xylem sap (Tiffin 1966; Brown and Chaney 1971). Recently, a certain form of Fe chelate has been identified in tomato xylem sap as a tri-Fe(III)-citrate (Rellán-Álvarez et al. 2010). An

Arabidopsis Ferric Chelate Reductase Defective 3 (FRD3) transporter that belongs to the multidrug and toxic compound extrusion (MATE) family appears to be responsible for citrate efflux into xylem sap (Rogers and Guerinot 2002; Green and Rogers 2004; Durrett et al. 2007). FRD3 is essential for iron homeostasis. The absence of FRD3 in the *frd3* mutant of *Arabidopsis* can lead to loss or inhibition of the capacity to load citrate into the xylem, resulting in root chlorosis with approximately 40 % less citrate in xylem. Importantly, the *frd3* mutant is allelic to *manganese accumulator 1 (man1)*. *frd3* can lead to abnormal distribution and accumulation of various metals. This suggests that FRD3 contributes crucial constitutive expression of iron uptake responses (Rogers and Guerinot 2002; Durrett et al. 2007). The FRD3-like gene in rice, OsFRDL1, which also encodes a citrate transporter responsible for citrate efflux (Yokosho et al. 2009), has been identified in root pericycle cells, which is similar to *Arabidopsis* FRD3 (Green and Rogers 2004; Inoue et al. 2004). Another novel transporter, PEZ1, which is located mainly in the plasma membrane of the stele root, can transport protocatechuic acid (PCA) and caffeic acid when expressed in *Xenopus laevis* oocytes (Yokosho et al. 2009). It is thought to be responsible for xylem loading of these phenolics.

Fe(II)-nicotianamine (Fe-NA) and Fe(III)-DMA are main influx forms transported in the phloem and cytoplasm. Among those influx transporters, YSL family members are widely involved in Fe translocation. ZmYS1, which encodes both in root and shoot, facilitates Fe(III)-DMA uptake from the rhizosphere (Curie et al. 2001). In rice, it is observed that Fe(II)-NA can be transported from roots to young leaves through phloem by OsYSL2 transporters (Tsukamoto et al. 2009; Ishimaru et al. 2010), whereas Fe(III)-DMA cannot (Koike et al. 2004). TOM1 expression can be observed both in root cells and leaf sheaths. As DMA is responsible for Fe accumulation from soil and translocation in plants, TOM1 appears to play an essential role in DMA chelation from the rhizosphere and efflux of DMA to the phloem and xylem for internal Fe transport (Nozoye et al. 2011). Besides being expressed in the epidermis and exodermis, OsIRT1 is also expressed in vascular tissue, implying a function of Fe^{2+} uptake into phloem cells (Ishimaru et al. 2006).

In comparison, there are few reports on Mn transport via phloem and xylem. Mn is taken up mainly in Mn(II) form. In barley, transpiration can make little effect on Mn translocation to younger leaves, whereas it crucially affects Mn translocation in older leaves (Tsukamoto et al. 2006), implying that major Mn transport via phloem to the young leaves is the main pathway. It has been suggested that Mn^{2+} is possibly exported from the cell via a Mn^{2+}/H^{+} antiport mechanism (Clarkson 1988; Shigaki et al. 2003). Transporter ECA1 isolated from *Arabidopsis*, which belongs to Ca^{2+}-ATPase subfamily, plays a role in manganese transport (Axelsen and Palmgren 2001). Besides being responsible with Fe translocation, OsYSL2 also appears to be involved in Mn-NA distribution into phloem for long-distance transport (Koike et al. 2004; Ishimaru et al. 2010).

3.4.2.3 Physiological Constraints Imposed by Excessive Mn and Fe Accumulation

Heavy metal toxicity is one of the major abiotic stress-causing hazardous effects in plants. Mineral elements such as copper (Cu), zinc (Zn), aluminium (Al), iron (Fe) and manganese (Mn) are, at certain concentrations, essential micronutrients that are critically involved in functional activities such as the activation of enzymes and proteins and development of living organisms (Hossain et al. 2012). It is proven that the concentration of elemental ions, such as Mn^{2+} and Fe^{2+}, could be significantly enhanced in a hypoxia or anoxia context, especially in waterlogged acidic soil (Khabaz-Saberi et al. 2006). However, elevated concentrations of those mineral elements in the soil can result in toxicity symptoms and inhibition of growth of most plants (Evans and Sorger 1966; Hall 2002). Elevated Mn levels can severely affect plant development in growth medium (Khabaz-Saberi et al. 2006). Excessive concentration of Mn can interfere with absorption, translocation and utilisation of other mineral elements, such as Ca, Mg, Fe and P, and alter the activities of enzymes and hormones, while essential Mn-requiring processes become less active or nonfunctional (Epstein 1961; Clark 1982; Horst 1988). Several factors determine Mn availability by plants, including soil PH, redox potential, soil moisture and microbiological activities in soils (Wang et al. 2002). Different from Al, excess Mn generally affects plant shoots more detrimentally than roots. Although the expression of Mn toxicity varies considerably among plant species, the typical symptom of Mn toxicity for many plants is marginal chlorosis and necrosis of leaves, while brown necrotic spots can be observed on older leaves in barley (Foy et al. 1978; El-Jaoual and Cox 1998). Additionally, those brown necrotic spots are reported to contain accumulations of oxidised Mn or precipitation of Mn compounds (Horst 1988). In some severe cases of Mn toxicity, high concentration of Mn causes root browning, and root browning also indicates the presence of oxidised Mn (Foy et al. 1978; Horiguchi 1987). High concentrations of Fe and Mn in waterlogging-affected acid soil are reported as crucial restrictions for intolerant wheat genotypes with 25–50 % decrease in shoot dry weight (Khabaz-Saberi and Rengel 2010) and two- to ten-fold increase in shoot Mn and Fe concentrations (Khabaz-Saberi et al. 2006).

ROS react with cellular constituents depending on properties such a chemical reactivity, redox potential, half-life and mobility within cellular compartments, and OH• radicals are the most reactive and short lived (1 ns) (Møller et al. 2007; Sharma and Dietz 2009). Mn and Fe toxicity can mediate H_2O_2 production or consumption and the oxidation of phenols in the leaf apoplast. Stimulating POD activity was observed in rice leaves while it absorbed excess Fe (Fang and Kao 2000). In cowpea, POD involving H_2O_2 consumption induced by the oxidation of Mn^{2+} has been seen to be a key reaction defending Mn stress, leading to Mn toxicity symptoms (Fecht-Christoffers et al. 2003).

Excessive Mn^{2+} and Fe^{2+} can also compete with other essential metal ions and cause nutrient ion deficiency. High Mn^{2+} concentration in the growth solution caused a significant decrease of K^+ influx in barley roots (Pang et al. 2007b). The displacement of one heavy metal ion by another generates the inhibition or loss of enzyme activities (Hossain et al. 2012). A high Mn availability in the soil can inhibit Fe accumulation and even alleviate Fe toxicity in plants (Fageria et al. 2008). Sequentially, this blocking of functional groups of metabolically important molecules, displacement of metal ions for biomolecules and membrane integrity attribute to altered plant metabolism and inhibition of photosynthesis respiration and alter activities of various key enzymes (Sharma and Dietz 2009; Hossain et al. 2012).

3.4.2.4 Mechanisms to Deal with Elemental Toxicities

Tolerance to heavy metals in plants may be stated as the ability to survive in soils that are toxic to other plants (Mark et al. 1999). However, the occurrence of multiple tolerance and co-tolerance combining several heavy metals remains to be resolved (Hall 2002). Although great genotypic variation in Mn toxicity tolerance was reported for some species such as rice, mechanisms of plant tolerance to Mn toxicity are fragmentary (Wang et al. 2002; Shabala 2011). Restriction of accumulation and transport of heavy metals is an essential avoidance strategy for plants, involving root interception of elemental ions and translocation to the shoot. The removal of metals from the cytoplasm, particularly segregation into the vacuole, is a crucial tolerance mechanism when accumulating excess metals. In Arabidopsis, several pathways have been identified for Mn import into vacuole. AtCAX2 and AtCAX4 cation/proton antiporters possess a capacity to transport Mn into the vacuole (Cheng et al. 2002; Schaaf et al. 2002; Koren'kov et al. 2007). Additionally, three metal tolerance proteins (MTPs) belonging to the cation diffusion facilitators (CDF) have been reported to be affiliated with Mn tolerance. The Mn-specific transport protein ShMTP8 is targeted to the vacuole membrane, conferring Mn tolerance by compartmentalisation of excess Mn when expressed in *Arabidopsis* (Delhaize et al. 2003). MTP11, another CDF family member which is expressed in pre-vacuolar or Golgi vesicles, has been revealed to be responsible for Mn tolerance (Delhaize et al. 2007). A great increase of Mn concentration is revealed in shoots and roots of the *mtp11* mutant, with 45–60 % higher concentration, implying MTP11 transporter can significantly activate segregation of excess Mn. In rice, OsMTP8.1 is expressed in cells of leaf blades and localised to the tonoplast. This transporter is also thought to play a role in sequestering Mn into vacuoles in the shoot (Chen et al. 2013). OsYSL6 is constitutively expressed at a high level both in roots and shoots under excess Mn or Fe conditions. Importantly, YSL6 is expressed unlike other YSL members whose expression is up- or down-regulated by Fe deficiency (Sasaki et al. 2011). NRAMP members have been involved in numerous functions including uptake, translocation, intracellular transport and detoxification of transition metals (Nevo and Nelson 2006). Overexpressed NRAPM1 in *Arabidopsis* is crucial to Fe homeostasis with excess Fe, indicating a tolerance to Fe toxicity (Curie et al. 2000).

3.4.3 Organic Phytotoxins in Flooded Soils

Another essential factor that affects overall flooding tolerance is the secondary metabolite toxicities that occur in waterlogged soils (Drew and Lynch 1980; Armstrong et al. 1996; Armstrong and Armstrong 1999, 2001, 2005; Shabala 2011; Setter et al. 2009). These substances may significantly affect plant physiological performance, both at the cellular and the whole-plant level. Toxic compounds can enter roots and move to shoots, with adverse effects on both organs (Baba et al. 1965; Armstrong and Armstrong 2005). Damage to plants, particularly in species well adapted to wetland conditions, has sometimes been attributed to the accumulation of harmful concentrations of organic and inorganic substances in the soil, rather than to the effects of oxygen deficiency per se (Drew et al. 1981).

3.4.3.1 Secondary Metabolite Toxins Produced under Flooded Conditions

In addition to inorganic phytotoxins such as Fe^{2+} and Mn^{2+}, a significant accumulation of organic substances also occurs in waterlogged soils. Numerous volatile (short-chain) fatty acids and phenolics accumulate in soils during prolonged waterlogging (Drew and Lynch 1980; Armstrong and Armstrong 1999). Organic acids are formed from plant residues and have been shown to reach rapidly phytotoxic concentrations (Drew and Lynch 1980). These are secondary metabolites, produced as a result of anaerobic metabolism in both plants (Lynch 1977; Armstrong and Armstrong 1999) and rhizosphere microorganisms (Drew and Lynch 1980). The type and amount of secondary metabolites produced depend upon the fermentative character of the microflora, the type and amount of organic materials added, the duration of waterlogging and physical and chemical changes associated with waterlogged soils.

As a consequence of microbial respiration, the redox potential (Eh) of the soil gradually declines (Drew and Lynch 1980). When Eh values fall below zero, SO_4^{2-} is reduced to H_2S, resulting consequently to methane production. Short-chain aliphatic acids (acetic, propionic and butyric) and aromatic acids including phenolics have been isolated from a range of soils under anaerobic conditions. With the volatile fatty acids, phenolic acids and nitrite (in equilibrium with nitrous acid), it is the undissociated acid form that is most active (Drew and Lynch 1980). The process of H_2S production is pH dependent as sulphate-reducing bacteria, e.g. *Desulfovibrio desulfuricans*, that produce sulphide require pH 5.5–9 and anaerobic conditions (Starkey 1966). H_2S produced in the reduction is readily soluble and above pH 7 dissociates to S^{2-} and HS^-. All three reduced S species are highly toxic to plants.

In addition, produced organic acids, as well as high CO_2, can impose "acid loads" on cells of roots in waterlogged soils (Greenway et al. 2006). Upon re-aeration after a period of O_2 deprivation, ethanol remaining in tissues will be converted into acetaldehyde that can induce post-anoxic cell injuries (Bailey-Serres and Voesenek 2008).

3.4.3.2 Whole-Plant and Cellular Responses to Phytotoxins

Flooding affects plants in two ways: while roots experience the direct effects of waterlogging, the shoots of plants in waterlogged soils suffer any consequences of root dysfunction. Phytotoxins have been shown to substantially affect nutrient uptake. Shoots can be damaged by deficiencies of mineral nutrients, particularly nitrogen, and by an influx of reduced soil toxins. For example, the application of submillimolar concentrations of cinnamic and benzoic acid derivatives caused 40–95 % reduction in both phosphorus (P) and potassium (K) reported in barley roots (Glass 1973, 1974). These effects were observed within minutes (Glass 1974) and were reversible (Glass 1973), suggesting a direct effect upon the cell membrane. Organic acids were also shown to reduce the uptake of other nutrients such as magnesium (Mg), calcium (Ca) and nitrogen (N) in a range of plant species (Booker et al. 1992; Armstrong and Armstrong 2001).

Plant performance under flood conditions largely depends on soil type, as was shown for different wheat genotypes (Setter et al. 2009). Armstrong and co-workers reported that the common reed, *Phragmites australis*, and rice responded to sulphide and the low molecular weight organic acids by the cessation of root growth and inhibition of lateral root emergence. The induction of blockages within the internal gas space system, lignification and suberisation in the normally permeable parts of the root system such as the superficial layers of fine laterals and of the apical regions of adventitious roots and blockages within the vascular system (Armstrong et al. 1996; Armstrong and Armstrong 1999, 2001, 2005). They also showed that ROL from adventitious root apices was greatly decreased in rice by acetic acid (15 mM as a single dose) and in *Phragmites* by mixtures of low molecular weight organic acids (Armstrong and Armstrong 2001). Damage to rice shoots attributed to hydrogen sulphide is known as browning, bronzing, brusome, akiochi, hie-imochi, suffocation and straight head diseases. Sulphide is a well-known inhibitor of oxidative metabolism in rice and other species (Armstrong and Armstrong 2005). Control of the disease is best achieved through increasing soil redox potential by the application of nitrate to prevent sulphide formation, rather than by promoting binding of sulphide.

It should also be stressed that many investigations analysed effects of only one of the potential toxins while there is possibility of synergistic interactions between toxins, particularly at limiting oxygen concentrations, thus calling for further attention.

3.4.3.3 Plant Adaptation to Organic Phytotoxins

Plants developed different survival approaches in the presence of high levels of phytotoxins and O_2 deprivation including minimisation of the adverse effects of phytotoxins to the roots and shoots and the repair of already damaged tissue. As mentioned earlier, some injuries caused by phytotoxins are attributed to inadequate uptake of inorganic nutrients and water. In order to compensate for the latter, plants developed physiological responses such as the slower growth rate of the shoot, the redistribution of inorganic nutrients to the younger leaves, senescence and abscission of the older leaves, epinasty and rapid stomatal closure (Drew and Lynch 1980; Colmer and Voesenek 2009). All these traits are designed to minimise nutrient requirements and water loss by the plant. Nitrogen is also redistributed from the older leaves to the younger (growing) leaves, thus compensating the lack of supply from the roots (Drew and Lynch 1980).

A number of mechanisms promote detoxification of reduced products in the anaerobic soil in the immediate vicinity of the root. These include the formation of root barriers to restrict entry of these compounds, as well as having some ROL from key sites (root tips and laterals) to re-oxidise these substances in the rhizosphere, regarded as important traits for minimising these toxicities (Armstrong 1979; Armstrong and Armstrong 2005). Oxidation of these compounds to less-toxic forms can occur either through the activities of aerobic microorganisms that colonise the oxygenated rhizospheres or by direct chemical oxidation (Begg et al. 1994; Kirk and Bajita 1995). In this context, radial oxygen loss (ROL) from root is considered to be essential for detoxification of phytotoxins (e.g. Fe^{2+}, Mn^{2+}, H_2S, S^{2-}, HS^-) and organic acids by direct oxidation or by the agencies of oxidising aerobic microorganisms from the rhizosphere (Armstrong and Armstrong 2005).

Finally, exposure to hypoxia for a short period of time prior to prolonged exposure to hypoxia allows cells to initiate processes that favour survival (Drew 1997; Chang et al. 2000; Bailey-Serres and Voesenek 2008). It would be of interest to assess whether species differ in their intrinsic tolerance to phytotoxins once the entry into the shoots and roots has already occurred.

3.5 Crop Breeding for Flooding Stress Tolerance

3.5.1 Transgenic Approach: Is This a Way Forward?

Genetically modified crops are plants used in agriculture, the DNA of which has been modified using genetic engineering techniques. In most cases, the aim is to introduce a new trait to the plant which does not occur naturally in the species. Genetic engineering promises to speed the process and broaden the scope of what can be done. According to the International Service for the Acquisition of Agri-biotech Applications (ISAAA), the area used for biotech crop increased by more than 100-fold from 1.7 million hectares in 1996 when first commercialisation started to over 175 million hectares in 2013 which are distributed in 27 countries. From 1996 to 2012, the use of biotech crops increased crop production valued at US$116.9 billion. In 2013, more countries, including Bangladesh, Indonesia and Panama approved biotech crop planting for commercialisation in 2014.

While the first generation of genetically engineered crops included mainly pest and herbicide tolerance (Ohkawa et al. 1999; Moellenbeck et al. 2001), the following research have targeted various quality traits (e.g. product shelf life; Bruening and Lyons 2000; Haroldsen et al. 2012; nutrition; Beyer et al. 2002) as well as a broad range of stress-tolerant traits such as cold (Kenward et al. 1999) and soil salinity (Zhang and Blumwald 2001). Attempts have also made to improve waterlogging/submergence tolerance. For example, *Arabidopsis* with transferred chimeric SAG12-ipt gene displayed higher-level of tolerance to waterlogging than wild-type plants (Zhang et al. 2000). The improved tolerance was explained by accelerated function of anti-oxidative system in transgenic plants.

However, the introduction of transgenic plants into agriculture has been vigorously opposed by both general public and authorities in many countries. Also, in addition to the potential risk of transgenes in commercial crops endangering native or nontarget species, there are some scientific issues. It is impossible to guide the insertion of the new gene, which leads to unpredictable effects. Also, gene expression is not necessarily directly correlated with the protein content so expression levels of the gene might not reflect expression of the phenotype (Zhou et al. 2013). Moreover, since genes do not work in isolation but in highly complex relationships which are not understood, any changes to the DNA at any point will affect it throughout its length in ways scientists cannot predict.

3.5.2 Marker-Assisted Selection (MAS) Approach to Plant Breeding

As a GMO approach is still being highly debated, conventional breeding remains the best way for variety improvement. Similar to other major abiotic stresses, little progress has been made in breeding for waterlogging tolerance because of its low heritability, variability among waterlogging treatments and the difficulty of screening a large number of lines in the field or under controlled conditions (Collako and Harrison 2005; Zhou 2010). Up to date, plant breeding for waterlogging tolerance was driven mainly by the availability of convenient screening tools. Agronomic characteristics represent the combined genetic and environmental effects on plant growth that have been widely used both in genetic studies

(Mano et al. 2005, 2006b; Li et al. 2008; Zhou et al. 2012; Xue et al. 2010) and in breeding programmes (Zhou 2010).

Marker-assisted selection (MAS) could be very effective in selecting the tolerance for various stresses which are easily affected by environment. Molecular markers give unambiguous, single site genetic differences that can easily be scored and mapped in most segregating populations (Kearsey 1998). However, QTL analysis depends on the close linkage between markers and the differences in the marker genotypes will be associated with different trait phenotypes (Kearsey 1998). Thus, the success of MAS depends on the development of reliable markers (accurate QTL location).

3.5.2.1 MAS for Agronomic Traits Linked to Waterlogging/ Submergence Tolerance

Many QTLs have been reported for waterlogging or submergence tolerance based on various agronomical and morphological characteristics. Some markers closely linked to the traits have been successfully used for many traits such as acid soil (http://www.theland.com.au/news/agriculture/cropping/grains/acidic-soils-options/2667100.aspx) and barley yellow dwarf virus (Western Australia breeding programme, pers. comm.) resistance in breeding programmes. In rice, a submergence tolerance locus has been successfully transferred to a submergence-intolerant Thai Jasmine rice variety KDML105 without changing any agronomically desirable traits through backcrossing using MAS (Siangliw et al. 2003). Using a population developed from a cross between an indica submergence tolerant line (IR40931-26) and a susceptible japonica line (PI543851), a major QTL was mapped to chromosome 9, designated as Sub1 (Xu and Mackill 1996). This QTL accounted for about 70 % of the phenotypic variation in submergence tolerance in the population studied. Nandi et al. (1997) confirmed the importance of Sub1 in submergence tolerance and identified four additional QTLs on four different chromosomes.

However, most of the traits used as the indices for waterlogging/submergence tolerance are not only easily affected by environment but controlled by many genes. For example, 34 different QTL-controlling waterlogging tolerance in maize were detected based on the analysis of agronomical traits such as plant height, root length and root and shoot dry weight; these QTLs were mapped to eight different chromosomes (Qiu et al. 2007). All the QTLs determined only a small proportion of phenotypic variation and it is not possible to conduct fine mapping in order to find molecular markers closely linked to the tolerance genes. Accurate phenotyping may narrow down the number to a few major QTLs, but it is still hard to make effective selection on many QTLs at the same time.

3.5.2.2 MAS for Major Physiological Traits Conferring Waterlogging Stress Tolerance

Plant adaptation to waterlogged conditions is physiologically complex and involves an orchestrated action of several contributing mechanisms. Linking tolerance with identifiable physiological traits may help guide evaluation and selection processes associated with breeding.

Minimising radial oxygen losses (ROL) from roots is one of the key features of waterlogging stress tolerance (Colmer and Voesenek 2009). Wetland species are able to form a barrier to ROL in the basal zones of roots with plenty of oxygen diffusion to the root apex (Abiko et al. 2012a). Many rice varieties grown in stagnant solutions also induced a tight barrier to ROL (Colmer 2003). The induced high root porosity and a barrier to ROL in wild relative *H. marinum* are the main contributors to waterlogging tolerance, with 84 % of the reduction in oxygen loss (Garthwaite et al. 2008). The wild relative of maize, *Z. nicaraguensis*, which is waterlogging tolerant, showed much greater ability to form a barrier to ROL in the basal zone of roots than less waterlogging-tolerant cultivated maize (Abiko et al. 2012a). Genetic variations in the formation of the barrier to ROL exist among varieties, and the formation of the barrier to ROL is correlated with waterlogging tolerance in rice (Colmer 2002), wheat (Malik et al. 2003), maize (Abiko et al. 2012a) and pasture species (Gibberd et al. 1999). However, the success of breeding waterlogging-tolerant varieties through ROL relies on the discovery of molecular markers linked to this trait since it's not practical to select a large number of individuals for ROL in a breeding programme.

Aerenchyma formation is considered to be one of the most crucial adaptive traits for waterlogging tolerance (Jackson and Armstrong 1999; Evans 2004; Colmer and Voesenek 2009). Waterlogging tolerance showed a strong correlation with aerenchyma formation in roots under waterlogging stress in barley (Colmer 2003; Pang et al. 2004; Broughton et al. 2015), wheat (Malik et al. 2003), maize (Abiko et al. 2012a), and pasture species (Gibberd et al. 2001). Wild relatives of maize are able to form constitutive lysigenous aerenchyma under non-waterlogging conditions. The constitutively formed aerenchyma allows cultivars to adapt to waterlogging conditions more rapidly than those without constitutive aerenchyma. QTLs were identified for aerenchyma formation (Mano et al. 2007) in maize. One major QTL on chromosome 1 was found in two populations, maize inbred B64×teosinte *Zea nicaraguensis* cross population and their advanced backcross population (Mano et al. 2007, 2008; Mano and Omori 2008, 2009). Nearly 50 % of the phenotypic variation was determined by this QTL (Mano et al. 2007). This QTL was also identified in other population B73×teosinte *Zea luxurians*, determining 36.3 % of the phenotypic variation (Mano et al. 2008). Six candidate genes were identified for constitutive aerenchyma formation in the region where the major QTL was located (Abiko et al. 2012b). In barley, a single major QTL was identified on 4H (Broughton et al. 2015), and this QTL was located in a similar position to a QTL for waterlogging tolerance (Zhou et al. 2012). This QTL explained 35.7 and 39.0 % of phenotypic variation in aerated and oxygen-deficient conditions, respectively. Comparative mapping revealed this QTL is syntenic with the Qaer1.02-3 QTL in maize (Mano et al. 2007) and the Sub1A-1 gene in rice (Xu and Mackill 1996), which are responsible for aerenchyma formation and submergence tolerance, respectively.

Superior ability to reduce detrimental effects of secondary metabolites and toxic ions such as Mn^{2+} was shown to be related to waterlogging tolerance (Pang et al. 2007a, b; Khabaz-Saberi and Rengel 2010; Khabaz-Saberi et al. 2012; Huang et al. 2014). Our preliminary study identified one major QTL and one minor QTL for Mn^{2+} tolerance in a DH population, and the minor QTL was in a similar position to one of the QTLs for waterlogging tolerance (Zhou 2011).

Many other physiological traits such as potassium retention in the cytosol, maintenance of a negative membrane potential and ROS tolerance may also contribute to waterlogging tolerance. However, no attempts have been undertaken in identifying QTL linked to these traits, which limits the direct use of these traits as selecting indices.

3.5.3 Limitations and Development of More Reliable Markers to Be Used in MAS

Many factors limit the progress of breeding for waterlogging/submergence tolerance. Since the direct selection on yield for waterlogging tolerance is not practical in a breeding programme, various agronomical and physiological traits were used as indirect selecting criteria. The relationships between these traits and waterlogging tolerance are crucial for the effectiveness of selection.

The mapping of quantitative trait loci (QTL) is an important step towards MAS. However, QTL mapping using segregating populations has limited resolution (Paterson et al. 1990) with most QTL covering a wide region of chromosome. Using QTL-derived markers for direct selection can often lead to disappointment as recombination at sites internal to the markers may disrupt the QTL. In most cases, several QTL were involved in the tolerance based on agronomic traits with each QTL controlling a relatively small proportion of phenotypic variance. This makes fine mapping of these QTL very hard. Physiological traits can be simply inherited (i.e. aerenchyma formation), and increased waterlogging tolerance can be achieved by pyramiding different traits. However, very few studies were conducted in identifying QTL for these traits, which is necessary for these traits to be used in breeding programmes.

To obtain markers that can be reliably used to follow a locus, additional experiments have to be conducted to fine map the region. One of the methods of resolving the map position of a QTL is by analysing a series of near-isogenic lines (NILs) that differ in markers flanking the QTL (Young et al. 1988; Paterson et al. 1990). NILs contain only a single introgression per line, which increases the power to detect small-effect QTL and conduct fine mapping and the study on molecular mechanisms of the individual QTL. In addition, because most of the genetic background is identical for all lines, NILs show more limited developmental and growth variation, increasing the homogeneity of growth stage within experiments. Thus these NILs can be used to (1) study the linkage between different indirect selection indices and yield under water logging conditions and (2) identify and fine map QTL for various agronomical and physiological traits linked to waterlogging tolerance, leading to more reliable markers in MAS.

3.6 Emerging Areas

3.6.1 Elucidating the Role of Membrane Transporters in Flooding Tolerance

Membrane transporters are central to perception and signalling of every known environmental factor (Ward et al. 1995; Zimmermann et al. 1999; Demidchik and Maathuis 2010). In *Arabidopsis*, 43 % of over 25,000 protein sequences have at least one transmembrane spanning domain, and 18 % of proteins have two such domains and thus are associated with cellular membranes (Ward 2001). Genetic manipulation with ion-permeable channels and transporters has resulted in improved tolerance to a range of key abiotic stresses such as salinity (Zhang et al. 2001), drought (Becker et al. 2003) and soil acidity (Ryan et al. 2011). Surprisingly, flooding tolerance is absent from this list. The role and the molecular identity of membrane transporters in signalling and adaptation to flooding stress remain largely unexplored and represent an under explored avenue for improving plant performance under stress conditions. From this perspective, at least three promising directions may be identified.

3.6.1.1 Hypoxia Sensing

The issue of what constitutes the oxygen sensor in plants still remains a hotly debated topic. Oxygen status might also be directly measured by the cell, for example, through sensor proteins or transcription factors whose activity is directly affected by oxygen levels (Banti et al. 2013). Recent studies in *Arabidopsis* (Gibbs et al. 2011; Licausi et al. 2011) showed that plant-specific group VII ethylene response factor (ERF) transcription factors may play this role. Under normoxia, ERF transcription factors are targets of proteasomal degradation via the N-end rule pathway. Under hypoxia, the N-degron is stabilised and ERF-VII transcription factors migrate to the nucleus to activate anaerobic gene expression (Licausi et al. 2013). Under anoxia, a mitochondrial imbalance or a nicotinamide adenine dinucleotide phosphate oxidase-mediated mechanism leads to the production of ROS, promoting the activation of genes involved in oxidative responses (Licausi et al. 2013; Bailey-Serres et al. 2012). However, by definition, sensing implies an *immediate* (e.g. microsecond timescale) response, while changes in ERF transcription factors operate in much slower timescale. Thus, it cannot be excluded that ERF factors represent an important downstream signalling component but *not the sensor per se*. In mammalian systems, a broad range of specialised K^+-, Ca^{2+}-, and Na^+-permeable channels were shown to act as oxygen sensors (reviewed by Kemp and Peers

2007), and it was suggested that such membrane-based oxygen sensors may also operate in plant systems (Pang and Shabala 2010; Shabala et al. 2014). If proven to be the case (to be elucidated in direct patch-clamp experiments), tempering with such oxygen sensor (a single protein transporter) will open exciting prospects of modifying plan flooding tolerance by genetic means.

3.6.1.2 Cytosolic pH Homeostasis

Most of cytosolic enzymes are "pH tuned" and operate at their optimum at around pH 7.0–7.2 (Ratcliffe 1997). Meanwhile, hypoxic stress results in significant cytosolic acidification. Within seconds after imposing hypoxia or anoxia to a cell, cytoplasmic pH rapidly drops by several tenths of a pH unit (Felle 2005). There may be a slow partial recovery of pH (e.g. in rice; Menegus et al. 1991), but for as long as the suboptimal oxygen conditions last, the pH remains below the normoxic set point (Felle 2005). Plasma membrane H^+-ATPase is thought to play a major role in cytoplasmic pH regulation (Greenway and Gibbs 2003; Felle 2005; Koizumi et al. 2011); however, reduced energy availability (see above) makes its operation questionable. The viable option is to switch to pyrophosphatase-driven H^+ transport. The H^+-pyrophosphatase pumps (termed V-PPase) exist at tonoplast membranes (Dietz et al. 2001), and it was argued that a switch from V-ATPase to V-PPase-driven H^+ transport will be beneficial to anoxically treated roots (Gibbs and Greenway 2003), both in the light of reduced ATP availability and due to the fact that hydrolysis of PPi is favoured to ATP when cytoplasmic pH drops as it does under anoxia (Felle 2005). Thus, transgenic crops with overexpressed PPase activity may be a possible way forward in developing flood-tolerant crops (Carystinos et al. 1995; Koizumi et al. 2011; Shabala et al. 2014).

3.6.1.3 Potassium Homeostasis

Potassium plays multiple roles in plants (Dreyer and Uozumi 2011; Anschütz et al. 2014; Shabala and Pottosin 2014), and the ability of roots to maintain a better cytosolic K^+ homeostasis and K^+ channel functionality was also named as an essential component of plant acclimation to hypoxia in this species (Mugnai et al. 2011; Barrett-Lennard and Shabala 2013). These traits may be, therefore, targeted in breeding programmes. It was argued, however, that reduction in the cytosolic K^+ pool may provide a mean by which plants "shut down" many energy-dependent processes such as protein synthesis, to survive the energy crisis (Shabala and Pottosin 2014; Shabala et al. 2014) and to redirect available energy to defence-related processes such as prevention of cytosolic acidification, ROS detoxification and production of molecular chaperones. Also, high cytosolic K^+ levels are also essential to suppress the activity of caspase-like proteases and endonucleases in plants (Shabala et al. 2007; Demidchik et al. 2010), and decrease in the cytosolic K^+ pool may result in the activation of these catabolic enzymes triggering PCD. It was argued (Shabala et al. 2014) that such a PCD-driven elimination of the cell may be beneficial if it occurs in the root cortex (for generating the air pocket and thus contributing to aerenchyma formation), but the same process in the root meristem (not capable of forming aerenchyma; Colmer 2003) may result in a complete arrest of growth and the eventual death of the root apex. Thus, the issue of potassium homeostasis and its regulation should be put in a strict time and tissue context.

3.6.2 Developing High-Throughput Technology Platforms for Fine QTL Mapping

Non-invasive microelectrode flux measurements have gradually emerged as a convenient tool to study membrane-transport processes in plants in situ (Shabala and Bose 2012). Some of their key features (non-invasiveness, high spatial and temporal resolution, etc.) allow us to establish and quantify causal links between membrane-transport processes and other metabolic or physiological processes in the cell. The MIFE technique developed in our laboratory (Shabala et al. 2006, 2012) has been successfully applied in the past to understand cellular mechanisms of waterlogging stress signalling and adaptation

(Pang et al. 2006, 2007a, b; Zeng et al. 2013; Teakle et al. 2013). We argue here that, if applied to the DH population, application of the MIFE technique may be useful for fine mapping QTL conferring key mechanisms contributing to flooding tolerance in crops.

It was shown in the past that K^+ leakage from root cells is mediated by two types of plasma membrane channels: depolarisation-activated outward-rectifying (GORK in *Arabidopsis*) and non-selective cation channels (NSCC) (Pang et al. 2007a, b). Thus, measurements of net K^+ fluxes from roots of DH plants contrasting in K^+ retention under either hypoxic or phytotoxin treatments may determine QTL conferring tolerance to these factors. The MIFE technique may be also used to map net O_2 fluxes from adventitious roots under hypoxic or anoxic conditions (Pang et al. 2006), determining QTL responsible for ROL. Finally, efficiency of H^+-ATPase activity operation may be determined by measuring the vanadate-sensitive component of H^+ efflux by H^+-selective microelectrodes (Shabala and Bose 2012). The fact that each of these physiological traits is mediated by a function of one (or perhaps a few), specific transport proteins gives hope that the detected QTL will be narrow and sharp and, hence, usable for practical purposes of pyramiding suitable traits.

3.6.3 Understanding ROS Signalling in Flooding Stress Tolerance

Plant survival of waterlogging or submergence also depends on their ability to limit or endure oxidative stress, which occurs during the transition from normoxia to anoxia as well as upon de-submergence (Bailey-Serres et al. 2012). H_2O_2 was identified as a second messenger in responses to low oxygen (Baxter-Burrell et al. 2002), and ROS is believed to be sensed by two-component sensors possessing a histidine kinase, subsequently inducing a mitogen-activated protein kinase (MAPK) signalling pathway (Voesenek et al. 2006). When produced in low quantities, ROS may have an important signalling role; when accumulated in large quantities, they may become cytotoxicants (Bailey-Serres and Voesenek 2008; Rodrigo-Moreno et al. 2013a, b). Thus, a better understanding of mechanisms of ROS production and signalling in flooded roots is needed. At least three possible mechanisms were suggested (Shabala et al. 2014): (1) by PM NADPH oxidase, (2) as a result of imbalance in mitochondrial functions, and (3) due to the accumulation of excess quantities of metal ions. Importantly, while H_2O_2 formation during flooding is principally caused by DPI-sensitive NADPH oxidase, the oxidative burst during recovery is mostly insensitive to DPI (Kumutha et al. 2009). High levels of antioxidant activity are essential to avoid damage during the reoxygenation period.

3.7 Conclusions and Prospectives

In light of the above, it is highly unlikely that progress in crop breeding for waterlogging stress tolerance will be achieved by targeting just one physiological mechanism. More likely, a thorough MAS-based pyramiding of physiological traits conferring efficient oxygen supply and energy maintenance, cytosolic ion homeostasis and ROS control and detoxification is required (Shabala et al. 2014). However, it appears that all the required techniques and methods seem to be in place, and high-throughput technology platforms for fine QTL mapping at the functional (protein) level are available for many traits. The next few years will show if these technological advances will be eventually implemented into flood-tolerant varieties in farmers' fields.

Acknowledgements This work was supported by the Australian Research Council Discovery (DP120101482) and Linkage (LP120200516) grants to Sergey Shabala.

References

Abiko T, Kotula L, Shiono K, Malik AI, Colmer TD, Nakazono M (2012a) Enhanced formation of aerenchyma and induction of a barrier to radial oxygen loss in adventitious roots of *Zea nicaraguensis* contribute to its waterlogging tolerance as compared with maize (*Zea mays* ssp. mays). Plant Cell Environ 35:1618–1630

Abiko T, Obara M, Abe F, Kawaguchi K, Oyanagi A, Yamauchi T, Nakazono M (2012b) Screening of candidate genes associated with constitutive aerenchyma formation in adventitious roots of the teosinte Zea nicaraguensis. Plant Roots 6:19–27

Agarwal S, Grover A (2006) Molecular biology, biotechnology and genomics of flooding-associated low O stress response in plants. Crc Cr Rev Plant Sci 25:1–21

Alscher RG, Erturk N, Heath LS (2002) Role of superoxide dismutases (SODs) in controlling oxidative stress in plants. J Exp Bot 53:1331–1341

Anschütz U, Becker D, Shabala S (2014) Going beyond nutrition: regulation of potassium homoeostasis as a common denominator of plant adaptive responses to environment. J Plant Physiol 171:670–687

Apel K, Hirt H (2004) Reactive oxygen species: metabolism, oxidative stress, and signal transduction. Annu Rev Plant Biol 55:373–399

Arbona V, Hossain Z, López-Climent MF, Pérez-Clemente RM, Gómez-Cadenas A (2008) Antioxidant enzymatic activity is linked to waterlogging stress tolerance in citrus. Physiol Plant 132:452–466

Armstrong W (1979) Aeration in higher plants. Adv Bot Res 7:225–332

Armstrong J, Armstrong W (1999) Phragmites die-back: toxic effects of propionic, butyric and caproic acids in relation to pH. New Phytol 142:201–217

Armstrong J, Armstrong W (2001) Rice and *Phragmites*: effects of organic acids on growth, root permeability, and radial oxygen loss to the rhizosphere. Am J Bot 88:1359–1370

Armstrong J, Armstrong W (2005) Rice: sulfide-induced barriers to root radial oxygen loss, Fe^{2+} and water uptake, and lateral root emergence. Ann Bot 96:625–638

Armstrong J, Armstrong W, Armstrong IB, Pittaway GR (1996) Senescence, and phytotoxin, insect, fungal and mechanical damage: factors reducing convective gasflows in *Phragmites australis*. Aquat Bot 54:211–226

Axelsen KB, Palmgren MG (2001) Inventory of the superfamily of P-type ion pumps in Arabidopsis. Plant Physiol 126:696–706

Baba I, Inada K, Takijima K (1965) Mineral nutrition and the occurrence of physiological diseases. In: Baltimore MD (ed) The mineral nutrition of the rice plant. Johns Hopkins University Press, Baltimore, pp 295–326

Bailey-Serres J, Voesenek L (2008) Flooding stress: acclimations and genetic diversity. Annu Rev Plant Biol 59:313–339

Bailey-Serres J, Fukao T, Gibbs DJ, Holdsworth MJ, Lee SC, Licausi F, Perata P, Voesenek LACJ, van Dongen JT (2012) Making sense of low oxygen sensing. Trends Plant Sci 17:129–138

Banti V, Mafessoni F, Loreti E, Alpi A, Perata P (2010) The heat-inducible transcription factor hsfa2 enhances anoxia tolerance in Arabidopsis. Plant Physiol 152:1471–1483

Banti V, Giuntoli B, Gonzali S, Loreti E, Magneschi L, Novi G, Paparelli E, Parlanti S, Pucciariello C, Santaniello A, Perata P (2013) Low oxygen response mechanisms in green organisms. Int J Mol Sci 14:4734–4761

Barret-Lennard EG, Shabala S (2013) The waterlogging/salinity interaction in higher plants revisited – focusing on the hypoxia-induced disturbance to K^+ homeostasis. Funct Plant Biol 40:872–882

Barrett-Lennard EG (2003) The interaction between waterlogging and salinity in higher plants: causes, consequences and implications. Plant Soil 253:35–54

Barrett-Lennard EG, Leighton PD, Buwalda F, Gibbs J, Armstrong W, Thomson CJ, Greenway H (1988) Effects of growing wheat in hypoxic nutrient solutions and of subsequent transfer to aerated solutions. I. Growth and carbohydrate status of shoots and roots. Funct Plant Biol 15:585–598

Baxter-Burrell A, Yang ZB, Springer PS, Bailey-Serres J (2002) RopGAP4-dependent Rop GTPase rheostat control of Arabidopsis oxygen deprivation tolerance. Science 296:2026–2028

Becker D, Hoth S, Ache P, Wenkel S, Roelfsema MRG, Meyerhoff O, Hartung W, Hedrich R (2003) Regulation of the ABA-sensitive Arabidopsis potassium channel gene GORK in response to water stress. FEBS Lett 554:119–126

Beckett PM, Armstrong W, Justin SHFW, Armstrong J (1988) On the relative importance of convective and diffusive gas flows in plant aeration. New Phytol 110:463–468

Begg CBM, Kirk GJD, Mackenzie AF, Neue H-U (1994) Root-induced iron oxidation and pH changes in the lowland rice rhizosphere. New Phytol 128:469–477

Benjamin LR, Greenway H (1979) Effects of a range of O_2 concentrations on porosity of barley roots and on their sugar and protein concentrations. Ann Bot 43:383–391

Beyer P, Al-Babili S, Ye XD, Lucca P, Schaub P, Welsch R, Potrykus I (2002) Golden rice: introducing the beta-carotene biosynthesis pathway into rice endosperm by genetic engineering to defeat vitamin A deficiency. J Nutr 132:506S–510S

Booker FL, Blum U, Fiscus EL (1992) Short-term effects of ferulic acid on ion uptake and water relations in cucumber seedlings. J Exp Bot 43:649–655

Branco-Price C, Kaiser KA, Jang CJH, Larive CK, Bailey-Serres J (2008) Selective mRNA translation coordinates energetic and metabolic adjustments to cellular oxygen deprivation and reoxygenation in *Arabidopsis thaliana*. Plant J 56:743–755

Broughton S, Zhou GF, Teakle LN, Matsuda R, Zhou MX, O'Leary AR, Colmer DT, Li CD (2015) Water logging tolerance is associated with root porosity in barley (Hordeum vulgare L.). Mol Breed 35:27. doi:10.1007/S11032-015-0243-3

Brown JC, Chaney RL (1971) Effect of iron on the transport of citrate into the xylem of soybeans and tomatoes. Plant Physiol 47:836–840

Bruening G, Lyons JM (2000) The case of the FLAVR SAVR tomato. Calif Agric 54:6–7

Burgos S, Stamp P, Schmid JE (2001) Agronomic and physiological study of cold and flooding tolerance of spelt (*Triticum spelta* L.) and wheat (*Triticum aestivum* L.). J Agron Crop Sci 187:195–202

Cailliatte R, Schikora A, Briat J-F, Mari S, Curie C (2010) High-affinity manganese uptake by the metal transporter NRAMP1 is essential for Arabidopsis growth in low manganese conditions. Plant Cell 22:904–917

Carystinos GD, MacDonald HR, Monroy AF, Dhindsa RS, Poole RJ (1995) Vacuolar H^+-translocating pyrophosphatase is induced by anoxia or chilling in seedlings of rice. Plant Physiol 108:641–649

Chang WW, Huang L, Shen M, Webster C, Burlingame AL, Roberts JK (2000) Patterns of protein synthesis and tolerance of anoxia in root tips of maize seedlings acclimated to a low-oxygen environment, and identification of proteins by mass spectrometry. Plant Physiol 122:295–318

Chen Z, Fujii Y, Yamaji N, Masuda S, Takemoto Y, Kamiya T, Yusuyin Y, Iwasaki K, S-i K, Maeshima M, Ma JF, Ueno D (2013) Mn tolerance in rice is mediated by MTP8.1, a member of the cation diffusion facilitator family. J Exp Bot 64:4375–4387

Cheng N-h, Pittman JK, Shigaki T, Hirschi KD (2002) Characterization of CAX4, an Arabidopsis H/cation antiporter. Plant Physiol 128:1245–1254

Clark RB (1982) Plant response to mineral element toxicity and deficiency. In: Christiansen MN, Lewis CF (eds) Breeding plants for less favorable environments. Wiley, New York

Clarkson DT (1988) The uptake and translocation of manganese by plant roots. In: Manganese in soils and plants. Springer

Colangelo EP, Guerinot ML (2004) The essential basic helix-loop-helix protein FIT1 is required for the iron deficiency response. Plant Cell 16:3400–3412

Collaku A, Harrison SA (2005) Heritability of waterlogging tolerance in wheat. Crop Sci 45:722–727

Colmer TD (2002) Aerenchyma and an inducible barrier to radial oxygen loss facilitate root aeration in upland, paddy and deep-water rice (*Oryza sativa* L.). Ann Bot 91:301–309

Colmer TD (2003) Long-distance transport of gases in plants: a perspective on internal aeration and radial oxygen loss from roots. Plant Cell Environ 26:17–36

Colmer TD, Greenway H (2010) Ion transport in seminal and adventitious roots of cereals during O_2 deficiency. J Exp Bot 62:39–57

Colmer TD, Pedersen O (2008) Under water photosynthesis and respiration in leaves of submerged wetland plants: gas films improve CO_2 and O_2 exchange. New Phytol 177:918–926

Colmer TD, Voesenek LACJ (2009) Flooding tolerance: suites of plant traits in variable environments. Funct Plant Biol 36:665–681

Colmer TD, Munns R, Flowers TJ (2005) Improving salt tolerance of wheat and barley: future prospects. Aust J Exp Agric 45:1425–1443

Colmer TD, Cox MCH, Voesenek LACJ (2006) Root aeration in rice (*Oryza sativa*): evaluation of oxygen, carbon dioxide, and ethylene as possible regulators of root acclimatization. New Phytol 170:767–778

Curie C, Alonso J, Le Jean M, Ecker J, Briat J (2000) Involvement of NRAMP1 from Arabidopsis thaliana in iron transport. Biochem J 347:749–755

Curie C, Panaviene Z, Loulergue C, Dellaporta SL, Briat J-F, Walker EL (2001) Maize yellow stripe1 encodes a membrane protein directly involved in Fe (III) uptake. Nature 409:346–349

Delhaize E, Gruber BD, Pittman JK, White RG, Leung H, Miao Y, Jiang L, Ryan PR, Richardson AE (2007) A role for the AtMTP11 gene of Arabidopsis in manganese transport and tolerance. Plant J 51:198–210

Delhaize E, Kataoka T, Hebb DM, White RG, Ryan PR (2003) Genes encoding proteins of the cation diffusion facilitator family that confer manganese tolerance. Plant Cell 15:1131–1142

Demidchik V, Maathuis FJM (2010) Ion channels and plant stress responses. Springer, Heidelberg, p 237

Demidchik V, Shabala SN, Davies JM (2007) Spatial variation in H_2O_2 response of *Arabidopsis thaliana* root epidermal Ca^{2+} flux and plasma membrane Ca^{2+} channels. Plant J 49:377–386

Demidchik V, Cuin TA, Svistunenko D, Smith SJ, Miller AJ, Shabala S, Sokolik A, Yurin V (2010) Arabidopsis root K^+-efflux conductance activated by hydroxyl radicals: single-channel properties, genetic basis and involvement in stress-induced cell death. J Cell Sci 123:1468–1479

Dietz KJ, Tavakoli N, Kluge C, Mimura T, Sharma SS, Harris GC, Chardonnens AN, Golldack D (2001) Significance of the V-type ATPase for the adaptation to stressful growth conditions and its regulation on the molecular and biochemical level. J Exp Bot 52:1969–1980

Drew MC (1997) Oxygen deficiency and root metabolism: injury and acclimation under hypoxia and anoxia. Annu Rev Plant Physiol Plant Mol Biol 48:223–250

Drew MC, Lynch JM (1980) Soil anaerobiosis, microorganisms, and root function. Annu Rev Phytopathol 18:37–66

Drew MC, Sisworo EJ (1977) Early effects of flooding on nitrogen deficiency and leaf chlorosis in barley. New Phytol 79:567–571

Drew MC, Jackson MB, Giffard SC, Campbell R (1981) Inhibition by silver ions of gas space (aerenchyma) formation in adventitious roots of *Zea mays L.* subjected to exogenous ethylene or to oxygen deficiency. Planta 153:217–224

Drew MC, Saglio PH, Pradet A (1985) Larger adenylate energy charge and ATP/ADP ratios in aerenchymatous roots of *Zea mays* in anaerobic media as a consequence of improved internal oxygen transport. Planta 165:51–58

Dreyer I, Uozumi N (2011) Potassium channels in plant cells. FEBS J 278:4293–4303

Durrett TP, Gassmann W, Rogers EE (2007) The FRD3-mediated efflux of citrate into the root vasculature is necessary for efficient iron translocation. Plant Physiol 144:197–205

El-Jaoual T, Cox D (1998) Manganese toxicity in plants. J Plant Nutr 21:353–386

Emdadul Haque M, Kawaguchi K, Komatsu S (2011) Analysis of proteins in aerenchymatous seminal roots of wheat grown in hypoxic soils under waterlogged conditions (supplementary material). Protein Pept Lett 18:912–924

Epstein E (1961) Mineral metabolism of halophytes. In: Rorison IH (ed) Eccological aspects of the mineral nutrition of plants. Blackwell, Oxford/Edinburg

Evans DE (2004) Aerenchyma formation. New Phytol 161:35–49

Evans HJ, Sorger GJ (1966) Role of mineral elements with emphasis on the univalent cations. Annu Rev Plant Physiol 17:47–76

Fageria N, Santos A, Barbosa Filho M, Guimaraes C (2008) Iron toxicity in lowland rice. J Plant Nutr 31:1676–1697

Fagerstedt KV (2010) Programmed cell death and aerenchyma formation under hypoxia. In: Mancuso S, Shabala S (eds) Waterlogging signalling and tolerance in plants. Springer, Heidelberg, pp 99–118

Fang W-C, Kao CH (2000) Enhanced peroxidase activity in rice leaves in response to excess iron, copper and zinc. Plant Sci 158:71–76

Fecht-Christoffers MM, Braun H-P, Lemaitre-Guillier C, VanDorsselaer A, Horst WJ (2003) Effect of manganese toxicity on the proteome of the leaf apoplast in cowpea. Plant Physiol 133:1935–1946

Felle HH (2005) pH regulation in anoxic plants. Ann Bot 96:519–532

Fleck AT, Nye T, Repenning C, Stahl F, Zahn M, Schenk MK (2011) Silicon enhances suberization and lignification in roots of rice (*Oryza sativa*). J Exp Bot 62:2001–2011

Foy CD, Chaney RL, White MC (1978) The physiology of metal toxicity in plants. Annu Rev Plant Physiol 29:511–566

Gambrell RP, Delaune RD, Patrick WH (1991) Redox processes in soils following oxygen depletion, plant life under oxygen deprivation. In: Jackson MB, Davies DD, Lambers H (eds) Ecology, physiology and biochemistry. SPB Academic, The Hague

Garthwaite AJ, von Bothmer R, Colmer TD (2003) Diversity in root aeration traits associated with waterlogging tolerance in the genus *Hordeum*. Funct Plant Biol 30:875–889

Garthwaite AJ, Steudle E, Colmer TD (2006) Water uptake by roots of *Hordeum marinum*: formation of a barrier to radial O_2 loss does not affect root hydraulic conductivity. J Exp Bot 57:655–664

Garthwaite AJ, Armstrong W, Colmer TD (2008) Assessment of O_2 diffusivity across the barrier to radial O_2 loss in adventitious roots of *Hordeum marinum*. New Phytol 179:405–416

Gechev TS, Van Breusegem F, Stone JM, Denev I, Laloi C (2006) Reactive oxygen species as signals that modulate plant stress responses and programmed cell death. Bioessays 28:1091–1101

Geisler-Lee J, Caldwell C, Gallie DR (2010) Expression of the ethylene biosynthetic machinery in maize roots is regulated in response to hypoxia. J Exp Bot 61:857–871

Gibberd MR, Colmer TD, Cocks PS (1999) Root porosity and oxygen movement in waterlogging-tolerant *Trifolium tomentosum* and -intolerant *Trifolium glomeratum*. Plant Cell Environ 22:1161–1168

Gibberd MR, Gray JD, Cocks PS, Colmer TD (2001) Waterlogging tolerance among a diverse range of *Trifolium* accessions is related to root porosity, lateral root formation and 'aerotropic rooting'. Ann Bot 88:579–589

Gibbs J, Greenway H (2003) Mechanisms of anoxia tolerance in plants. I. Growth, survival and anaerobic catabolism. Funct Plant Biol 30:1–47

Gibbs DJ, Lee SC, Isa NM, Gramuglia S, Fukao T, Bassel GW, Correia CS, Corbineau F, Theodoulou FL, Bailey-Serres J, Holdsworth MJ (2011) Homeostatic response to hypoxia is regulated by the N-end rule pathway in plants. Nature 479:415–418

Glass ADM (1973) Influence of phenolic acids on ion uptake. 1. Inhibition of phosphate uptake. Plant Physiol 51:1037–1041

Glass ADM (1974) Influence of phenolic acids upon ion uptake. 3. Inhibition of potassium absorption. J Exp Bot 25:1104–1113

Green LS, Rogers EE (2004) FRD3 controls iron localization in Arabidopsis. Plant Physiol 136:2523–2531

Greenway H, Gibbs J (2003) Mechanisms of anoxia tolerance in plants. II. Energy requirements for maintenance and energy distribution to essential processes. Funct Plant Biol 30:999–1036

Greenway H, Armstrong W, Colmer TD (2006) Conditions leading to high CO2 (>5 kPa) in waterlogged-flooded soils and possible effects on root growth and metabolism. Ann Bot 98:9–32

Guerinot ML, Yi Y (1994) Iron: nutritious, noxious, and not readily available. Plant Physiol 104:815

Hall JL (2002) Cellular mechanisms for heavy metal detoxification and tolerance. J Exp Bot 53:1–11

Haroldsen VM, Paulino G, Chi-Ham CL, Bennett AB (2012) Research and adoption of biotechnology strategies could improve California fruit and nut crops. Calif Agric 66:62–69

Hattori Y, Nagai K, Ashikari M (2010) Rice growth adapting to deep-water. Curr Opin Plant Biol 14:100–105

He C-J, Morgan PW, Drew MC (1992) Enhanced sensitivity to ethylene in nitrogen- or phosphate-starved roots of *Zea mays* L. during aerenchyma formation. Plant Physiol 98:137–142

He C-J, Finlayson SA, Drew MC, Jordan WR, Morgan PW (1996a) Ethylene biosynthesis during aerenchyma formation in roots of maize subjected to mechanical impedance and hypoxia. Plant Physiol 112:1679–1685

He C-J, Morgan PW, Drew MC (1996b) Transduction of an ethylene signal is required for cell death and lysis in the root cortex of maize during aerenchyma formation induced by hypoxia. Plant Physiol 112:463–472

Hell R, Stephan UW (2003) Iron uptake, trafficking and homeostasis in plants. Planta 216:541–551

Henriques R, Jásik J, Klein M, Martinoia E, Feller U, Schell J, Pais MS, Koncz C (2002) Knock-out of Arabidopsis metal transporter gene IRT1 results in iron deficiency accompanied by cell differentiation defects. Plant Mol Biol 50:587–597

He Y, Zhang J, Ma H, Tu Y, Chen J, Chen F, Xu H, Song Z, Cai D, Xue X (2012) Aerenchyma formation and increased accumulation of free proline in roots of xerophytic *Aloe vera* L. cultured in nutrient solutions. J Med Plants 6:243–252

Horiguchi T (1987) Mechanism of manganese toxicity and tolerance of plant. II. Deposition of oxidized manganese in plant tissues. Soil Sci Plant Nutr 33:595–606

Horst WJ (1988) The physiology of Mn toxicity. In: Graham RD, Hannam RJ, Uren NC (eds) Manganese in soils and plants. Kluwer Academic, Dordrecht

Hossain MA, Piyatida P, da Silva JAT, Fujita M (2012) Molecular mechanism of heavy metal toxicity and tolerance in plants: central role of glutathione in detoxification of reactive oxygen species and methylglyoxal and in heavy metal chelation. J Bot 2012

Huang X, Shabala S, Shabala L, Rengel Z, Wu XJ, Zhang GP, Zhou MX (2014) Linking waterlogging tolerance with Mn^{2+} toxicity: a case study for barley. Plant Biol 17:26-33

Inoue H, Mizuno D, Takahashi M, Nakanishi H, Mori S, Nishizawa NK (2004) A rice FRD3-like (OsFRDL1) gene is expressed in the cells involved in long-distance transport. Soil Sci Plant Nutr 50:1133–1140

Ishimaru Y, Suzuki M, Tsukamoto T, Suzuki K, Nakazono M, Kobayashi T, Wada Y, Watanabe S, Matsuhashi S, Takahashi M, Nakanishi H, Mori S, Nishizawa NK (2006) Rice plants take up iron as an Fe -phytosiderophore and as Fe. Plant J 45:335–346

Ishimaru Y, Masuda H, Bashir K, Inoue H, Tsukamoto T, Takahashi M, Nakanishi H, Aoki N, Hirose T, Ohsugi R, Nishizawa NK (2010) Rice metalnicotianamine transporter, OsYSL2, is required for the long-distance transport of iron and manganese. Plant J 62:379–390

Ishimaru Y, Bashir K, Nakanishi H, Nishizawa NK (2012a) OsNRAMP5, a major player for constitutive iron and manganese uptake in rice. Plant Signal Behav 7:763–766

Ishimaru Y, Takahashi R, Bashir K, Shimo H, Senoura T, Sugimoto K, Ono K, Yano M, Ishikawa S, Arao T (2012b) Characterizing the role of rice NRAMP5 in manganese, iron and cadmium transport. Scientific reports 2

Izaguirre-Mayoral ML, Sinclair TR (2005) Variation in manganese and iron accumulation among soybean genotypes growing on hydroponic solutions of differing manganese and nitrate concentrations. J Plant Nutr 28:521–535

Jackson MB (1985) Ethylene and responses of plants to soil waterlogging and submergence. Annu Rev Plant Biol 36:145–174

Jackson MB (2002) Long-distance signalling from roots to shoots assessed: the flooding story. J Exp Bot 53:175–181

Jackson MB, Armstrong W (1999) Formation of aerenchyma and the processes of plant ventilation in relation to soil flooding and submergence. Plant Biol 1:274–287

Jackson MB, Hall KC (1987) Early stomatal closure in waterlogged pea plants is mediated by abscisic acid in the absence of foliar water deficits. Plant Cell Environ 10:121–130

Jackson M, Dobson C, Herman B, Merryweather A (1984) Modification of 3,5-diiodo-4-hydroxybenzoic acid (DIHB) activity and stimulation of ethylene production by small concentrations of oxygen in the root environment. Plant Growth Regul 2:251–262

Jiang Z, Song X-F, Zhou Z-Q, Wang L-K, Li J-W, Deng X-Y, Fan H-Y (2010) Aerenchyma formation: programmed cell death in adventitious roots of winter wheat (*Triticum aestivum*) under waterlogging. Funct Plant Biol 37:748–755

John CD (1977) The structure of rice roots grown in aerobic and anaerobic environments. Plant Soil 47:269–274

Joshi R, Kumar P (2012) Lysigenous aerenchyma formation involves non-apoptotic programmed cell death in rice (*Oryza sativa* L.) roots. Physiol Mol Biol Plants 18:1–9

Joshi R, Shukla A, Mani SC, Kumar P (2010) Hypoxia induced non-apoptotic cellular changes during aerenchyma formation in rice (*Oryza sativa* L.) roots. Physiol Mol Biol Plants 16:99–106

Jung J, Lee S, Choi H-K (2008) Anatomical patterns of aerenchyma in aquatic and wetland plants. J Plant Biol 51:428–439

Justin SHFW, Armstrong W (1987) The anatomical characteristics of roots and plant response to soil flooding. New Phytol 465–495

Karuppanapandian T, Moon J-C, Kim C, Manoharan K, Kim W (2011) Reactive oxygen species in plants: their generation, signal transduction, and scavenging mechanisms. Aust J Crop Sci 5:709–725

Kawase M, Whitmoyer RE (1980) Aerenchyma development in waterlogged plants. Am J Bot 67:18–22

Kearsey MJ (1998) The principles of QTL analysis (a minimal mathematics approach). J Exp Bot 49:1619–1623

Kemp PJ, Peers C (2007) Oxygen sensing by ion channels. In: Peers C (ed) Oxygen sensing and hypoxia-induced responses, vol 43, Essays in biochemistry. Portland Press, London, pp 77–90

Kenward KD, Brandle J, McPherson J, Davies PL (1999) Type II fish antifreeze protein accumulation in transgenic tobacco does not confer frost resistance. Transgenic Res 8:105–117

Khabaz-Saberi H, Setter TL, Waters I (2006) Waterlogging induces high to toxic concentrations of iron, aluminum, and manganese in wheat varieties on acidic soil. J Plant Nutr 29:899–911

Khabaz-Saberi H, Rengel Z (2010) Aluminum, manganese, and iron tolerance improves performance of wheat genotypes in waterlogged acidic soils. J Plant Nutr Soil Sci 173:461–468

Khabaz-Saberi H, Barker SJ, Rengel Z (2012) Tolerance to ion toxicities enhances wheat (*Triticum aestivum* L.) grain yield in waterlogged acidic soils. Plant Soil 354:371–381

Kirk GJD, Bajita JB (1995) Root-induced iron oxidation, pH changes and zinc solubilization in the rhizosphere of lowland rice. New Phytol 131:129–137

Kirk G, Solivas J, Alberto M (2003) Effects of flooding and redox conditions on solute diffusion in soil. Eur J Soil Sci 54:617–624

Koike S, Inoue H, Mizuno D, Takahashi M, Nakanishi H, Mori S, Nishizawa NK (2004) OsYSL2 is a rice metal-nicotianamine transporter that is regulated by iron and expressed in the phloem. Plant J 39:415–424

Koizumi Y, Hara Y, Yazaki Y, Sakano K, Ishizawa K (2011) Involvement of plasma membrane H^+-ATPase in anoxic elongation of stems in pondweed (*Potamogeton distinctus*) turions. New Phytol 190:421–430

Koren'kov V, Park S, Cheng N-H, Sreevidya C, Lachmansingh J, Morris J, Hirschi K, Wagner G (2007) Enhanced Cd -selective root-tonoplasttransport in tobaccos expressing Arabidopsis cation exchangers. Planta 225:403–411

Kotula L, Steudle E (2009) Measurements of oxygen permeability coefficients of rice (*Oryza sativa* L.) roots using a new perfusion technique. J Exp Bot 60:567–580

Kotula L, Ranathunge K, Schreiber L, Steudle E (2009a) Functional and chemical comparison of apoplastic barriers to radial oxygen loss in roots of rice (*Oryza sativa* L.) grown in aerated or deoxygenated solution. J Exp Bot 60:2155–2167

Kotula L, Ranathunge K, Steudle E (2009b) Apoplastic barriers effectively block oxygen permeability across outer cell layers of rice roots under deoxygenated conditions: roles of apoplastic pores and of respiration. New Phytol 184:909–917

Kotula L, Colmer TD, Nakazono M (2014) Effects of organic acids on the formation of the barrier to radial oxygen loss in roots of *Hordeum marinum*. Funct Plant Biol 41:187–202

Kumutha D, Ezhilmathi K, Sairam RK, Srivastava GC, Deshmukh PS, Meena RC (2009) Waterlogging induced oxidative stress and antioxidant activity in pigeon pea genotypes. Biol Plant 53:75–84

Lachavisit A, Davies JM (2007) The gas that opens gates: calcium channel activation by ethylene. New Phytol 174:470–473

Laurie S, Tancock N, McGrath S, Sanders J (1995) Influence of EDTA complexation on plant uptake of manganese (II). Plant Sci 109:231–235

Licausi F (2011) Regulation of the molecular response to oxygen limitations in plants. New Phytol 190:550–555

Licausi F, Kosmacz M, Weits DA, Giuntoli B, Giorgi FM, Voesenek L, Perata P, van Dongen JT (2011) Oxygen sensing in plants is mediated by an N-end rule pathway for protein destabilization. Nature 479:419–422

Licausi F, Pucciariello C, Perata P (2013) New role for an old rule: N-end rule-mediated degradation of ethylene responsive factor proteins governs low oxygen response in plants. J Int Plant Biol 55:31–39

Li H, Vaillancourt R, Mendham N, Zhou M (2008) Comparative mapping of quantitative trait loci associated with waterlogging tolerance in barley (Hordeum vulgare L.). BMC Genomics 9:401

Lynch JM (1977) Phytotoxicity of acetic-acid produced in anaerobic decomposition of wheat straw. J Appl Bacteriol 42:81–87

MacEwan RJ, Gardner WK, Ellington A, Hopkins DG, Bakker AC (1992) Tile and mole drainage for control of waterlogging in duplex soils of southeastern Australia. Aust J Exp Agric 32:865–878

Mark M, Gavin T, Susanne S (1999) The genetics of metal tolerance and accumulation in higher plants. In: Phytoremediation of contaminated soil and water. CRC Press

Malik AI, Colmer TD, Lambers H, Schortemeyer M (2001) Changes in physiological and morphological traits of roots and shoots of wheat in response to different depths of waterlogging. Aust J Plant Physiol 28:1121–1131

Malik AI, Colmer TD, Lambers H, Setter TL, Schortemeyer M (2002) Short-term waterlogging has long-term effects on the growth and physiology of wheat. New Phytol 153:225–236

Malik AI, Colmer TD, Lambers H, Schortemeyer M (2003) Aerenchyma formation and radial O_2 loss along adventitious roots of wheat with only the apical root portion exposed to O_2 deficiency. Plant Cell Environ 26:1713–1722

Malik AI, English JP, Colmer TD (2009) Tolerance of Hordeum marinum accessions to O_2 deficiency, salinity and these stresses combined. Ann Bot 103:237–248

Mano Y, Omori F (2008) Verification of QTL controlling root aerenchyma formation in a maize x teosinte "*Zea nicaraguensis*" advanced backcross population. Breed Sci 58:217–223

Mano Y, Omori F (2009) High-density linkage map around the root aerenchyma locus Qaer1.06 in the backcross populations of maize Mi29 x teosinte "*Zea nicaraguensis*". Breed Sci 59:427–433

Mano Y, Omori F (2013a) Flooding tolerance in interspecific introgression lines containing chromosome segments from teosinte (*Zea nicaraguensis*) in maize (*Zea mays* subsp. *mays*). Ann Bot 112:1125–1139

Mano Y, Omori F (2013b) Relationship between constitutive root aerenchyma formation and flooding tolerance in *Zea nicaraguensis*. Plant Soil 370:447–460

Mano Y, Muraki M, Fujimori M, Takamizo T, Kindiger B (2005) Identification of QTL controlling adventitious root formation during flooding conditions in teosinte (*Zea mays* ssp. *huehuetenangensis*) seedlings. Euphytica 142:33–42

Mano Y, Muraki M, Takamizo T (2006a) Identification of QTL controlling flooding tolerance in reducing soil conditions in maize (*Zea mays* L.) seedlings. Plant Prod Sci 9:176–181

Mano Y, Omori F, Takamizo T, Kindiger B, Bird RM, Loaisiga CH (2006b) Variation for root aerenchyma formation in flooded and non-flooded maize and teosinte seedlings. Plant Soil 281:269–279

Mano Y, Omori F, Takamizo T, Kindiger B, Bird RM, Loaisiga CH, Takahashi H (2007) QTL mapping of root aerenchyma formation in seedlings of a maize x rare teosinte "*Zea nicaraguensis*" cross. Plant Soil 295:103–113

Mano Y, Omori F, Kindiger B, Takahashi H (2008) A linkage map of maize x teosinte Zea luxurians and identification of QTLs controlling root aerenchyma formation. Mol Breed 21:327–337

Marschner H (1995) Mineral nutrition of higher plants. Academic, London

Mcdonald MP, Galwey NW, Colmer TD (2001) Waterlogging tolerance in the tribe Triticeae: the adventitious roots of *Critesion marinum* have a relatively high porosity and a barrier to radial oxygen loss. Plant Cell Environ 24:585–596

McDonald MP, Galwey NW, Colmer TD (2002) Similarity and diversity in adventitious root anatomy as related to root aeration among a range of wetland and dryland grass species. Plant Cell Environ 25:441–451

Menegus F, Cattaruzza L, Mattana M, Beffagna N, Ragg E (1991) Response to anoxia in rice and wheat seedlings. Plant Physiol 95:760–767

Mino Y, Ishida T, Ota N, Inoue M, Nomoto K, Takemoto T, Tanaka H, Sugiura Y (1983) Mugineic acid-iron (III) complex and its structurally analogous cobalt (III) complex: characterization and implication for absorption and transport of iron in gramineous plants. J Am Chem Soc 105:4671–4676

Moellenbeck DJ, Peters ML, Bing JW, Rouse JR, Higgins LS, Sims L, Nevshemal T, Marshall L, Ellis RT, Bystrak PG, Lang BA, Stewart JL, Kouba K, Sondag V, Gustafson V, Nour K, Xu DP, Swenson J, Zhang J, Czapla T, Schwab G, Jayne S, Stockhoff BA, Narva K, Schnepf HE, Stelman SJ, Poutre C, Koziel M, Duck N (2001) Insecticidal proteins from *Bacillus thuringiensis* protect corn from corn rootworms. Nat Biotechnol 19:668–672

Møller IM, Jensen PE, Hansson A (2007) Oxidative modifications to cellular components in plants. Annu Rev Plant Biol 58:459–481

Mugnai S, Marras AM, Mancuso S (2011) Effect of hypoxic acclimation on anoxia tolerance in *Vitis* roots: response of metabolic activity and K^+ fluxes. Plant Cell Physiol 52:1107–1116

Muhlenbock P, Plaszczyca M, Plaszczyca M, Mellerowicz E, Karpinski S (2007) Lysigenous aerenchyma formation in Arabidopsis is controlled by *LESION SIMULATING DISEASE1*. Plant Cell 19:3819–3830

Murata Y, Ma JF, Yamaji N, Ueno D, Nomoto K, Iwashita T (2006) A specific transporter for iron (III)–phytosiderophore in barley roots. Plant J 46:563–572

Nandi S, Subudhi PK, Senadhira D, Manigbas NL, SenMandi S, Huang N (1997) Mapping QTLs for submergence tolerance in rice by AFLP analysis and selective genotyping. Mol Gen Genet 255:1–8

Nevo Y, Nelson N (2006) The NRAMP family of metal-ion transporters. BBA-Mol Cell Res 1763:609–620

Nishiuchi S, Yamauchi T, Takahashi H, Kotula L, Nakazono M (2012) Mechanisms for coping with submergence and waterlogging in rice. Rice 5:2

Nozoye T, Nagasaka S, Kobayashi T, Takahashi M, Sato Y, Sato Y, Uozumi N, Nakanishi H, Nishizawa NK (2011) Phytosiderophore efflux transporters are crucial for iron acquisition in graminaceous plants. J Biol Chem 286:5446–5454

Ohkawa H, Tsujii H, Ohkawa Y (1999) The use of cytochrome P450 genes to introduce herbicide tolerance in crops: a review. Pest Sci 55:867–874

Overmyer K, Brosche M, Kangasjarvi J (2003) Reactive oxygen species and hormonal control of cell death. Trends Plant Sci 8:335–342

Pang J, Shabala S (2010) Membrane transporters and waterlogging tolerance. In: Mancuso S, Shabala S (eds) Waterlogging signalling and tolerance in plants. Springer, Heidelberg, pp 197–219

Pang J, Zhou M, Mendham N, Shabala S (2004) Growth and physiological responses of six barley genotypes to waterlogging and subsequent recovery. Aust J Agr Res 55:895–906

Pang JY, Newman I, Mendham N, Zhou M, Shabala S (2006) Microelectrode ion and O_2 fluxes measurements reveal differential sensitivity of barley root tissues to hypoxia. Plant Cell Environ 29:1107–1121

Pang J, Cuin T, Shabala L, Zhou M, Mendham N, Shabala S (2007a) Effect of secondary metabolites associated with anaerobic soil conditions on ion fluxes and electrophysiology in barley roots. Plant Physiol 145:266–276

Pang J, Ross J, Zhou M, Mendham N, Shabala S (2007b) Amelioration of detrimental effects of waterlogging by foliar nutrient sprays in barley. Funct Plant Biol 34:221–227

Paterson AH, Deverna JW, Lanini B, Tanksley SD (1990) Fine mapping of quantitative trait loci using selected overlapping recombinant chromosomes, in an interspecies cross of tomato. Genetics 124:735–742

Pedas P, Ytting CK, Fuglsang AT, Jahn TP, Schjoerring JK, Husted S (2008) Manganese efficiency in barley: identification and characterization of the metal ion transporter HvIRT1. Plant Physiol 148:455–466

Pedersen O, Rich SM, Colmer TD (2009) Surviving floods: leaf gas films improve O_2 and CO_2 exchange, root aeration, and growth of completely submerged rice. Plant J 58:147–156

Perata P, Voesenek L (2007) Submergence tolerance in rice requires Sub1A, an ethylene-response-factor-like gene. Trends Plant Sci 12:43–46

Peschke VM, Sachs MM (1994) Characterization and expression of transcripts induced by oxygen deprivation in maize (*Zea mays* L.). Plant Physiol 104:387–394

Pezeshki S (2001) Wetland plant responses to soil flooding. Environ Exp Bot 46:299–312

Pezeshki S, DeLaune R (1998) Responses of seedlings of selected woody species to soil oxidation-reduction conditions. Environ Exp Bot 40:123–133

Pittman JK (2005) Managing the manganese: molecular mechanisms of manganese transport and homeostasis. New Phytol 167:733–742

Qiu F, Zheng Y, Zhang Z, Xu S (2007) Mapping of QTL associated with waterlogging tolerance during the seedling stage in maize. Ann Bot 99:1067–1081

Rajhi I, Yamauchi T, Takahashi H, Nishiuchi S, Shiono K, Watanabe R, Mliki A, Nagamura Y, Tsutsumi N, Nishizawa NK, Nakazono M (2011) Identification of genes expressed in maize root cortical cells during lysigenous aerenchyma formation using laser microdissection and microarray analyses. New Phytol 190:351–368

Ranathunge K, Kotula L, Steudle E, Lafitte R (2004) Water permeability and reflection coefficient of the outer part of young rice roots are differently affected by closure of water channels (aquaporins) or blockage of apoplastic pores. J Exp Bot 55:433–447

Ranathunge K, Lin J, Steudle E, Schreiber L (2011) Stagnant deoxygenated growth enhances root suberization and lignifications, but differentially affects water and NaCl permeabilities in rice (*Oryza sativa* L.) roots. Plant Cell Environ 34:1223–1240

Ratcliffe RG (1997) In vivo NMR studies of the metabolic response of plant tissues to anoxia. Ann Bot 79:39–48

Raven JA (2008) Not drowning but photosynthesizing: probing plant plastrons. New Phytol 177:841–845

Rellán-Álvarez R, Giner-Martínez-Sierra J, Orduna J, Orera I, Rodríguez-Castrillón JÁ, García-Alonso JI, Abadía J, Álvarez-Fernández A (2010) Identification of a tri-iron (III), tri-citrate complex in the xylem sap of iron-deficient tomato resupplied with iron: new insights into plant iron long-distance transport. Plant Cell Physiol 51:91–102

Rengel Z (2000) Manganese uptake and transport in plants. Met Ions Biol Syst 37:57–87

Rivetta A, Negrini N, Cocucci M (1997) Involvement of Ca -calmodulin in Cd toxicity during the early phases of radish (Raphanus sativus L.) seed germination. Plant Cell Environ 20:600–608

Robinson NJ, Procter CM, Connolly EL, Guerinot ML (1999) A ferric-chelate reductase for iron uptake from soils. Nature 397:694–697

Rodrigo-Moreno A, Andreas-Colas N, Poschenrieder C, Gunse B, Pellarrubia L, Shabala S (2013a) Calcium- and potassium-permeable plasma membrane transporters are activated by copper in Arabidopsis root tips: linking copper transport with cytosolic hydroxyl radical production. Plant Cell Environ 36:844–855

Rodrigo-Moreno A, Poschenrieder C, Shabala S (2013b) Transition metals: a double edge sward in ROS generation and signaling. Plant Signal Behav 8:e23425.23421–e23425.23425

Rogers EE, Guerinot ML (2002) FRD3, a member of the multidrug and toxin efflux family, controls iron deficiency responses in Arabidopsis. Plant Cell 14:1787–1799

Ryan PR, Tyerman SD, Sasaki T, Furuichi T, Yamamoto Y, Zhang WH, Delhaize E (2011) The identification of aluminium-resistance genes provides opportunities for enhancing crop production on acid soils. J Exp Bot 62:9–20

Saab IN, Sachs MM (1996) A flooding-induced xyloglucan *Endo*-transglycosylase homolog in maize is responsive to ethylene and associated with aerenchyma. Plant Physiol 112:385–391

Sairam RK, Kumutha D, Ezhilmathi K, Deshmukh PS, Srivastava GC (2008) Physiology and biochemistry of waterlogging tolerance in plants. Biol Plant 52:401–412

Samad A, Meisner C, Saifuzzaman M, van Ginkel M (2001) Waterlogging tolerance. application of physiology in wheat breeding. In: Reynolds MP, Ortiz-Monasterio JI, McNab A (eds) Application of physiology in wheat breeding. CIMMYT, Mexico, pp 136–144

Sasaki A, Yamaji N, Xia J, Ma JF (2011) OsYSL6 is involved in the detoxification of excess manganese in rice. Plant Physiol 157:1832–1840

Sasaki A, Yamaji N, Yokosho K, Ma JF (2012) Nramp5 is a major transporter responsible for manganese and cadmium uptake in rice. Plant Cell 24:2155–2167

Schaaf G, Catoni E, Fitz M, Schwacke R, Schneider A, Nv W, Frommer W (2002) A putative role for the vacuolar calcium/manganese proton antiporter AtCAX2 in heavy metal detoxification. Plant Biol 4:612–618

Schaaf G, Erenoglu BE, von Wirén N (2004) Physiological and biochemical characterization of metal-phytosiderophore transport in graminaceous species. Soil Sci Plant Nutr 50:989–995

Seago J, Marsh L, Stevens K, Soukup A, Votrubova O, Enstone DE (2005) A re-examination of the root cortex in wetland flowering plants with respect to aerenchyma. Ann Bot 96:565–579

Setter T, Waters I (2003) Review of prospects for germplasm improvement for waterlogging tolerance in wheat, barley and oats. Plant Soil 253:1–34

Setter TL, Burgess P, Waters I, Kuo J (1999) Genetic diversity of barley and wheat for waterlogging tolerance in Western Australia. In: 9th Australian Barley technical symposium, Melbourne, Australia, pp 2.17.1–2.17.7

Setter TL, Waters I, Sharma SK, Singh KN, Kulshreshtha N, Yaduvanshi NPS, Ram PC, Singh BN, Rane J, McDonald G et al (2009) Review of wheat improvement for waterlogging tolerance in Australia and India: the importance of anaerobiosis and element toxicities associated with different soils. Ann Bot 103:221–235

Shabala S (2011) Physiological and cellular aspects of phytotoxicity tolerance in plants: the role of membrane transporters and implications for crop breeding for waterlogging tolerance. New Phytol 190:289–298

Shabala S, Bose J (2012) Application of non-invasive microelectrode flux measurements in plant stress physiology. In: Volkov AG (ed) Plant electrophysiology: methods and cell electrophysiology. Springer, Berlin/Heidelberg, pp 91–126

Shabala S, Pottosin I (2014) Regulation of potassium transport in plants under hostile conditions: implications for abiotic and biotic stress tolerance. Physiol Plant 151:257–279. doi:10.1111/ppl.12165

Shabala L, Ross T, McMeekin T, Shabala S (2006) Non-invasive microelectrode ion flux measurements to study adaptive responses of microorganisms to the environment. FEMS Microbiol Rev 30:472–486

Shabala S, Cuin TA, Prismall L, Nemchinov LG (2007) Expression of animal CED-9 anti-apoptotic gene in tobacco modifies plasma membrane ion fluxes in response to salinity and oxidative stress. Planta 227:189–197

Shabala S, Shabala L, Newman IA (2012) Studying membrane transport processes by non-invasive microelectrodes: basic principles and methods. In: Volkov AG (ed) Plant electrophysiology: methods and cell electrophysiology. Springer, Berlin/Heidelberg, pp 167–186

Shabala S, Shabala L, Barcelo J, Poschenrieder C (2014) Membrane transporters mediating root signalling and adaptive responses to oxygen deprivation and soil flooding. Plant Cell Environ 37(10):2216–2233

Sharma SS, Dietz K-J (2009) The relationship between metal toxicity and cellular redox imbalance. Trends Plant Sci 14:43–50

Shaw RE, Meyer WS, McNeill A, Tyerman SD (2013) Waterlogging in Australian agricultural landscapes: a review of plant responses and crop models. Crop Pasture Sci 64:549–562

Shigaki T, Pittman JK, Hirschi KD (2003) Manganese specificity determinants in the Arabidopsis metal/H antiporter CAX2. J Biol Chem 278:6610–6617

Shimamura S, Yoshida S, Mochizuki T (2007) Cortical aerenchyma formation in hypocotyl and adventitious roots of *Luffa cylindrica* subjected to soil flooding. Ann Bot 100:1431–1439

Shimamura S, Yamamoto R, Nakamura T, Shimada S, Komatsu S (2010) Stem hypertrophic lenticels and secondary aerenchyma enable oxygen transport to roots of soybean in flooded soil. Ann Bot 106:277–284

Shiono K, Takahashi H, Colmer TD, Nakazono M (2008) Role of ethylene in acclimations to promote oxygen transport in roots of plants in waterlogged soils. Plant Sci 175:52–58

Shiono K, Ogawa S, Yamazaki S, Isoda H, Fujimura T, Nakazono M, Colmer TD (2011) Contrasting dynamics of radial O_2-loss barrier induction and aerenchyma formation in rice roots of two lengths. Ann Bot 107:89–99

Siangliw M, Toojinda T, Tragoonrung S, Vanavichit A (2003) Thai jasmine rice carrying QTLch9 (SubQTL) is submergence tolerant. Ann Bot 91:255–261

Starkey RL (1966) Oxidation and reduction of sulphur compounds in soils. Soil Sci 101:297–306

Steffens B, Geske T, Sauter M (2010) Aerenchyma formation in the rice stem and its promotion by H_2O_2. New Phytol 190:369–378

Striker GG, Insausti P, Grimoldi AA, Vega AS (2007) Trade-off between root porosity and mechanical strength in species with different types of aerenchyma. Plant Cell Environ 30:580–589

Suralta RR, Yamauchi A (2008) Root growth, aerenchyma development, and oxygen transport in rice genotypes subjected to drought and waterlogging. Environ Exp Bot 64:75–82

Teakle NL, Armstrong J, Barrett-Lennard EG, Colmer TD (2011) Aerenchymatous phellem in hypocotyl and roots enables O_2 transport in *Melilotus siculus*. New Phytol 190:340–350

Teakle NL, Bazihizina N, Shabala S, Colmer TD, Barrett-Lennard EG, Rodrigo-Moreno A, Läuchli AE (2013) Differential tolerance to combined salinity and O_2 deficiency in the halophytic grasses *Puccinellia ciliata* and *Thinopyrum ponticum*: the importance of K^+ retention in roots. Environ Exp Bot 87:69–78

Thomas AL, Guerreiro SMC, Sodek L (2005) Aerenchyma formation and recovery from hypoxia of the flooded root system of nodulated soybean. Ann Bot 96:1191–1198

Thomson CJ, Armstrong W, Waters I, Greenway H (1990) Aerenchyma formation and associated oxygen movement in seminal and nodal roots of wheat. Plant Cell Environ 13:395–403

Tiffin LO (1966) Iron translocation I. Plant culture, exudate sampling, iron-citrate analysis. Plant Physiol 41:515–518

Torres MA, Dangl JL (2005) Functions of the respiratory burst oxidase in biotic interactions, abiotic stress and development. Curr Opin Plant Biol 8:397–403

Trought MCT, Drew MC (1980) The development of waterlogging damage in wheat seedlings (*Triticum aestivum* L.). Plant Soil 54:77–94

Tsukamoto T, Nakanishi H, Kiyomiya S, Watanabe S, Matsuhashi S, Nishizawa NK, Mori S (2006) 52Mn translocation in barley monitored using a positron-emitting tracer imaging system. Soil Sci Plant Nutr 52:717–725

Tsukamoto T, Nakanishi H, Uchida H, Watanabe S, Matsuhashi S, Mori S, Nishizawa NK (2009) 52Fe translocation in barley as monitored by a positron-emitting tracer imaging system (PETIS): evidence for the direct translocation of Fe from roots to young leaves via phloem. Plant Cell Physiol 50:48–57

Varotto C, Maiwald D, Pesaresi P, Jahns P, Salamini F, Leister D (2002) The metal ion transporter IRT1 is necessary for iron homeostasis and efficient photosynthesis in Arabidopsis thaliana. Plant J 31:589–599

Vert G, Barberon M, Zelazny E, Séguéla M, Briat J-F, Curie C (2009) Arabidopsis IRT2 cooperates with the high-affinity iron uptake system to maintain iron homeostasis in root epidermal cells. Planta 229:1171–1179

Vert G, Briat JF, Curie C (2001) Arabidopsis IRT2 gene encodes a root-periphery iron transporter. Plant J 26:181–189

Vert G, Grotz N, Dédaldéchamp F, Gaymard F, Guerinot ML, Briat J-F, Curie C (2002) IRT1, an Arabidopsis transporter essential for iron uptake from the soil and for plant growth. Plant Cell 14:1223–1233

Visser EJW, Bögemann GM (2003) Measurement of porosity in very small samples of plant tissue. Plant Soil 253:81–90

Visser EJW, Colmer TD, Blom C, Voesenek L (2000) Changes in growth, porosity, and radial oxygen loss from adventitious roots of selected mono- and dicotyledonous wetland species with contrasting types of aerenchyma. Plant Cell Environ 23:1237–1245

Voesenek L, Sasidharan R (2013) Ethylene – and oxygen signalling – drive plant survival during flooding. Plant Biol 15:426–435

Voesenek LACJ, Armstrong W, Bogemann GM, McDonald MP, Colmer TD (1999) A lack of aerenchyma and high rates of radial oxygen loss from the root base contribute to the waterlogging intolerance of *Brassica napus*. Aust J Plant Physiol 26:87–93

Voesenek L, Colmer TD, Pierik R, Millenaar FF, Peeters AJM (2006) How plants cope with complete submergence. New Phytol 170:213–226

Wang YX, Wu P, Wu YR, Yan XL (2002) Molecular marker analysis of manganese toxicity tolerance in rice under greenhouse conditions. Plant Soil 238:227–233

Ward JM (2001) Identification of novel families of membrane proteins from the model plant Arabidopsis thaliana. Bioinformatics 17:560–563

Ward JM, Pei ZM, Schroeder JI (1995) Roles of ion channels in initiation of signal-transduction in higher-plants. Plant Cell 7:833–844

Watson ER, Lapins P, Barron RJW (1976) Effect of waterlogging on the growth, grain and straw yield of wheat, barley and oats. Aust J Exp Agric 16:114–122

Webb J, Jackson MB (1986) A transmission and cryo-scanning electron microscopy study of the formation of aerenchyma (cortical gas-filled space) in adventitious roots of rice (*Oryza sativa*). J Exp Bot 37:832–841

Wegner LH (2010) Oxygen transport in waterlogged plants. In: Mancuso S, Shabala S (eds) Waterlogging signalling and tolerance in plants. Springer, Heidelberg, pp 3–22

Wong HL, Sakamoto T, Kawasaki T, Umemura K, Shimamoto K (2004) Down-regulation of metallothionein, a reactive oxygen scavenger, by the small GTPase OsRac1 in rice. Plant Physiol 135:1447–1456

Xu K, Mackill D (1996) A major locus for submergence tolerance mapped on rice chromosome 9. Mol Breed 2:219–224

Xu K, Xu X, Fukao T, Canlas P, Maghirang-Rodriguez R, Heuer S, Ismail AM, Bailey-Serres J, Ronald PC, Mackill DJ (2006) *Sub1A* is an ethylene-response-factor-like gene that confers submergence tolerance to rice. Nature 442:705–708

Xue DW, Zhou MX, Zhang XQ, Chen S, Wei K, Zeng FR, Mao Y, Wu FB, Zhang GP (2010) Identification of QTLs for yield and yield components of barley under different growth conditions. J Zhejiang Univ Sci B 11:169–176

Yamauchi T, Shimamura S, Nakazono M, Mochizuki T (2013) Aerenchyma formation in crop species: a review. Field Crop Res 152:8–16

Yamauchi T, Watanabe K, Fukazawa A, Mori H, Abe F, Kawaguchi K, Oyanagi A, Nakazono M (2014) Ethylene and reactive oxygen species are involved in root aerenchyma formation and adaptation of wheat seedlings to oxygen-deficient conditions. J Exp Bot 65:261–273

Yan B, Dai Q, Liu X, Huang S, Wang Z (1996) Flooding-induced membrane damage, lipid oxidation and activated oxygen generation in corn leaves. Plant Soil 179:261–268

Yang M, Zhang W, Dong H, Zhang Y, Lv K, Wang D, Lian X (2013) OsNRAMP3 is a vascular bundles-specific manganese transporter that is responsible for manganese distribution in rice. PLoS One 8, e83990

Yin D, Chen S, Chen F, Jiang J (2013) Ethylene promotes induction of aerenchyma formation and ethanolic fermentation in waterlogged roots of *Dendranthema* spp. Mol Biol Rep 40:4581–4590

Yokosho K, Yamaji N, Ueno D, Mitani N, Ma JF (2009) OsFRDL1 is a citrate transporter required for efficient translocation of iron in rice. Plant Physiol 149:297–305

Yordanova RY, Popova LP (2001) Photosynthetic response of barley plants to soil flooding. Photosynthetica 39:515–520

Yordanova RY, Uzunova AN, Popova LP (2005) Effects of short-term soil flooding on stomata behaviour and leaf gas exchange in barley plants. Biol Plant 49:317–3199

Young ND, Zamir D, Ganal MW, Tanksley SD (1988) Use of isogenic lines and simultaneous probing to identify dna markers tightly linked to the tm-2-alpha gene in tomato. Genetics 120:579–585

Yu Q, Rengel Z (1999) Waterlogging influences plant growth and activities of superoxide dismutases in narrow-leafed lupin and transgenic tobacco plants. J Plant Physiol 155:431–438

Zaharieva T, Römheld V (2000) Specific Fe^{2+} uptake system in strategy I plants inducible under Fe deficiency. J Plant Nutr 23:1733–1744

Zeng F, Shabala L, Zhou M, Zhang GP, Shabala S (2013) Barley responses to combined waterlogging and salinity stress: separating effects of oxygen deprivation and elemental toxicity. Front Plant Physiol 4:313. doi:10.3389/fpls.2013.00313

Zhang HX, Blumwald E (2001) Transgenic salt-tolerant tomato plants accumulate salt in foliage but not in fruit. Nat Biotechnol 19:765–768

Zhang J, Zhang X (1994) Can early wilting of old leaves account for much of the ABA accumulation in flooded pea plants? J Exp Bot 45:1335–1342

Zhang J, Van Toai T, Huynh L, Preiszner J (2000) Development of flooding-tolerant Arabidopsis thaliana by autoregulated cytokinin production. Mol Breed 6:135–144

Zhang HX, Hodson JN, Williams JP, Blumwald E (2001) Engineering salt-tolerant Brassica plants: characterization of yield and seed oil quality in transgenic plants with increased vacuolar sodium accumulation. Proc Natl Acad Sci U S A 98:12832–12836

Zhao M-G, Tian Q-Y, Zhang W-H (2007) Ethylene activates a plasma membrane Ca^{2+}-permeable channel in tobacco suspension cells. New Phytol 174:507–515

Zhou MX (2010) Improvement of plant waterlogging tolerance. In: Mancuso S, Shabala S (eds) Waterlogging signalling and tolerance in plants. Springer, Berlin/Heidelberg, pp 267–286

Zhou M (2011) Accurate phenotyping reveals better QTL for waterlogging tolerance in barley. Plant Breed 130:203–208

Zhou W, Zhao D, Lin X (1997) Effects of waterlogging on nitrogen accumulation and alleviation of waterlogging damage by application of nitrogen fertilizer and mixtalol in winter rape (*Brassica napus* L.). J Plant Growth Regul 16:47–53

Zhou M, Johnson P, Zhou G, Li C, Lance R (2012) Quantitative trait loci for waterlogging tolerance in a barley cross of Franklin × YuYaoXiangTian Erleng and the relationship between waterlogging and salinity tolerance. Crop Sci 52:2082–2088

Zhou MX, Li HB, Mendham NJ (2007) Combining ability of waterlogging tolerance in barley. Crop Sci 47:278–284

Zhou G, Delhaize E, Zhou M, Ryan PR (2013) The barley MATE gene, HvAACT1, increases citrate efflux and Al-3 tolerance when expressed in wheat and barley. Ann Bot 112:603–612

Zimmermann S, Ehrhardt T, Plesch G, Muller-Rober B (1999) Ion channels in plant signalling. Cell Mol Life Sci 55:183–203

4 Genetic Improvement of Drought Resistance in Rice

Jun You and Lizhong Xiong

Abstract

Among multiple abiotic stresses, drought or water deficit is considered the most important constraint in rice production in many rice-growing areas. Therefore, development of drought-resistant rice varieties is urgently needed. Through identification of drought-resistant germplasms, numerous genetic resources with drought resistance traits have been introduced to commercialized paddy rice cultivars through conventional breeding. Many drought resistance quantitative trait loci (QTLs) were also identified based on root-, leaf-, and yield-related traits. Especially, some QTLs that showed a large and consistent effect in a wide range of environments are of special interest for improving drought resistance of elite rice cultivars following marker-assisted breeding. Using transgenic techniques, many drought-resistance transgenic rice lines have been produced by changing the expression of endogenous drought-responsive genes or by introducing exogenous genes. Some transgenic rice plants that harbored the drought-responsive genes driven by appropriate promoters showed significantly improved drought resistance without phenotypic changes or yield penalty in the field conditions, suggesting of potential promise of genetic engineering in genetic improvement of drought resistance in rice. All these genetic resources, combined with the development of new strategies and drought screening methods, provide great opportunities to improve drought resistance in rice. This article provides an overview and perspective of the strategies, resources, progresses, and challenges toward the improvement of drought resistance in rice.

J. You • L. Xiong (✉)
National Key Laboratory of Crop Genetic Improvement and National Center of Plant Gene Research (Wuhan), Huazhong Agricultural University, Wuhan, China
e-mail: lizhongx@mail.hzau.edu.cn

P.K. Jaiwal et al. (eds.), *Genetic Manipulation in Plants for Mitigation of Climate Change*,
DOI 10.1007/978-81-322-2662-8_4

Keywords
Abiotic stress • Drought tolerance • Genetic engineering • *Oryza sativa*

4.1 Introduction

Rice is one of the major staple food crops in the world. In some developing countries, rice productivity is closely associated with food security and is even related to political security. The continuous increase of the human population, together with the loss of agricultural land which is mainly due to urbanization, industrialization, and desertification, poses serious challenges to world food security. To meet the Declaration of the World Summit on Food Security (http://www.fao.org/wsfs/forum2050/wsfs-forum/en/) target of 70 % more food by 2050, an average annual increase in production of 44 million metric tons per year is required (Tester and Langridge 2010). This goal is even more challenging in light of the simultaneous occurrence of different abiotic stresses, as well as the increased frequency of extreme weather conditions (Mittler and Blumwald 2010).

Among multiple abiotic stresses, drought or water deficit is considered the most important constraint in rice production in many rice-growing areas due to variation in the rainfall patterns from one year to another and the uneven distribution of rainfall in the rice-growing season (Zhang 2007). Drought conditions can affect the plant at any growth stage, but drought stress during the reproductive stage directly results in yield losses of 24–84 % (Venuprasad et al. 2007). To cope with the challenges of sustainable rice production, a Green Super Rice Project was put forward to develop environment-friendly cultivated rice (Zhang 2007). Water conservation and drought resistance are indispensable features of the Green Super Rice. Increasing the understanding of the drought resistance mechanism in rice, identification of drought-resistant rice germplasm, development of new strategies and selection methods, identification of drought-tolerant quantitative trait loci (QTLs) and genes, and international workshops provide great opportunities to improve drought tolerance in rice. This article provides a perspective of the strategies, resources, current status, and challenges toward the improvement of drought tolerance in rice.

4.2 Evaluation of Drought Resistance in Rice

For crops, drought resistance is not only the ability to survive or grow in a water-deficient environment, but also encompasses minimum loss of yield resulting from the stress. Drought resistance of rice can be defined as the survival ability and production capacity under drought conditions (Luo 2010). Drought resistance is a quantitative trait and is often indicated by relative values (the values under drought stress relative to the values under normal growth conditions) of various traits. Drought resistance in rice consists of four mechanisms: drought escape, drought avoidance, drought tolerance, and drought recovery (Yue et al. 2006; Luo 2010; Lawlor 2013; Fukai and Cooper 1995). Drought escape (DE) refers to escape away from stress by a short life cycle or developmental plasticity. In agricultural production, DE means avoiding the seasonal or climatic drought by adjusting the planting time or by the application of short-duration varieties. Drought avoidance (DA) is characterized mainly by the plant capacity in sustaining high water status under drought conditions via enhanced water uptake and reduced water loss. For example, DA can be achieved through the development of a dense and deep root system to exploit water or through the closure of stomata, leaf rolling, or a non-permeable leaf cuticle to reduce transpiration. Drought tolerance (DT) is defined as the capacity of plant cells to maintain function under water-deficit status through reducing stress-induced damage by regulation of gene expression and metabolic pathways. DT is often associated with the accumulation of osmotic adjustment (OA)-related molecules (such as proline) in plant cells for turgor maintenance. DT is also commonly related to the removal of harmful substances accumulated in plants such as reactive oxygen species (ROS). Drought recovery (DR) refers to the recovery capability of a plant after a period of severe drought stress, which causes the complete loss of turgor and leaf desiccation.

Drought resistance is a complex trait reflected by changes at the morphological, physiological, biochemical, and molecular levels. It is important to identify the main criteria or indices to evaluate the drought resistance for different crops. Drought resistance indices in rice can be divided into three categories. (1) DA-related indices. The main criteria of DA include root morphological traits (such as root length, root diameter, root volume, etc.) and physiological traits (such as stomatal conductance, leaf water potential, leaf relative water content, water loss rate, photosynthetic rate, canopy temperature, etc). (2) DT-related indices. This category includes several physiological traits related to OA (such as proline and soluble sugar content), abscisic acid (ABA) content and response, and oxidative stress protection (peroxidase or superoxide dismutase activity, chlorophyll content, etc). (3) Composite indices. These indices include the traits related to biological yield (survival rate, fresh weight, dry weight, green leaf area, and dead leaf scoring) or economic yield (seed setting rate and grain yield) under drought stress conditions.

Comprehensive evaluation of rice drought resistance is very hard. Actually, researchers from different disciplines typically choose part of the above indices to evaluate drought resistance according to their purposes. Although it is difficult to apply the comprehensive index system to reveal the mechanism of drought resistance, some of the composite indices (especially yield-related traits) are preferable and effective for drought-resistance breeding by agronomists and breeders.

4.3 Conventional Breeding of Drought Resistance in Rice

Conventional breeding for drought tolerance in rice has made little progress in the twentieth century (Fukai and Cooper 1995). This is mainly due to the polygenic nature of the phenomenon with low heritability and high G×E interaction (Ingram and Bartels 1996; Fukai and Cooper 1995; Pantuwan et al. 2002; Lafitte et al. 2007). Indirect selection of secondary traits, which has poor correlation with grain yield under stress, may also lead to the slow progress in breeding for improved drought tolerance in rice (Guan et al. 2010; Bernier et al. 2008). Furthermore, the long-time practice of breeding for higher yield under well-watered conditions has narrowed the genetic spectrum of current cultivars with respect to their ability to tolerate drought stress (Yang et al. 2010).

Rice is rich in germplasm resources. The International Rice Genebank holds >105,000 types of Asian and African cultivated rice and ≈ 5000 ecotypes of wild relatives. In addition, many major rice-producing countries have established national germplasm banks (Zhang 2007). In recent years, drought resistance as well as the major agronomic characteristics, physiological traits, and leaf anatomical and root traits of the drought-resistant varieties originating from Asia and Africa were examined by a field screen facility, and several accessions showed a high level of drought resistance (Luo 2010; Lafitte et al. 2006).

A large-scale backcross breeding project was initiated at the International Rice Research Institute (IRRI) to improve drought tolerance in rice. Three elite lines were used as the recurrent parents: two broadly adapted, high-yielding lowland *indica* cultivars (IR64 and Teqing) and the tropical *japonica* new plant type breeding line IR68552-55-3-2. A total of 163 varieties of diverse origins, which represented the breadth of cultivated rice germplasm, were used as donors (Yu et al. 2003). A total of 322 crosses were made between the recurrent parents and donors, and 325 BC2F2 bulk populations were screened in drought lowland or upland nurseries (Lafitte et al. 2006). The high frequency of BC progeny showing apparent transgressive performance for yield in the severe stress of the selection nurseries over the parental lines in most populations was found. This indicates substantial genetic drought tolerance in the donors (Lafitte et al. 2006). Further analysis of IR64-derived superior lines indicated that improving drought tolerance in a background of a semidwarf modern variety is not necessarily linked to losses in yield potential (Lafitte et al. 2007). Dual goals of good yield potential in flooded environments and yield under drought stress conditions were also not incompatible in these materials (Lafitte et al. 2007). A new and promising strategy combining BC breeding with designed QTL pyramiding has been practiced at the IRRI and in China, in which exploiting useful genetic diversity for drought tolerance from the primary gene pool of rice by BC breeding and developing drought tolerance introgression lines in elite genetic backgrounds, discovery, allelic mining, and characterization of QTL networks for DT/ST and directed trait improvement by designed QTL pyramiding are well designed and integrated (Li and Xu 2007). For an instance, 48 pyramiding lines (PLs) developed by two rounds of QTL pyramiding in the IR64 background were evaluated over 2 years for yield performance and related traits under severe drought stress and irrigated conditions, and all of these PLs had significantly improved drought tolerance (Guan et al. 2010). In addition, 17 PLs had higher yield potential than IR64, and the remaining 31 PLs had a yield potential similar to IR64 under the irrigated control. The results indicated that selection for yield plus some secondary traits under appropriately managed stress and nonstress conditions similar to the target environments are critically important for improving drought resistance without yield penalties in rice (Guan et al. 2010).

Upland rice is a special ecotype of cultivated rice, which has low yield but high potential in water conservation and drought resistance (Luo 2010). One strategy to improve the drought resistance of paddy rice (also called irrigated or lowland rice) is to introduce the water conservation and drought resistance capacity of the upland rice into the commercialized paddy rice cultivars. Using upland rice IRAT 109 as a drought resistance donor, large-scale hybridization and backcrossing with the paddy rice varieties was done at the Shanghai Agrobiological Gene Center (SAGC). After selection in a water-limited environment, a drought-resistant variety Huhan 3 was developed and registered at the national level and released to farmers in 2004 (Luo 2010). The drought resistance of Huhan 3 was also introduced to Hanfong B, a paddy rice possessing the ability to maintain the cytoplasm male sterility of *japonica*, which led to the release of Huhan 2B which retained the same yield potential as Hanfong B and the same drought resistance as Huhan 3 (Luo 2010). By this strategy, SAGC released a batch of drought-resistant varieties, including the conventional varieties, such as Zhonghan 3, Huhan 3, and Huhan 15 and hybrid rice such as Hanyou 2 and Hanyou 3. These varieties performed better in terms of drought resistance and/or water conservation properties and have been applied for cultivation in low and moderate yielding fields (Luo 2010).

4.4 MAS for Drought Resistance in Rice

Since the development of molecular markers allows the construction of saturated linkage maps, QTL analysis could be employed to analyze the genetics of complex traits (Price et al. 2002b). With the development of DNA markers, marker-assisted selection (MAS) provides an opportunity for examining the usefulness of introducing special genomic regions from drought-resistant germplasms into elite rice cultivars for genetic improvement of drought resistance in rice.

The root system is very important to drought resistance because plants obtain water and nutrients through roots. The ability to access water in deep soil layers through an effective rooting system has been considered as an important DA mechanism for upland rice (Price and Courtois 1999). Many genetic studies have been conducted on the root system, and numerous root-related QTLs were identified, including root length and thickness (Champoux et al. 1995; Yadav et al. 1997; Price et al. 2002a; MacMillan et al. 2006; Qu et al. 2008), root volume (Hemamalini et al. 2000; Price and Tomos 1997; Yue et al. 2006), and root penetration ability (Ray et al. 1996; Ali et al. 2000; Price et al. 2000; Zheng et al. 2000) (Table 4.1). Yue et al. (2006) detected 38 QTLs for seven root traits under drought stress conditions (maximum root depth, drought-induced root growth in depth, root growth rate in depth, root volume, deep root rate in volume, root growth rate in volume, deep root rate in volume induced by drought conditions) using a

Table 4.1 QTLs associated with drought tolerance in rice

Trait	Cross	Population	Number of QTLs	Reference
Root system trait				
Root thickness, root-shoot ratio, root dry weight per tiller, deep root dry weight per tiller	CO39/Moroberekan	RILs	56	Champoux et al. (1995)
Number of penetrated roots, total root number, root penetration index	CO39/Moroberekan	RILs	29	Ray et al. (1996)
Maximum root length, root thickness, total root weight, deep root weight, deep root weight per tiller, deep root to shoot ratio	IR64/Azucena	DH	39	Yadav et al. (1997)
Total root number, penetrated root number, root penetration index, penetrated root thickness, penetrated root length	IR58821/IR52561	RILs F7	28	Ali et al. (2000)
Root length, total root number, root volume, root thickness, root dry weight	IR64/Azucena	DH	28	Hemamalini et al. (2000)
Root penetration ability, root thickness, penetrated root number, total root number	IR64/Azucena	DH	12	Zheng et al. (2000)
Maximum root length, root thickness, root dry weight	IAC165 × Co39	RILs F7	29	Courtois et al. (2003)
Maximum root length, root volume, adventitious root thickness, root cell length	Bala/Azucena	F2	24	Price and Tomos (1997)
Number of roots, number of penetrated roots, ratio of penetrated to total roots	Bala/Azucena	RILs F6	17	Price et al. (2000)
Maximum root length, root thickness, dry weight of roots, number of roots, root to shoot ratio	Bala/Azucena	RILs F6	49	Price et al. (2002a)

(continued)

Table 4.1 (continued)

Trait	Cross	Population	Number of QTLs	Reference
Maximum root length, maximum root thickness, root mass below 50 cm, % root mass	Bala/Azucena	RILs F6	51	MacMillan et al. (2006)
Root axis length, branching index, root dry weight	Akihikari/IRAT109	BC1F5	9	Horii et al. (2006)
Deep root mass, deep root ratio, deep root per tiller, rooting depth, root thickness	IR58821/IR52561	RILs F7	31	Kamoshita et al. (2002)
Root penetration index, root thickness, root pulling force, root dry weight, penetrated root length	CT9993/IR62266	DH	36	Zhang et al. (2001a)
Root pulling resistance, root dry weigh	CT9993/IR62266	DH	3	Kumar et al. (2007)
Maximum root length, root dry weight, total dry weight, root-shoot ratio	ZS97/Minghui63	RILs F10	20	Xu et al. (2004)
Maximum root depth, drought-induced root growth in depth, root growth rate in depth, root volume, deep root rate in volume, root growth rate in volume, deep root rate in volume induced by drought conditions	ZS97/IRAT109	RILs F9	74	Yue et al. (2006)
Seminal root length	IR1552/Azucena	RILs F10	2	Zhang et al. (2001b)
Seminal root length, adventitious root number, lateral root length, lateral root number	IR1552/Azucena	RILs F10	23	Zheng et al. (2003)
Basal root thickness, total root number, maximum root length, root fresh weight, root dry weight, ratio of root fresh weight to shoot fresh weight, ratio of root dry weight to shoot dry weight	IRAT109/Yuefu	DH	19	Li et al. (2005)

(continued)

Table 4.1 (continued)

Trait	Cross	Population	Number of QTLs	Reference
Basal root thickness, root number, maximum root length, root fresh weight, root dry weight, root volume	Yuefu/IRAT109	RILs	84	Qu et al. (2008)
Basal root thickness	Yuefu/IRAT109	BC5F3	22	Li et al. (2011b)
Leaf trait related to drought resistance				
Osmotic adjustment, dehydration tolerance	Mor/CO39	RIL F7	7	Lilley et al. (1996)
Osmotic adjustment	CT9993/IR62266	DH	5	Zhang et al. (2001a)
Osmotic adjustment	IR62266/IR60080	BC3F3	14	Robin et al. (2003b)
ABA content	IR20/63-83	F2	10	Quarrie et al. (1997)
Cell membrane stability	CT9993/IR62266	DH	9	Tripathy et al. (2000)
Canopy temperature, relative water content, leaf rolling, leaf drying	CT9993/IR62266	DH	9	Babu et al. (2003)
Canopy temperature, leaf rolling, leaf drying	ZS97/IRAT109	RILs F9	17	Yue et al. (2005)
Leaf drying, leaf rolling time	ZS97/IRAT109	RILs F9	10	Yue et al. (2006)
Leaf rolling, leaf drying, relative water content	Bala/Azucena	RILs F6	40	Price et al. (2002c)
Leaf rolling, leaf drying, relative water content	IR64/Azucena	DH	32	Courtois et al. (2000)
Leaf rolling, drought score	IR64/Azucena	DH	3	Hemamalini et al. (2000)
Leaf rolling	Guichao2/Dongxiang	ILs	12	Zhang et al. (2006)
Stomatal density, leaf rolling time	Nipponbare/Kasalath	BC1F7	5	Ishimaru et al. (2001a)
Stomatal density, stomatal size	IR69093/IR72	RILs	14	Laza et al. (2010)
Water use efficiency	Nipponbare/Kasalath	BC1F5	5	Xu et al. (2009)
Water use efficiency	Nipponbare/Kasalath	BC1F7	6	Ishimaru et al. (2001b)
Water use efficiency	Bala/Azucena	RILs F6	9	Price et al. (2002b)
Net photosynthesis rate, stomatal conductance, transpiration rate, quantum yield of PSII, proportion of open PSII, chlorophyll fluorescence	Shennong265/Haogelao	BC3F6	15	Gu et al. (2012)

(continued)

Table 4.1 (continued)

Trait	Cross	Population	Number of QTLs	Reference
Growth and developmental traits				
Relative growth rate	IR64/Azucena	DH	10	Courtois et al. (2000)
Plant height, tiller	IR64/Azucena	DH	7	Hemamalini et al. (2000)
Plant height, biomass, heading date	CT9993/IR62266	DH	10	Babu et al. (2003)
Plant height, biomass, flowering date	CT9993/IR62266	DH	24	Lanceras et al. (2004)
Biomass, flowering date, flowering delay	CT9993/IR62266	DH	8	Kumar et al. (2007)
Plant height, biomass, flowering date, flowering delay	Bala/Azucena	RILs F6	28	Lafitte et al. (2004)
Plant height, heading date	Lemont/Teqing	ILs	24	Xu et al. (2005)
Relative biomass	ZS97/IRAT109	RILs F9	4	Yue et al. (2006)
Biomass	Bala/Azucena	RILs F6	8	Price et al. (2002a)
Plant height, relative plant height, flag leaf length, flag leaf width, relative leaf length, relative leaf width, panicle exsertion, difference of panicle exsertion	ZS97/IRAT109	RILs F9	53	Yue et al. (2008)
Plant height, biomass, flowering date, panicle length	IR62266/Norungan	RILs F7	23	Suji et al. (2012)
Plant height, biomass, flowering date	N22/MTU1010, N22/Swarna, N22/IR64	F3:4	7	Vikram et al. (2011)
Plant height, maximum root length, shoot fresh weight, root fresh weight, number of roots, root-shoot ratio (seedling stage)	ZS97/Minghui63	RILs F12	34	Cui et al. (2008)
Adventitious root lengths, shoot height, shoot biomass, root to shoot dry weight ratio (seedling stage)	IR64/Azucena	DH	16	Zheng et al. (2008)
Yield-related traits				
Grain yield, relative yield, grains per panicle, harvest index	CT9993/IR62266	DH	9	Babu et al. (2003)

(continued)

Table 4.1 (continued)

Trait	Cross	Population	Number of QTLs	Reference
Grain yield, grain weight, grains per panicle, fraction sterile panicles, panicle number, spikelet fertility, harvest index	Bala/Azucena	RILs F6	49	Lafitte et al. (2004)
Grain yield, grain weight, spikelet fertility, panicle number, spikelet number per panicle	ZS97/IRAT109	RILs F9	14	Zou et al. (2005)
Grain yield, grain weight, panicle number	Lemont/Teqing	ILs	10	Xu et al. (2005)
Index of drought resistance	IRAT109/Yuefu	DH	4	Li et al. (2005)
Relative yield, relative spikelet fertility, drought response index	ZS97/IRAT109	RILs F9	16	Yue et al. (2005)
Relative yield, relative grain weight, relative spikelet fertility, relative rate of fertile panicles, relative harvest index, relative number of spikelets per panicle	ZS97/IRAT109	RILs F9	23	Yue et al. (2006)
Grain number per plant, panicle neck diameter, panicle length, primary branch number, second branch number	ZS97/IRAT109	RILs F9	11	Liu et al. (2008)
Grain yield, grain number per plant, primary branch number, second branch number, spikelet density, panicle length	ZS97/IRAT109	RILs F9	14	Liu et al. (2010)
Grain yield, spikelet number, spikelet sterility, panicle number, harvest index	CT9993/IR62266	DH	53	Lanceras et al. (2004)
Grain yield, relative grain yield, spikelet fertility, harvest index	IR62266/Norungan	RILs F7	17	Suji et al. (2012)
Grain yield	Apo/Swarna	BC1F4:5	2	Venuprasad et al. (2009)
Grain yield	Vandana/Way Rarem	F3	1	Bernier et al. (2007)

(continued)

Table 4.1 (continued)

Trait	Cross	Population	Number of QTLs	Reference
Grain yield	CT9993/IR62266	DH	1	Kumar et al. (2007)
Grain yield, harvest index	N22/MTU1010, N22/Swarna, N22/IR64	F3:4	7	Vikram et al. (2011)
Grain yield	IR74371-46-1-1/Sabitri	BC1F3:5	1	Mishra et al. (2013)

recombinant inbred line (RIL) population. Most QTLs have a low genetic effect and could not be detected in different populations or different environments. For example, 29, 26, 14, and 10 QTLs related to root penetration ability were identified by Ray et al. (1996), Ali et al. (2000), Price et al. (2000), and Zheng et al. (2000) using different mapping populations with different genetic backgrounds, respectively. Only the QTL region for root penetration ability on chromosome 2 was revealed in all four mapping populations (Ali et al. 2000; Zheng et al. 2000; Price et al. 2002b).

Drought resistance is associated with leaf traits of rice, and numerous QTLs have been reported for drought resistance-related leaf traits. For example, QTLs have been identified for leaf rolling (Babu et al. 2003; Yue et al. 2005, 2006; Price et al. 2002c; Courtois et al. 2000; Hemamalini et al. 2000), which is considered to be a DA mechanism since leaf rolling can reduce the water loss in addition to reducing the leaf area exposed to heat and light radiation (Price and Courtois 1999). QTLs for other DA-related traits, such as canopy temperature (Babu et al. 2003; Yue et al. 2005), relative water content (Babu et al. 2003; Courtois et al. 2000; Price et al. 2002c), and water use efficiency (Xu et al. 2009; Ishimaru et al. 2001b; Price et al. 2002b), were also identified. In another study, Gu et al. (2012) detected 15 QTLs for six photosynthesis parameters in leaves under drought stress conditions. Many QTLs for DT-related traits such as osmotic adjustment (Lilley et al. 1996; Zhang et al. 2001a; Robin et al. 2003a), ABA content (Quarrie et al. 1997), and cell membrane stability (Tripathy et al. 2000) were also identified.

Drought stress coinciding with the rice booting to heading stage affects the development of the panicle and causes significant yield loss. The traits related to drought resistance at the reproductive stages include heading date or flowering time, plant height, and biomass (Babu et al. 2003; Lanceras et al. 2004; Kumar et al. 2007; Lafitte et al. 2004; Xu et al. 2005) and yield-related traits (Babu et al. 2003; Lafitte et al. 2004; Zou et al. 2005; Xu et al. 2005; Li et al. 2005, 2010; Yue et al. 2006; Suji et al. 2012) (Table 4.1). Using a subset of 154 doubled haploid lines and a line source sprinkler irrigation system, 53 QTLs related to grain yield and yield-related traits under different water limited regimes were identified (Lanceras et al. 2004). Yue et al. (2006) analyzed the genetic basis of DT and DA at the reproductive stage in rice using an RIL population. The RIL plants were grown individually in PVC pipes, and two cycles of drought stress were applied to individual plants at the same panicle development stage. Little correlation of relative yield traits with potential yield, plant size, and root traits was detected, suggesting that DT and DA were detected separately in the experiment. A total of 23 QTLs were detected for six traits of relative performance of fitness and yield, and 74 QTLs were detected for root traits. Only a small portion of QTLs for fitness and yield-related traits overlapped with QTLs for root traits, indicating that DT and DA had distinct genetic bases (Yue et al. 2006). A large-effect QTL, *qtl12.1* (explaining 51 % of the genetic variance), was identified for grain yield under drought stress conditions in upland situations in 2 years by screening a population of 436 F_3 lines derived from a cross between two upland rice cultivars, Vandana and Way Rarem (Bernier et al. 2007). Two QTLs, *qDTY1.1* and *qDTY3.1*, were identified under lowland drought stress conditions. The *qDTY1.1* showed a consistent effect in the background of high-yielding mega-varieties Swarna, IR64, and MTU1010, explaining up to 16.9 % of the phenotypic variance (Vikram et al. 2011). The *qDTY3.1* was identified in an Apo/Swarna population by bulk-segregant analysis and explained 31 % of the genetic variance (Venuprasad et al. 2009).

Although a large number of QTLs for drought tolerance have been identified in rice, few of them have been validated in advanced backcross lines and further used in marker-assisted breeding programs. Shen et al. (2001) used MAS to transfer four QTLs for deep root traits from the *japonica* upland cultivar "Azucena" to the lowland *indica* variety "IR64," and some BC_3F_3 near-isogenic lines (NILs) showed improvement on the root depth trait. Similarly, Steele et al. (2006) conducted a marker-assisted backcrossing (MABC) breeding program to improve the root traits of the Indian upland rice variety, Kalinga III. Cultivar Azucena was used as a donor parent. Four segments carrying QTLs related to root morphological traits (root length and thickness) from cultivar Azucena were selected for introgression. Twenty-two NILs were evaluated for root traits in five field experiments, and the target segment on chromosome 9 (RM242-RM201) significantly increased root length under both irrigated and drought stress treatments. Steele et al. (2007) further tested some key agronomic traits in NILs derived from the previous study (Steele et al. 2006) in a field trial. All four NILs excelled Kalinga III in terms of grain and straw yield under drought-prone conditions. All of these efforts resulted in the release of the first upland rice variety, Birsa Vikas Dhan 111 (PY 84), with early maturing, high drought tolerance, and high grain yield with good grain quality bred using MAS (Steele 2009).

In another study conducted at the IRRI, Bernier et al. (2007) detected a QTL (*qtl12.1*) on chromosome 12 with a large effect on grain yield under stress conditions. The large effect of this QTL on grain yield under drought stress was further confirmed under a wider range of environments, stress intensities, and stress timing, and also in the target environment of Eastern India (Bernier et al. 2009). The QTL *qtl12.1* improved grain yield in nine out of ten direct-seeded upland trials where drought stress was severe or moderate, indicating that *qtl12.1* has a large and consistent effect on grain yield under upland drought stress conditions, in a wide range of environments (Bernier et al. 2009). Recently, *qtl12.1* (*qDTY12.1*) also showed a consistent effect across environments for high grain yield under lowland reproductive-stage drought stress in the background of the popular high-yielding but drought-susceptible recipient variety Sabitri (Mishra et al. 2013). *qtl12.1* is the only QTL reported so far in rice to have shown a large effect against multiple recipient genetic backgrounds, as well as under highly diverse upland and lowland rice ecosystems. Therefore, this QTL can be successfully introgressed to improve grain yield of popular high-yielding but drought-susceptible lowland as well as upland-adapted varieties under drought stress conditions following marker-assisted breeding (Mishra et al. 2013).

4.5 Transgenic Techniques for Drought Tolerance in Rice

For about 30 years, the techniques referred to as transgenic or genetic engineering have offered the prospect of directly altering the genomes of higher plants to change their metabolisms and improve growth and yield under adverse environmental conditions to better serve human requirements (Lawlor 2013). In contrast to classical breeding, the transgenic approach allows the incorporation of specific gene(s) into different organisms (Ashraf 2010).

Using microarray technology, a large number of genes responding to drought stress have been identified. The functions of many drought-responsive genes have been tested by the reverse genetic approach in plants, particularly in the model species *Arabidopsis thaliana*. Meanwhile, many transgenic rice lines with altered expression of drought-responsive genes have been produced and tested for drought tolerance (Table 4.2). These genes are involved in particular aspects of cellular homeostasis such as osmotic adjustment, chaperones, or antioxidants. Alternatively, the overexpression or suppression

Table 4.2 Genes improve drought resistance in rice

Functional category	Gene name	Protein function	Origin	Materials, expression	Testing condition and stage	Parameters	Negative effects on growth	Negative effects on yield	Reference
Signaling factors									
Protein kinase (PK)									
MAPK cascade	*OsMAPK5*	MAPK	*O. sativa*	OE, *CaMV35SP*	GH; VS	Survivability	N	ND	Xiong and Yang (2003)
	NPK1	MAPKKK	*N. tabacum*	OE, rice *Actin1P/LEA3-1P* (Stress inducible)	FD; RS	Seed setting rate	N	N	Xiao et al. (2009)
	DSM1	MAPKKK	*O. sativa*	OE, Maize *Ubi1P*	HO; VS	Plant growth	Y	ND	Ning et al. 2010
CIPK	*OsCIPK12*	CBL-interacting PK	*O. sativa*	OE, *CaMV35SP*	GH; VS	Survivability	N	ND	Xiang et al. (2007)
CDPK	*OsCDPK7*	Calcium-dependent PK	*O. sativa*	OE, *CaMV35SP*	GH; VS	Plant growth; Fv/Fm	N	ND	Saijo et al. (2000)
Other kinase	*OsSIK1*	Receptor-like kinase	*O. sativa*	OE, *CaMV35SP*	GH; VS	Survivability	N	ND	Ouyang et al. (2010)
	SOS2	Ser/Thr kinase	*A. thaliana*	OE, Rice *LEA3-1P*	FD; RS	Seed setting rate	N	N	Xiao et al. (2009)
	OsGSK1	Orthologue of BIN2		KM	DH; VS	Fv/Fm	N	ND	Koh et al. (2007)
Transcription factors									
AP2/ERF	*OsDREB1A, 1B*	DREB1/CBF	*O. sativa*	OE, *CaMV35SP*/ maize *Ubi1P*	DH/GH; VS	Survivability	Y	ND	Ito et al. (2006)
	DREB1A, 1B, 1C	DREB1/CBF	*A. thaliana*	OE, maize *Ubi1P*	DH/GH; VS	Survivability	Y	ND	Ito et al. (2006)
	DREB1A	DREB1/CBF	*A. thaliana*	OE, maize *Ubi1P*	GH; VS	Survivability	N	N	Oh et al. (2005)
	DREB1A	DREB1/CBF	*A. thaliana*	OE, *Arabidopsis RD29AP*	GH; VS/RS	Survivability; yield	N	N	Datta et al. (2012)
	DREB1A	DREB1/CBF	*A. thaliana*	OE, rice *LEA3-1P*	FD; RS	Yield; seed setting rate	N	N	Xiao et al. (2009)
	DREB1C	DREB1/CBF	*A. thaliana*	OE, rice *LIP9P* (stress-inducible)	HO/GH; VS/RS	Dry weights; spikelets per plant; seed setting rate	N	N	Ishizaki et al. (2013)
	ZmCBF3	DREB1/CBF	*Z. mays*	OE, maize *Ubi1P*	GH; VS	Survivability	N	N	Xu et al. (2011)

	HvCBF4	DREB1/CBF	*H. vulgare*	OE, maize *Ubi1P*	GH; VS	Survivability	N	N	Oh et al. (2007)
	OsDREB1F	DREB1/CBF	*O. sativa*	OE, *CaMV35SP*	DH; VS	Survivability	N	ND	Wang et al. (2008)
	OsDREB1G,2B	DREB1/CBF, DREB2	*O. sativa*	OE, *CaMV35SP*	GH; VS	Survivability	N	ND	Chen et al. (2008)
	OsDREB2A	DREB2	*O. sativa*	OE, ARPC (stress inducible)	HO/GH; VS	Survivability	N	ND	Cui et al. (2011)
	ARAG1	DREB	*O. sativa*	OE, *CaMV35SP*	DH; VS	Water concentrations	N	ND	Zhao et al. (2010)
	SUB1A	ERF	*O. sativa*	Isogenic line	GH; VS	Survivability; fresh weight	N	ND	Fukao et al. (2011)
	OsWR1	ERF	*O. sativa*	OE, *CaMV35SP*	GH; VS	Survivability	N	ND	Wang et al. (2012)
	OsDERF1	ERF		RI	GH; VS	Survivability	N	ND	Wan et al. (2011)
	TSRF1	ERF	*L. esculentum*	OE, *CaMV35SP*	HO/GH; VS	Survivability; fresh weight	N	ND	Quan et al. (2010)
	JERF1	ERF	*L. esculentum*	OE, *CaMV35SP*	GH; VS	Survivability; fresh weight	N	N	Zhang et al. (2010b)
	JERF3	ERF	*L. esculentum*	OE, *CaMV35SP*	HO/GH; VS	Plant growth; survivability	N	ND	Zhang et al. (2010a)
	TERF1	ERF	*L. esculentum*	OE, *CaMV35SP*	GH; VS	Survivability	N	ND	Gao et al. (2008)
	OsAP37	AP2	*O. sativa*	OE, rice *OsCc1P* (constitutive expression)	GH/FD; VS/RS	Survivability; Fv/Fm; seed setting rate; yield	N	N	Oh et al. (2009)
	OsAP59	AP2	*O. sativa*	OE, rice *OsCc1P*	GH; VS	Survivability; Fv/Fm	N	Y	Oh et al. (2009)
	HARDY	AP2/ERF-like	*A. thaliana*	OE, *CaMV35SP*	GH; VS	Survivability; WUE; photosynthesis	N	N	Karaba et al. (2007)
bZIP	*OsbZIP23*	bZIP	*O. sativa*	OE, maize *Ubi1P*	GH; VS/RS	Relative yield	N	ND	Xiang et al. (2008)
	OsbZIP46 (constitutive active form)	bZIP	*O. sativa*	OE, maize *Ubi1P*	HO/GH; VS/RS	Survivability; relative yield	N	ND	Tang et al. (2012)
	OsbZIP72	bZIP	*O. sativa*	OE, *CaMV35SP*	DH; VS	Survivability	N	ND	Lu et al. (2009)

(continued)

Table 4.2 (continued)

Functional category	Gene name	Protein function	Origin	Materials, expression	Testing condition and stage	Parameters	Negative effects on growth	Negative effects on yield	Reference
	OsbZIP16	bZIP	*O. sativa*	OE, rice *Actin1P*	DH/GH; VS	Survivability	N	ND	Chen et al. (2012)
	OsABI5	bZIP		RI	HO; VS	Survivability	Y	Y	Zou et al. (2008)
	ABF3	bZIP	*A. thaliana*	OE, maize *Ubi1P*	GH; VS	Survivability; Fv/Fm	N	N	Oh et al. (2005)
NAC	*SNAC1/OsNAC9*	NAC	*O. sativa*	OE, *CaMV35SP*	GH/FD; VS/RS	Survivability; seed setting rate	N	N	Hu et al. (2006)
	SNAC1/OsNAC9	NAC	*O. sativa*	OE, rice *RCc3P* (root-specific expression)	GH/FD; VS/RS	Survivability; Fv/Fm; seed setting rate; yield	N	N	Redillas et al. (2012)
	OsNAC10	NAC	*O. sativa*	OE, rice *RCc3P*	GH/FD; VS/RS	Survivability; Fv/Fm; seed setting rate; yield	N	N	Jeong et al. (2010)
	OsNAC5	NAC	*O. sativa*	OE, maize *Ubi1P*	GH; VS	Survivability	N	ND	Takasaki et al. (2010) and Song et al. (2011)
	OsNAC5	NAC	*O. sativa*	OE, rice *RCc3P*	GH/FD; VS/RS	Survivability; Fv/Fm; seed setting rate; yield	N	N	Jeong et al. (2013)
	ONAC045	NAC	*O. sativa*	OE, *CaMV35SP*	DH; VS	Survivability	N	ND	Zheng et al. (2009)
	OsNAC6/SNAC2	NAC	*O. sativa*	OE, maize *Ubi1P*	DH; VS	Survivability	Y	Y	Nakashima et al. (2007)
	OsNAC6/SNAC2	NAC	*O. sativa*	OE, maize *Ubi1P*	HO; VS	Plant growth	N	N	Hu et al. (2008)
Zinc finger	*DST*	C2H2 zinc finger		EMS mutant	HO/GH; VS	Survivability	N	Y	Huang et al. (2009b)
	ZFP182	TFIIIA-type zinc finger	*O. sativa*	OE, *CaMV35SP*	GH; VS	Survivability	N	ND	Huang et al. (2012)
	ZFP252	C2H2 zinc finger	*O. sativa*	OE, *CaMV35SP*	GH; VS	Survivability	N	ND	Xu et al. (2008)

	ZFP245	C2H2 zinc finger	*O. sativa*	OE, *CaMV35SP*	GH; VS	Survivability	N	ND	Huang et al. (2009a)
	OsiSAP8	A20/AN1-type ZF	*O. sativa*	OE, maize *Ubi1P*	HO/GH; VS/RS	Plant growth; relative yield	Y	Y	Kanneganti and Gupta (2008)
	OsCOIN	RING finger	*O. sativa*	OE, maize *Ubi1P*	HO; VS	Survivability	N	ND	Liu et al. (2007)
	Zat10	C2H2-EAR zinc finger	*A. thaliana*	OE, Rice *Actin1P*/*LEA3-1P*	FD; RS	Yield; seed setting rate	N	N	Xiao et al. (2009)
	AlSAP	Stress-associated protein	*A. littoralis*	OE, *CaMV35SP*	HO/GH; GS/VS/RS	Germination rate; survivability; yield	N	N	Ben Saad et al. (2012)
Other TF	*OsbHLH148*	bHLH	*O. sativa*	OE, rice *OsCc1P*	GH; VS	Survivability; Fv/Fm	N	ND	Seo et al. (2011)
	OsMYB2	MYB	*O. sativa*	OE, maize *Ubi1P*	HO/GH; VS	Survivability	N	ND	Yang et al. (2012)
	OsWRKY11	WRKY	*O. sativa*	OE, rice *HSP101P*	GH; VS	Survivability	N	ND	Wu et al. (2009)
	OsWRKY30	WRKY	*O. sativa*	OE, *CaMV35SP*	HO; VS	Survivability	N	ND	Shen et al. (2012)
	Oshox22	HD-Zip		KM	GH; VS	Survivability	N	ND	Zhang et al. (2012b)
	OsTIFY11a	TIFY	*O. sativa*	OE, rice *LEA3-1P*	HO; VS	Plant growth	Y	ND	Ye et al. (2009)
Protein degradation	*OsDIS1*	E3 ubiquitin ligase		RI	GH; VS	Survivability	N	ND	Ning et al. (2011)
	OsDSG1	E3 ubiquitin ligase		KM	GH; VS	Fresh weight	Y	ND	Park et al. (2010)
	OsSDIR1	E3 ubiquitin ligase	*O. sativa*	OE, maize *Ubi1P*	GH; VS	Survivability	N	ND	Gao et al. (2011)
	OsRDCP1	E3 ubiquitin ligase	*O. sativa*	OE, *CaMV35SP*	GH; VS	Survivability	N	ND	Bae et al. (2011)
Protein modification	*SQS1*	Farnesyltransferase/squalene synthase		RI	GH; VS/RS	Survivability; relative yield	N	N	Manavalan et al. (2012)
Other nuclear proteins	*OsSKIPa*	Ski-interaction protein	*O. sativa*	OE, *CaMV35SP*	GH; VS/RS	Survivability; yield	N	N	Hou et al. (2009)
	OSRIP18	Ribosome-inactivating protein	*O. sativa*	OE, *CaMV35SP*	HO/GH; VS/RS	Survivability	N	N	Jiang et al. (2012)
	ZmGF14-6	14-3-3 proteins	*Z. mays*	OE, maize *Ubi1P*	GH; VS	Survivability	Y	ND	Campo et al. (2012)
Other	*PHYB*	Phytochrome B		EMS mutant	GH; VS	Survivability	N	N	Liu et al. (2012)
Functional proteins									

(continued)

Table 4.2 (continued)

Functional category	Gene name	Protein function	Origin	Materials, expression	Testing condition and stage	Parameters	Negative effects on growth	Negative effects on yield	Reference
ABA metabolism	*DSM2*	Carotene hydroxylase	*O. sativa*	OE, *CaMV35SP*	GH; VS/RS	Survivability; seed setting rate	N	N	Du et al. (2010)
	LOS5	Molybdenum cofactor sulfurase	*A. thaliana*	OE, rice *Actin1P/LEA3-1P*	FD; RS	Yield; seed setting rate	N	N	Xiao et al. (2009)
Metabolism of other hormones	*TLD1/OsGH3.13*	IAA-AA synthetase		Activation mutant	GH; VS	Survivability	Y	Y	Zhang et al. (2009)
	IPT	Isopentenyltransferase	*A. tumefaciens*	OE, bean *SAPKP* (maturation and stress inducible)	GH; RS	Yield; biomass	N	N	Peleg et al. (2011)
Osmotic adjustment									
Trehalose	*OsTPS1*	Trehalose-6-phosphate synthase	*O. sativa*	OE, rice *Actin1P*	DH/HO; VS	Survivability	N	ND	Li et al. (2011a)
	TPSP (otsA + otsB)	Trehalose-6-phosphate synthase/phosphatase	*E. coli*	OE, ARPC/rice rbcSP (mesophyll-specific expression)	GH; VS	Plant growth; Fv/Fm	N	N	Garg et al. (2002)
	TPSP (otsA + otsB)	Trehalose-6-phosphate synthase/phosphatase	*E. coli*	OE, maize *Ubi1P*	GH; VS	Plant growth; Fv/Fm	N	N	Jang et al. (2003)
Proline	*P5CS*	Δ1-pyrroline-5-carboxylate synthetase	*P. aconitifolius*	OE, ARPC	GH; VS	Fresh weight	N	ND	Zhu et al. (1998)
Spermine	*ADC*	Arginine decarboxylase	*D. stramonium*	OE, maize *Ubi1P*	HO (in soil); VS	Plant growth	N	ND	Capell et al. (2004)
	SAMDC	S-adenosylmethionine decarboxylase	*D. stramonium*	OE, maize *Ubi1P*	HO (in soil); VS	Survivability	N	ND	Peremarti et al. (2009)
Dehydrin/LEA	*OsLEA3-1*	LEA protein	*O. sativa*	OE, *CaMV35SP*/rice *LEA3-1P*	GH/FD; RS	Yield; seed setting rate	N	N	Xiao et al. (2007)
	OsLEA3-2	LEA protein	*O. sativa*	OE, *CaMV35SP*	GH; VS	Survivability; grains per spike	N	N	Duan and Cai (2012)

	HVA1	LEA protein	*H. vulgare*	OE, rice *Actin1P*	GH; VS	Plant growth; survivability; RWC	N	ND	Xu et al. (1996) and Chandra Babu et al. (2004)
Heat or cold shock proteins	*OsHsp17.0, 23.7*	Heat shock proteins	*O. sativa*	OE, *CaMV35SP*	DH/HO; GS/VS	Germination rate; survivability	N	ND	Zou et al. (2012)
	sHSP17.7	Small heat shock protein	*O. sativa*	OE, *CaMV35SP*	HO/GH; VS	Survivability	ND	ND	Sato and Yokoya (2008)
	CspB	Cold shock proteins	*B. subtilis*	OE, *CaMV35SP*	GH; VS	Plant growth	ND	ND	Castiglioni et al. (2008)
Aquaporins	*RWC3*	Water channel protein	*O. sativa*	OE, sweet potato *SWPA2P* (stress inducible)	HO; VS	Root osmotic hydraulic conductivity (Lp); leaf water potential	N	ND	Lian et al. (2004)
Other transporters	*AtNHX1*	Na⁺/H⁺ antiporters	*A. thaliana*	OE, rice *Actin1P*	FD; RS	Seed setting rate	N	N	Xiao et al. (2009)
	AtNHX5	Na⁺/H⁺ antiporters	*A. thaliana*	OE, *CaMV35SP*	HO/GH; VS	Plant growth; survivability	N	ND	Li et al. (2011c)
	OsPIN3t	Auxin efflux carrier	*O. sativa*	OE, *CaMV35SP*	HO/GH; VS	Plant growth; Survivability	Y	N	Zhang et al. (2012a)
ROS scavenging	*Sod1*	Cytosolic copper zinc superoxide dismutase	*A. marina*	OE, maize *Ubi1P*	GH; VS	Survivability	N	ND	Prashanth et al. (2008)
	MnSOD	Manganese superoxide dismutase	*P. sativum*	OE, sweet potato *SWPA2P*	HO; VS	Relative ion leakage; relative photosynthetic rate	N	ND	(Wang et al. 2005)
Other defense-related proteins	*OsMT1a*	Type 1 metallothionein	*O. sativa*	OE, rice *Actin1P*	HO; VS	Survivability	N	N	Yang et al. (2009)
	OCPI1	Chymotrypsin inhibitor	*O. sativa*	OE, *CaMV35SP*	FD; RS	Yield; seed setting rate	N	N	Huang et al. (2007)
	OsPR4a	Pathogenesis-related 4	*O. sativa*	OE, rice *LEA3-1P*	GH/FD; VS/RS	Survivability; relative yield	N	N	Wang et al. (2011)

(continued)

Table 4.2 (continued)

Functional category	Gene name	Protein function	Origin	Materials, expression	Testing condition and stage	Parameters	Negative effects on growth	Negative effects on yield	Reference
	hrf1	Harpin proteins	*X. oryzae*	OE, *CaMV35SP*	GH; VS	Survivability	N	ND	Zhang et al. (2011)
Pyrimidine nucleotide biosynthesis	*OsDHODH1*	Dihydroorotate dehydrogenase	*O. sativa*	OE, *CaMV35SP*	GH; VS	Survivability; fresh weight	N	ND	Liu et al. (2009)
Porphyrin biosynthesis	*PPO*	Protoporphyrinogen oxidase	*M. xanthus*	OE, maize *Ubi1P*	GH; VS	Survivability; water potential	N	ND	Phung et al. (2011)
Amino acid metabolism	*OsOAT*	Ornithine δ-aminotransferase	*O. sativa*	OE, maize *Ubi1P*	GH/FD; VS/RS	Survivability; relative seed setting rate	N	ND	You et al. (2012)
Myoinositol metabolism	*OsMIOX*	Myoinositol oxygenase	*O. sativa*	OE, *CaMV35SP*	HO; VS	Plant growth; survivability	N	ND	Duan et al. (2012)
Unknown function	*TaSTRG*	Salt tolerance-related gene	*T. aestivum*	OE, maize *Ubi1P*	HO; VS	Survivability; fresh weight	N	ND	Zhou et al. (2009)

Notes: Testing conditions: *DH* dehydration, *HO* high osmotic medium supplied with PEG or mannitol, *GH* greenhouse or growth chamber (with plants grown in pots containing soil), *FD* field. Testing stage: *GS* germination stage, *VS* vegetative or seedling stage, *RS* reproductive or flowering stage. Abbreviations: *OE* overexpression, *RI* RNAi repression, *KM* knockout mutant, *Y* yes, *N* no, *ND* no data of some regulatory genes, which may potentially activate multiple mechanisms of stress tolerance simultaneously, also resulted in drought tolerance (Deikman et al. 2012).

Transcriptional factors (TFs) are a kind of important regulatory proteins involved in abiotic stress responses. Members of the dehydration-responsive element-binding factors (DREB), basic leucine zipper (bZIP), zinc finger and NAM (no apical meristem), ATAF1-2, and CUC2 (cup-shaped cotyledon) (NAC) families have been characterized with roles in the regulation of plant drought responses (Yamaguchi-Shinozaki and Shinozaki 2006; Ariel et al. 2007; Fang et al. 2008; Ciftci-Yilmaz and Mittler 2008), and some of them have been modified to improve drought resistance in rice (Hu et al. 2006; Xiao et al. 2009). The members of the AP2/ERF (APETALA2/ethylene response factor) TF family, including DREB/CBF TFs, are especially important as they regulate many genes involved in drought, salinity, and freezing responses based on the studies conducted in *Arabidopsis* (Mizoi et al. 2012). To test the effect of these well-documented DREB genes on improving stress resistance in crop plants, the *Arabidopsis DREB1A* was overexpressed under the control of the rice *LEA3-1* promoter in rice. Transgenic rice plants showed significantly improved seed setting rate and yield under drought conditions in the field (Xiao et al. 2009), indicating that the stress tolerance genes identified in non-crop species are important gene resources for drought resistance engineering in crop plant species. Overexpression of AP2 family gene *OsAP37* in rice, driven by a *OsCc1* constitutive promoter significantly enhanced drought tolerance in the field and increased grain yield by 16–57 % more than the control under severe drought conditions, yet exhibited no significant difference under normal growth conditions (Oh et al. 2009). Some bZIPs have been well characterized for their roles in ABA signaling during dehydration and seed maturation (Jakoby et al. 2002). Overexpression of *OsbZIP23* and *OsbZIP46CA1* (constitutively active form) in rice also resulted in significant improvement in drought tolerance and sensitivity to ABA (Tang et al. 2012; Xiang et al. 2008). NACs are plant-specific TFs with various roles in plant development and stress response (Puranik et al. 2012). *Stress-responsive NAC1* (*SNAC1*) is predominantly induced in guard cells by drought. Overexpression of *SNAC1* significantly enhanced drought resistance in transgenic rice (with 22–34 % more seed setting than the control) in the field under severe drought stress conditions at the reproductive stage, while showing no phenotypic changes or yield penalty under normal growth conditions (Hu et al. 2006). Compared with the wild type, the transgenic rice was more sensitive to ABA and lost water more slowly by closing more stomatal pores, yet displayed no significant difference in the rate of photosynthesis (Hu et al. 2006). In two additional studies, overexpression of *OsNAC5* and *OsNAC10* in rice under the control of the root-specific promoter *RCc3* resulted in significantly enhanced drought tolerance at the reproductive stage and increased grain yield in the field under both drought stress and normal conditions (Jeong et al. 2010, 2013). Zinc finger proteins play a critical role in many cellular functions, including transcriptional regulation, RNA binding, regulation of apoptosis, and protein-protein interactions (Ciftci-Yilmaz and Mittler 2008). Overexpression of an *Arabidopsis* zinc finger protein gene *ZAT10* in rice improved seed setting rate and yield under drought stress in field conditions (Xiao et al. 2009). AlSAP, a zinc finger protein isolated from the halophytic grass *Aeluropus littoralis*, also has a positive effect on the drought tolerance of rice (Ben Saad et al. 2012). In addition to TFs, some transcriptional coregulators also have functions in drought resistance in rice. For example, OsSKIPa, a rice homolog of the human Ski-interacting protein (SKIP), has been studied for its effect on stress resistance. The *OsSKIPa*-overexpressing rice exhibited significantly improved drought resistance at both the seedling and reproductive stages by increasing the ROS scavenging ability and transcript levels of many stress-related genes (Hou et al. 2009).

Various signal transduction systems, involving protein phosphorylation and dephosphorylation, calcium sensing, protein degradation and modification, phospholipid metabolism, and so on, have very important roles in abiotic stress responses (Umezawa et al. 2006). Although these complex signaling processes are not yet fully understood, several genes encoding signaling factors that function in the drought response have been identified (Umezawa et al. 2006; Xiao et al. 2009). Reversible protein phosphorylation mediated by protein kinases and protein phosphatases is a major event in signal transduction, regulating virtually all cellular activities in eukaryotic systems (Luan 2003; DeLong 2006; Becraft 2002). Protein kinases related to osmotic adaptation include mitogen-activated protein kinases (MAPKs), calcium-dependent protein kinases (CDPKs), CBL (calcineurin B-like)-interacting protein kinase (CIPK), and sucrose non-fermenting protein (SNF1)-related kinase 2 (SnRK2). The tobacco MAPKKK gene *NPK1* is known to activate an oxidative stress-response signaling cascade and enhanced tolerance to freezing, heat, and salt stress in transgenic tobacco (Kovtun et al. 2000). When *NPK1* was expressed under the control of the drought-inducible *LEA3-1* promoter in rice, the overexpression lines showed enhanced drought tolerance and maintained significantly increased spikelet fertility and yield under field drought conditions (Xiao et al. 2009). A rice MAPK gene *OsMAPK5* and MAPK kinase kinase (MAPKKK) gene *DSM1* enhanced drought resistance in rice at the seedling stage (Xiong and Yang 2003; Ning et al. 2010). The *Arabidopsis SOS2* gene encodes a Ser/Thr protein kinase that is required for salt tolerance (Liu et al. 2000). Overexpression of *SOS2* in rice under the control of the drought-inducible *LEA3-1* promoter showed enhanced drought tolerance and significantly increased spikelet fertility than wild type in the field under drought stress conditions (Xiao et al. 2009). The functions of other types of protein kinases such as OsCIPK12, OsCDPK7, OsSIK1, and OsGSK1 in drought resistance at the seedling stage have been characterized (Xiang et al. 2007; Saijo et al. 2000; Ouyang et al. 2010; Koh et al. 2007). Farnesylation is a post-translational protein modification that mediates the COOH-terminal lipidation of specific cellular signaling proteins. Farnesylated proteins function to regulate many different developmental and stress response processes (Galichet and Gruissem 2003). Manavalan et al. (2012) demonstrated that RNAi-mediated disruption of a rice farnesyltransferase/squalene synthase (SQS) improves drought tolerance at both the vegetative and reproductive stages. When subjected to drought stress, the transgenic plants had higher yield and lost water more slowly compared with the wild type, through reduced stomatal conductance and the retention of high leaf relative water content (RWC) (Manavalan et al. 2012). The ubiquitination/26S proteasome-dependent protein degradation pathway, consisting of ubiquitin-activating enzymes (E1), ubiquitin-conjugating enzymes (E2), and ubiquitin ligases (E3), is one of the most prominent mechanisms that plants use to control growth and development and to respond to biotic and abiotic stresses (Smalle and Vierstra 2004). The functions of four E3 ubiquitin ligases (OsDIS1, OsDSG1, OsSDIR1, and OsRDCP1) in drought resistance have been demonstrated in rice (Ning et al. 2011; Park et al. 2010; Gao et al. 2011; Bae et al. 2011).

Phytohormones, especially ABA, play important roles in the drought response. In response to drought stress, plants synthesize ABA, which triggers ABA-inducible gene expression and closing of stomatal pores to reduce water loss (Yamaguchi-Shinozaki and Shinozaki 2006; Schroeder et al. 2001). The ABA biosynthesis and metabolism pathways have been well elucidated in *Arabidopsis* by utilizing different mutants of these pathways as well as genetic approaches (Nambara and Marion-Poll 2005). Some studies reported manipulation of ABA metabolism to enhance drought tolerance in rice. *Arabidopsis LOS5* encodes the molybdenum cofactor (MoCo) sulfurase, which is a key enzyme function in the last step of ABA biosynthesis. The *LOS5* mutant exhibited reduced tolerance to drought, salt, and cold stresses (Xiong et al. 2001). The constitutive or drought-inducible overexpression of *LOS5* in rice significantly increased the spikelet fertility and yield of transgenic plants under field drought conditions, with no yield decrease observed under normal irrigation growth conditions (Xiao et al. 2009).

However, 9-cis-epoxycarotenoid dioxygenase 2 (NCED2), a rate-limiting enzyme for ABA biosynthesis from *Arabidopsis*, has no effect on drought resistance when controlled by either constitutive or drought-inducible promoters in transgenic rice plants (Xiao et al. 2009). *DSM2*, a gene encoding β-carotene hydroxylase which is predicted to have a role in the biosynthesis of zeaxanthin (carotenoid precursor of ABA), was isolated in rice by screening drought-hypersensitive mutants (Du et al. 2010). Overexpression of *DSM2* in rice resulted in significantly increased resistance to drought and oxidative stresses by control of the xanthophyll cycle and ABA synthesis, both of which play critical roles in the establishment of drought resistance in rice (Du et al. 2010). Cytokinin (CK) is involved in responses to adverse environmental stimuli (Hare et al. 1997) and is also involved in delaying stress-induced senescence (Rivero et al. 2007). Overexpression of the isopentenyl transferase (a critical enzyme for CK synthesis) gene *IPT*, driven by a stress-induced promoter, exhibited a delayed response to stress with significantly higher grain yield when compared to wild-type plants (Peleg et al. 2011). *IPT-transgenic* rice displayed changes in the expression of many genes associated with hormone synthesis and homeostasis and modification of sink-source relationships and enhanced drought tolerance (Peleg et al. 2011). The rice *GH3.13* gene encodes indole-3-acetic acid (IAA)-amido synthetase, which is dramatically induced by drought stress. The activation of *OsGH3.13* in *tld1-D* mutant rice resulted in an IAA deficiency and dramatic changes in architecture; however, it also resulted in enhanced drought tolerance (Zhang et al. 2009).

The biosynthesis and accumulation of compatible solutes, such as trehalose, proline, and spermine, are important adaptive mechanisms that enable protection of cell turgor and restoration of the water status of cells by maintaining cellular water potential as well as stabilizing membranes and/or scavenging ROS (Reguera et al. 2012). Many genes encoding the synthesis of such organic solutes have been engineered to overproduce these solutes in transgenic plants to improve drought tolerance. To increase the biosynthesis of trehalose in rice, a trehalose-6-phosphate synthase/phosphatase (TPSP) fusion gene that includes the coding regions of the *Escherichia coli* trehalose biosynthetic genes (*otsA* and *otsB*) was engineered into rice under the control of either tissue-specific, stress-responsive, or constitutive promoters (Jang et al. 2003; Garg et al. 2002). During abiotic stress, transgenic plants accumulated increased amounts of trehalose and showed high levels of tolerance to salt, drought, and low-temperature stresses, as compared with the non-transformed plants (Jang et al. 2003; Garg et al. 2002). Overexpression of the rice trehalose biosynthetic gene *OsTPS1* also improved the tolerance of rice seedlings to cold, high salinity, and drought stresses (Li et al. 2011a). Proline accumulation has been demonstrated to be correlated with tolerance to drought and salt stress in plants (Delauney and Verma 2002). Overexpression of Δ1-pyrroline-5-carboxylate synthetase (P5CS), a key enzyme in proline biosynthesis, increased tolerance to drought stress in rice (Zhu et al. 1998). Polyamines (spermidine and spermine) are derived from arginine via enzymes such as arginine decarboxylase (ADC) and S-adenosylmethionine decarboxylase (SAMDC) in plants. The modification of polyamine levels by the overexpression of genes such *ADC* and *SAMDC* from *Datura* was reported, and transgenic rice containing these genes showed enhanced tolerance to drought stress (Capell et al. 2004; Peremarti et al. 2009).

LEA proteins are low molecular weight proteins that play crucial roles in cellular dehydration tolerance by preventing protein aggregation during desiccation or water stress, and they may have antioxidant capacity together with a possible role as chaperones (Reguera et al. 2012; Hand et al. 2011). Transgenic rice overexpressing *OsLEA3-1* under the control of constitutive and stress-inducible promoters had a significantly higher relative yield (yield under drought stress treatment/yield under normal growth conditions) than the wild type under drought stress conditions, while it showed no yield penalties under control conditions (Xiao et al. 2007). Another rice LEA gene, *OsLEA3-2*, and the LEA gene *HVA1* from barley were also engineered in rice and confirmed to confer drought tolerance (Xu et al. 1996; Chandra Babu et al. 2004; Duan and Cai 2012).

Some defense-related, metabolic-related, and transporter proteins were reported to have a significant effect on improving drought tolerance in rice. *Oryza sativa chymotrypsin inhibitor-like 1* (*OCPI1*) was strongly induced by dehydration stresses and ABA, and *OCPI1*-overexpressing transgenic plants had significantly higher grain yields and seed setting rates than the negative transgenic control under severe drought stress conditions (Huang et al. 2007). PR4 proteins constitute a pathogenesis-related (PR) protein family with a conserved Barwin domain, and five PR4-homologous genes were identified in rice (Wang et al. 2011). Transgenic rice overexpressing *OsPR4a* showed enhanced tolerance to drought at both the seedling and reproductive stages (Wang et al. 2011). Ornithine δ-aminotransferase (δ-OAT) is a pyridoxal-5′-phosphate-dependent enzyme that has been proposed to be involved in proline and arginine metabolism. *OsOAT*-overexpressing rice showed significantly increased δ-OAT activity and proline levels under normal growth conditions and enhanced drought, osmotic, and oxidative stress tolerance (You et al. 2012). AtNHX1, the most abundant vacuolar Na^+/H^+ antiporter in *Arabidopsis*, contributes to salt tolerance by transporting Na^+ and K^+ into the vacuole (Apse et al. 1999). Interestingly, overexpression of *AtNHX1* in rice under the control of a constitutive promoter resulted in enhanced drought resistance (Xiao et al. 2009).

There is evidence that many of the other genes that are listed in Table 4.2 are involved in drought tolerance, although they are not described individually due to the limited space of this chapter.

4.6 Conclusions and Perspectives

Compared to conventional breeding, MAS offers the potential to assemble target traits in the same genotype more precisely, with fewer unintentional losses in fewer selection cycles (Xu and Crouch 2008). Successful cases of genetic improvement for yield under drought stress have been obtained by selecting for secondary traits related to drought tolerance. For example, the silk-tassel interval was identified as an indicative secondary trait for drought resistance in maize, which showed a strong and consistent genetic correlation to grain yield under drought stress conditions (Bolanos and Edmeades 1996). However, indirect selection for secondary traits was used in several cases with little success in improvement of rice drought tolerance because of the low broad-sense heritability, a lack of effective selection criteria, and the poor correlation of secondary traits with yield under stress (Atlin and Lafitte 2000; Serraj et al. 2009; Bernier et al. 2008). Recently, direct selection for yield under stress by backcrossing breeding populations has been shown to be promising for improving drought tolerance in rice (Guan et al. 2010; Steele et al. 2007; Lafitte et al. 2006). Numerous QTLs for traits associated with drought tolerance were identified in rice. However, few of them have been successfully used in drought-tolerance rice breeding mainly due to the large genotype × environment, QTL × environment, and QTL × recipient genetic background interactions, as well as the absence of QTLs with a large and consistent effect against high-yielding but drought-susceptible varieties (Mishra et al. 2013). Therefore, a strategy of screening in different environments, particularly in the target population of environments, is advocated for developing high-yield varieties with broader adaptation (Mishra et al. 2013; Fukai and Cooper 1995; Cooper 1999).

With the completion of the rice genome sequencing project, there have been rapid developments in functional genomic resources, including a large number of mutants and full-length cDNA libraries and whole-genome expression profiling datasets. Using these resources in combination with rice germplasm, a large number of drought-responsive genes have been identified by forward and/or reverse genetic approaches (Hadiarto and Tran 2011; Ashraf 2010). The functions of many drought-responsive genes have been precisely characterized in transgenic plants. However, our knowledge of drought tolerance in rice remains fragmental, which is an important subject especially for finding solutions to dramatically improve drought tolerance without yield penalty in elite rice varieties under field conditions. New methods (such as association mapping) and new regulatory mechanisms (such as epigenetic regulation and small RNAs) should be intensively adopted to discover novel genes and/or pathways for the drought tolerance improvement of rice. Meanwhile, most of the previous genetic engineering efforts in rice for improving drought tolerance used constitutive promoters (such as CaMV35S and ubiquitin). Although these promoters have been effective in improving the drought tolerance of transgenic plants, the constitutive expression of candidate genes is not always desirable because of the negative effects on growth and development under normal conditions. The application of stress-inducible or tissue-specific promoters has been successfully used in many studies to overcome the negative effects of genes, such as *DREB1A* and *IPT* (Datta et al. 2012; Peleg et al. 2011). Most of the transgenic rice plants were tested under controlled laboratory or greenhouse conditions at the vegetative stage, and they showed a remarkable performance (recovery or death) under drought conditions. Fewer studies have reported the drought resistance of transgenic rice at the reproductive or flowering stage showing good performance based on yield and/or setting rate, and very few of these tests were conducted under drought-prone field conditions. In addition, the drought stresses in some of the reported experiments were too severe, and the yield was too low to be referenced for agricultural practice. A big challenge ahead is to incorporate multiple necessary genes into the genetic backgrounds of elite cultivars or hybrids and to evaluate their performance under real agricultural field conditions (Zhang 2007).

With a long-term goal to improve the drought resistance of rice, more and more QTLs with large and consistent effects and key regulators of plant stress responses need to be identified. The pyramiding or stacking of several well-characterized QTLs and/or key genes by MAS and/or transgenic approaches in elite cultivars targeting for a specific zone of breeding may be a strategy to create new cultivars with high-yield stability in drought-prone environments.

Acknowledgment This work was supported by grants from the National Program for Basic Research of China (2012CB114305) and the National Program on High Technology Development (2012AA10A303).

References

Ali M, Pathan M, Zhang J, Bai G, Sarkarung S, Nguyen H (2000) Mapping QTLs for root traits in a recombinant inbred population from two *indica* ecotypes in rice. Theor Appl Genet 101(5):756–766

Apse MP, Aharon GS, Snedden WA, Blumwald E (1999) Salt tolerance conferred by overexpression of a vacuolar Na^+/H^+ antiport in Arabidopsis. Science 285(5431):1256–1258

Ariel FD, Manavella PA, Dezar CA, Chan RL (2007) The true story of the HD-Zip family. Trends Plant Sci 12(9):419–426

Ashraf M (2010) Inducing drought tolerance in plants: recent advances. Biotechnol Adv 28(1):169–183

Atlin G, Lafitte H (2000) Marker-assisted breeding versus direct selection for drought tolerance in rice. In: Saxena N, O'Toole J (eds) Field screening for drought tolerance in crop plants with emphasis on rice: Proceedings of an international workshop on field screening for drought tolerance in rice, 11–14 Dec 2000. ICRISAT/The Rockefeller Foundation, Patancheru/New York, pp 71–81

Babu RC, Nguyen BD, Chamarerk V, Shanmugasundaram P, Chezhian P, Jeyaprakash P, Ganesh S, Palchamy A, Sadasivam S, Sarkarung S (2003) Genetic analysis of drought resistance in rice by molecular markers. Crop Sci 43(4):1457–1469

Bae H, Kim SK, Cho SK, Kang BG, Kim WT (2011) Overexpression of OsRDCP1, a rice RING domain-containing E3 ubiquitin ligase, increased tolerance to drought stress in rice (*Oryza sativa* L.). Plant Sci 180(6):775–782

Becraft PW (2002) Receptor kinase signaling in plant development. Annu Rev Cell Dev Biol 18:163–192

Ben Saad R, Fabre D, Mieulet D, Meynard D, Dingkuhn M, Al-Doss A, Guiderdoni E, Hassairi A (2012) Expression of the *Aeluropus littoralis AlSAP* gene in rice confers broad tolerance to abiotic stresses through maintenance of photosynthesis. Plant Cell Environ 35(3):626–643

Bernier J, Kumar A, Ramaiah V, Spaner D, Atlin G (2007) A large-effect QTL for grain yield under reproductive-stage drought stress in upland rice. Crop Sci 47(2):507–516

Bernier J, Atlin GN, Serraj R, Kumar A, Spaner D (2008) Breeding upland rice for drought resistance. J Sci Food Agric 88(6):927–939

Bernier J, Kumar A, Venuprasad R, Spaner D, Verulkar S, Mandal NP, Sinha PK, Peeraju P, Dongre PR, Mahto R (2009) Characterization of the effect of a QTL for drought resistance in rice, *qtl12.1*, over a range of environments in the Philippines and eastern India. Euphytica 166(2):207–217

Bolanos J, Edmeades G (1996) The importance of the anthesis-silking interval in breeding for drought tolerance in tropical maize. Field Crop Res 48(1):65–80

Campo S, Peris-Peris C, Montesinos L, Penas G, Messeguer J, San Segundo B (2012) Expression of the maize *ZmGF14-6* gene in rice confers tolerance to drought stress while enhancing susceptibility to pathogen infection. J Exp Bot 63(2):983–999

Capell T, Bassie L, Christou P (2004) Modulation of the polyamine biosynthetic pathway in transgenic rice confers tolerance to drought stress. Proc Natl Acad Sci U S A 101(26):9909–9914

Castiglioni P, Warner D, Bensen RJ, Anstrom DC, Harrison J, Stoecker M, Abad M, Kumar G, Salvador S, D'Ordine R (2008) Bacterial RNA chaperones confer abiotic stress tolerance in plants and improved grain yield in maize under water-limited conditions. Plant Physiol 147(2):446–455

Champoux M, Wang G, Sarkarung S, Mackill D, O'Toole J, Huang N, McCouch S (1995) Locating genes associated with root morphology and drought avoidance in rice via linkage to molecular markers. Theor Appl Genet 90(7):969–981

Chandra Babu R, Zhang J, Blum A, David Ho T-H, Wu R, Nguyen H (2004) *HVA1*, a LEA gene from barley confers dehydration tolerance in transgenic rice (*Oryza sativa* L.) via cell membrane protection. Plant Sci 166(4):855–862

Chen JQ, Meng XP, Zhang Y, Xia M, Wang XP (2008) Over-expression of *OsDREB* genes lead to enhanced drought tolerance in rice. Biotechnol Lett 30(12):2191–2198

Chen H, Chen W, Zhou J, He H, Chen L, Deng XW (2012) Basic leucine zipper transcription factor OsbZIP16 positively regulates drought resistance in rice. Plant Sci 193–194:8–17

Ciftci-Yilmaz S, Mittler R (2008) The zinc finger network of plants. Cell Mol Life Sci 65(7–8):1150–1160

Cooper M (1999) Concepts and strategies for plant adaptation research in rainfed lowland rice. Field Crop Res 64(1):13–34

Courtois B, McLaren G, Sinha P, Prasad K, Yadav R, Shen L (2000) Mapping QTLs associated with drought avoidance in upland rice. Mol Breed 6(1):55–66

Courtois B, Shen L, Petalcorin W, Carandang S, Mauleon R, Li Z (2003) Locating QTLs controlling constitutive root traits in the rice population IAC 165× Co39. Euphytica 134(3):335–345

Cui K, Huang J, Xing Y, Yu S, Xu C, Peng S (2008) Mapping QTLs for seedling characteristics under different water supply conditions in rice (*Oryza sativa*). Physiol Plant 132(1):53–68

Cui M, Zhang W, Zhang Q, Xu Z, Zhu Z, Duan F, Wu R (2011) Induced over-expression of the transcription factor *OsDREB2A* improves drought tolerance in rice. Plant Physiol Biochem 49(12):1384–1391

Datta K, Baisakh N, Ganguly M, Krishnan S, Yamaguchi Shinozaki K, Datta SK (2012) Overexpression of Arabidopsis and rice stress genes' inducible transcription factor confers drought and salinity tolerance to rice. Plant Biotechnol J 10(5):579–586

Deikman J, Petracek M, Heard JE (2012) Drought tolerance through biotechnology: improving translation from the laboratory to farmers' fields. Curr Opin Biotechnol 23(2):243–250

Delauney AJ, Verma DPS (2002) Proline biosynthesis and osmoregulation in plants. Plant J 4(2):215–223

DeLong A (2006) Switching the flip: protein phosphatase roles in signaling pathways. Curr Opin Plant Biol 9(5):470–477

Du H, Wang N, Cui F, Li X, Xiao J, Xiong L (2010) Characterization of the beta-carotene hydroxylase gene *DSM2* conferring drought and oxidative stress resistance by increasing xanthophylls and abscisic acid synthesis in rice. Plant Physiol 154(3):1304–1318

Duan J, Cai W (2012) *OsLEA3-2*, an abiotic stress induced gene of rice plays a key role in salt and drought tolerance. PLoS One 7(9):e45117

Duan J, Zhang M, Zhang H, Xiong H, Liu P, Ali J, Li J, Li Z (2012) *OsMIOX*, a myo-inositol oxygenase gene, improves drought tolerance through scavenging of reactive oxygen species in rice (*Oryza sativa* L.). Plant Sci 196:143–151

Fang Y, You J, Xie K, Xie W, Xiong L (2008) Systematic sequence analysis and identification of tissue-specific or stress-responsive genes of NAC transcription factor family in rice. Mol Genet Genomics 280(6):547–563

Fukai S, Cooper M (1995) Development of drought-resistant cultivars using physiomorphological traits in rice. Field Crop Res 40(2):67–86

Fukao T, Yeung E, Bailey-Serres J (2011) The submergence tolerance regulator SUB1A mediates crosstalk between submergence and drought tolerance in rice. Plant Cell 23(1):412–427

Galichet A, Gruissem W (2003) Protein farnesylation in plants – conserved mechanisms but different targets. Curr Opin Plant Biol 6(6):530–535

Gao S, Zhang H, Tian Y, Li F, Zhang Z, Lu X, Chen X, Huang R (2008) Expression of *TERF1* in rice regulates expression of stress-responsive genes and enhances tolerance to drought and high-salinity. Plant Cell Rep 27(11):1787–1795

Gao T, Wu Y, Zhang Y, Liu L, Ning Y, Wang D, Tong H, Chen S, Chu C, Xie Q (2011) *OsSDIR1* overexpression greatly improves drought tolerance in transgenic rice. Plant Mol Biol 76(1–2):145–156

Garg AK, Kim JK, Owens TG, Ranwala AP, Choi YD, Kochian LV, Wu RJ (2002) Trehalose accumulation in rice plants confers high tolerance levels to different abiotic stresses. Proc Natl Acad Sci U S A 99(25):15898–15903

Gu J, Yin X, Struik PC, Stomph TJ, Wang H (2012) Using chromosome introgression lines to map quantitative trait loci for photosynthesis parameters in rice (*Oryza sativa* L.) leaves under drought and well-watered field conditions. J Exp Bot 63(1):455–469

Guan YS, Serraj R, Liu SH, Xu JL, Ali J, Wang WS, Venus E, Zhu LH, Li ZK (2010) Simultaneously improving yield under drought stress and non-stress conditions: a case study of rice (*Oryza sativa* L.). J Exp Bot 61(15):4145–4156

Hadiarto T, Tran LS (2011) Progress studies of drought-responsive genes in rice. Plant Cell Rep 30(3):297–310

Hand SC, Menze MA, Toner M, Boswell L, Moore D (2011) LEA proteins during water stress: not just for plants anymore. Annu Rev Physiol 73:115–134

Hare P, Cress W, Van Staden J (1997) The involvement of cytokinins in plant responses to environmental stress. Plant Growth Regul 23(1):79–103

Hemamalini G, Shashidhar H, Hittalmani S (2000) Molecular marker assisted tagging of morphological and physiological traits under two contrasting moisture regimes at peak vegetative stage in rice (*Oryza sativa* L.). Euphytica 112(1):69–78

Horii H, Nemoto K, Miyamoto N, Harada J (2006) Quantitative trait loci for adventitious and lateral roots in rice. Plant Breed 125(2):198–200

Hou X, Xie K, Yao J, Qi Z, Xiong L (2009) A homolog of human ski-interacting protein in rice positively regulates cell viability and stress tolerance. Proc Natl Acad Sci U S A 106(15):6410–6415

Hu H, Dai M, Yao J, Xiao B, Li X, Zhang Q, Xiong L (2006) Overexpressing a NAM, ATAF, and CUC (NAC) transcription factor enhances drought resistance and salt tolerance in rice. Proc Natl Acad Sci U S A 103(35):12987–12992

Hu H, You J, Fang Y, Zhu X, Qi Z, Xiong L (2008) Characterization of transcription factor gene *SNAC2* conferring cold and salt tolerance in rice. Plant Mol Biol 67(1–2):169–181

Huang Y, Xiao B, Xiong L (2007) Characterization of a stress responsive proteinase inhibitor gene with positive effect in improving drought resistance in rice. Planta 226(1):73–85

Huang J, Sun SJ, Xu DQ, Yang X, Bao YM, Wang ZF, Tang HJ, Zhang H (2009a) Increased tolerance of rice to cold, drought and oxidative stresses mediated by the overexpression of a gene that encodes the zinc finger protein ZFP245. Biochem Biophys Res Commun 389(3):556–561

Huang XY, Chao DY, Gao JP, Zhu MZ, Shi M, Lin HX (2009b) A previously unknown zinc finger protein, DST, regulates drought and salt tolerance in rice via stomatal aperture control. Genes Dev 23(15):1805–1817

Huang J, Sun S, Xu D, Lan H, Sun H, Wang Z, Bao Y, Wang J, Tang H, Zhang H (2012) A TFIIIA-type zinc finger protein confers multiple abiotic stress tolerances in transgenic rice (*Oryza sativa* L.). Plant Mol Biol 80(3):337–350

Ingram J, Bartels D (1996) The molecular basis of dehydration tolerance in plants. Annu Rev Plant Physiol Plant Mol Biol 47:377–403

Ishimaru K, Shirota K, Higa M, Kawamitsu Y (2001a) Identification of quantitative trait loci for adaxial and abaxial stomatal frequencies in *Oryza sativa*. Plant Physiol Biochem 39(2):173–177

Ishimaru K, Yano M, Aoki N, Ono K, Hirose T, Lin S, Monna L, Sasaki T, Ohsugi R (2001b) Toward the mapping of physiological and agronomic characters on a rice function map: QTL analysis and comparison between QTLs and expressed sequence tags. Theor Appl Genet 102(6):793–800

Ishizaki T, Maruyama K, Obara M, Fukutani A, Yamaguchi-Shinozaki K, Ito Y, Kumashiro T (2013) Expression of *Arabidopsis DREB1C* improves survival, growth, and yield of upland New Rice for Africa (NERICA) under drought. Mol Breed 31(1):255–264

Ito Y, Katsura K, Maruyama K, Taji T, Kobayashi M, Seki M, Shinozaki K, Yamaguchi-Shinozaki K (2006) Functional analysis of rice DREB1/CBF-type transcription factors involved in cold-responsive gene expression in transgenic rice. Plant Cell Physiol 47(1):141–153

Jakoby M, Weisshaar B, Droge-Laser W, Vicente-Carbajosa J, Tiedemann J, Kroj T, Parcy F, b ZIPRG (2002) bZIP transcription factors in Arabidopsis. Trends Plant Sci 7(3):106–111

Jang IC, Oh SJ, Seo JS, Choi WB, Song SI, Kim CH, Kim YS, Seo HS, Choi YD, Nahm BH, Kim JK (2003) Expression of a bifunctional fusion of the *Escherichia coli* genes for trehalose-6-phosphate synthase and trehalose-6-phosphate phosphatase in transgenic rice plants increases trehalose accumulation and abiotic stress tolerance without stunting growth. Plant Physiol 131(2):516–524

Jeong JS, Kim YS, Baek KH, Jung H, Ha SH, Do Choi Y, Kim M, Reuzeau C, Kim JK (2010) Root-specific expression of *OsNAC10* improves drought tolerance and grain yield in rice under field drought conditions. Plant Physiol 153(1):185–197

Jeong JS, Kim YS, Redillas MC, Jang G, Jung H, Bang SW, Choi YD, Ha SH, Reuzeau C, Kim JK (2013) *OsNAC5* overexpression enlarges root diameter in rice plants leading to enhanced drought tolerance and increased grain yield in the field. Plant Biotechnol J 11(1):101–114

Jiang SY, Bhalla R, Ramamoorthy R, Luan HF, Venkatesh PN, Cai M, Ramachandran S (2012) Over-expression of *OSRIP18* increases drought and salt tolerance in transgenic rice plants. Transgenic Res 21(4):785–795

Kamoshita A, Wade L, Ali M, Pathan M, Zhang J, Sarkarung S, Nguyen H (2002) Mapping QTLs for root morphology of a rice population adapted to rainfed lowland conditions. Theor Appl Genet 104(5):880–893

Kanneganti V, Gupta AK (2008) Overexpression of *OsiSAP8*, a member of stress associated protein (SAP) gene family of rice confers tolerance to salt, drought and cold stress in transgenic tobacco and rice. Plant Mol Biol 66(5):445–462

Karaba A, Dixit S, Greco R, Aharoni A, Trijatmiko KR, Marsch-Martinez N, Krishnan A, Nataraja KN, Udayakumar M, Pereira A (2007) Improvement of water use efficiency in rice by expression of *HARDY*, an Arabidopsis drought and salt tolerance gene. Proc Natl Acad Sci U S A 104(39):15270–15275

Koh S, Lee SC, Kim MK, Koh JH, Lee S, An G, Choe S, Kim SR (2007) T-DNA tagged knockout mutation of rice *OsGSK1*, an orthologue of Arabidopsis *BIN2*, with enhanced tolerance to various abiotic stresses. Plant Mol Biol 65(4):453–466

Kovtun Y, Chiu WL, Tena G, Sheen J (2000) Functional analysis of oxidative stress-activated mitogen-activated protein kinase cascade in plants. Proc Natl Acad Sci U S A 97(6):2940–2945

Kumar R, Venuprasad R, Atlin G (2007) Genetic analysis of rainfed lowland rice drought tolerance under naturally-occurring stress in eastern India: heritability and QTL effects. Field Crop Res 103(1):42–52

Lafitte HR, Price AH, Courtois B (2004) Yield response to water deficit in an upland rice mapping population: associations among traits and genetic markers. Theor Appl Genet 109(6):1237–1246

Lafitte H, Li Z, Vijayakumar C, Gao Y, Shi Y, Xu J, Fu B, Yu S, Ali A, Domingo J (2006) Improvement of rice drought tolerance through backcross breeding: evaluation of donors and selection in drought nurseries. Field Crop Res 97(1):77–86

Lafitte HR, Yongsheng G, Yan S, Li ZK (2007) Whole plant responses, key processes, and adaptation to drought stress: the case of rice. J Exp Bot 58(2):169–175

Lanceras JC, Pantuwan G, Jongdee B, Toojinda T (2004) Quantitative trait loci associated with drought tolerance at reproductive stage in rice. Plant Physiol 135(1):384–399

Lawlor DW (2013) Genetic engineering to improve plant performance under drought: physiological evaluation of achievements, limitations, and possibilities. J Exp Bot 64(1):83–108

Laza MRC, Kondo M, Ideta O, Barlaan E, Imbe T (2010) Quantitative trait loci for stomatal density and size in lowland rice. Euphytica 172(2):149–158

Li ZK, Xu JL (2007) Breeding for drought and salt tolerant rice (*Oryza sativa* L.): progress and perspectives. In: Jenks MA, Hasegawa PM, Jain SM (eds) Advances in molecular breeding toward drought and salt tolerant crops. Springer, Dordrecht, pp 531–564

Li Z, Mu P, Li C, Zhang H, Gao Y, Wang X (2005) QTL mapping of root traits in a doubled haploid population from a cross between upland and lowland japonica rice in three environments. Theor Appl Genet 110(7):1244–1252

Li HW, Zang BS, Deng XW, Wang XP (2011a) Overexpression of the trehalose-6-phosphate synthase gene *OsTPS1* enhances abiotic stress tolerance in rice. Planta 234(5):1007–1018

Li J, Wang D, Xie Y, Zhang H, Hu G, Dai A, Liu L, Li Z (2011b) Development of upland rice introgression lines and identification of QTLs for basal root thickness under different water regimes. J Genet Genomics 38(11):547–556

Li M, Lin X, Li H, Pan X, Wu G (2011c) Overexpression of *AtNHX5* improves tolerance to both salt and water stress in rice (*Oryza sativa* L.). Plant Cell Tissue Org Cult 107(2):283–293

Lian HL, Yu X, Ye Q, Ding X, Kitagawa Y, Kwak SS, Su WA, Tang ZC (2004) The role of aquaporin RWC3 in drought avoidance in rice. Plant Cell Physiol 45(4):481–489

Lilley J, Ludlow M, McCouch S, O'Toole J (1996) Locating QTL for osmotic adjustment and dehydration tolerance in rice. J Exp Bot 47(9):1427–1436

Liu J, Ishitani M, Halfter U, Kim CS, Zhu JK (2000) The *Arabidopsis thaliana SOS2* gene encodes a protein kinase that is required for salt tolerance. Proc Natl Acad Sci U S A 97(7):3730–3734

Liu K, Wang L, Xu Y, Chen N, Ma Q, Li F, Chong K (2007) Overexpression of OsCOIN, a putative cold inducible zinc finger protein, increased tolerance to chilling, salt and drought, and enhanced proline level in rice. Planta 226(4):1007–1016

Liu G, Mei H, Yu X, Zou G, Liu H, Hu S, Li M, Wu J, Chen L, Luo L (2008) QTL analysis of panicle neck diameter, a trait highly correlated with panicle size, under well-watered and drought conditions in rice (*Oryza sativa* L.). Plant Sci 174(1):71–77

Liu WY, Wang MM, Huang J, Tang HJ, Lan HX, Zhang HS (2009) The *OsDHODH1* gene is involved in salt and drought tolerance in rice. J Integr Plant Biol 51(9):825–833

Liu G, Mei H, Liu H, Yu X, Zou G, Luo L (2010) Sensitivities of rice grain yield and other panicle characters to late-stage drought stress revealed by phenotypic correlation and QTL analysis. Mol Breed 25(4):603–613

Liu J, Zhang F, Zhou J, Chen F, Wang B, Xie X (2012) Phytochrome B control of total leaf area and stomatal density affects drought tolerance in rice. Plant Mol Biol 78(3):289–300

Lu G, Gao C, Zheng X, Han B (2009) Identification of OsbZIP72 as a positive regulator of ABA response and drought tolerance in rice. Planta 229(3):605–615

Luan S (2003) Protein phosphatases in plants. Annu Rev Plant Biol 54:63–92

Luo LJ (2010) Breeding for water-saving and drought-resistance rice (WDR) in China. J Exp Bot 61(13):3509–3517

MacMillan K, Emrich K, Piepho HP, Mullins CE, Price AH (2006) Assessing the importance of genotype x environment interaction for root traits in rice using a mapping population II: conventional QTL analysis. Theor Appl Genet 113(5):953–964

Manavalan LP, Chen X, Clarke J, Salmeron J, Nguyen HT (2012) RNAi-mediated disruption of squalene synthase improves drought tolerance and yield in rice. J Exp Bot 63(1):163–175

Mishra KK, Vikram P, Yadaw RB, Swamy BM, Dixit S, Cruz MT, Maturan P, Marker S, Kumar A (2013) *qDTY12.1*: a locus with a consistent effect on grain yield under drought in rice. BMC Genet 14(1):12

Mittler R, Blumwald E (2010) Genetic engineering for modern agriculture: challenges and perspectives. Annu Rev Plant Biol 61:443–462

Mizoi J, Shinozaki K, Yamaguchi-Shinozaki K (2012) AP2/ERF family transcription factors in plant abiotic stress responses. Biochim Biophys Acta 1819(2): 86–96

Nakashima K, Tran LS, Van Nguyen D, Fujita M, Maruyama K, Todaka D, Ito Y, Hayashi N, Shinozaki K, Yamaguchi-Shinozaki K (2007) Functional analysis of a NAC-type transcription factor OsNAC6 involved in abiotic and biotic stress-responsive gene expression in rice. Plant J 51(4):617–630

Nambara E, Marion-Poll A (2005) Abscisic acid biosynthesis and catabolism. Annu Rev Plant Biol 56:165–185

Ning J, Li X, Hicks LM, Xiong L (2010) A Raf-like MAPKKK gene *DSM1* mediates drought resistance through reactive oxygen species scavenging in rice. Plant Physiol 152(2):876–890

Ning Y, Jantasuriyarat C, Zhao Q, Zhang H, Chen S, Liu J, Liu L, Tang S, Park CH, Wang X, Liu X, Dai L, Xie Q, Wang GL (2011) The SINA E3 ligase OsDIS1 negatively regulates drought response in rice. Plant Physiol 157(1):242–255

Oh SJ, Song SI, Kim YS, Jang HJ, Kim SY, Kim M, Kim YK, Nahm BH, Kim JK (2005) Arabidopsis CBF3/DREB1A and ABF3 in transgenic rice increased tolerance to abiotic stress without stunting growth. Plant Physiol 138(1):341–351

Oh SJ, Kwon CW, Choi DW, Song SI, Kim JK (2007) Expression of barley HvCBF4 enhances tolerance to abiotic stress in transgenic rice. Plant Biotechnol J 5(5):646–656

Oh SJ, Kim YS, Kwon CW, Park HK, Jeong JS, Kim JK (2009) Overexpression of the transcription factor AP37 in rice improves grain yield under drought conditions. Plant Physiol 150(3):1368–1379

Ouyang SQ, Liu YF, Liu P, Lei G, He SJ, Ma B, Zhang WK, Zhang JS, Chen SY (2010) Receptor-like kinase OsSIK1 improves drought and salt stress tolerance in rice (*Oryza sativa*) plants. Plant J 62(2):316–329

Pantuwan G, Fukai S, Cooper M, Rajatasereekul S, O'Toole J (2002) Yield response of rice (*Oryza sativa* L.) genotypes to different types of drought under rainfed lowlands: Part 1. Grain yield and yield components. Field Crop Res 73(2):153–168

Park GG, Park JJ, Yoon J, Yu SN, An G (2010) A RING finger E3 ligase gene, *Oryza sativa* Delayed Seed Germination 1 (OsDSG1), controls seed germination and stress responses in rice. Plant Mol Biol 74(4–5):467–478

Peleg Z, Reguera M, Tumimbang E, Walia H, Blumwald E (2011) Cytokinin-mediated source/sink modifications improve drought tolerance and increase grain yield in rice under water-stress. Plant Biotechnol J 9(7):747–758

Peremarti A, Bassie L, Christou P, Capell T (2009) Spermine facilitates recovery from drought but does not confer drought tolerance in transgenic rice plants expressing *Datura stramonium* S-adenosylmethionine decarboxylase. Plant Mol Biol 70(3):253–264

Phung TH, Jung HI, Park JH, Kim JG, Back K, Jung S (2011) Porphyrin biosynthesis control under water stress: sustained porphyrin status correlates with drought tolerance in transgenic rice. Plant Physiol 157(4):1746–1764

Prashanth SR, Sadhasivam V, Parida A (2008) Over expression of cytosolic copper/zinc superoxide dismutase from a mangrove plant *Avicennia marina* in *indica* rice var Pusa Basmati-1 confers abiotic stress tolerance. Transgenic Res 17(2):281–291

Price A, Courtois B (1999) Mapping QTLs associated with drought resistance in rice: progress, problems and prospects. Plant Growth Regul 29(1):123–133

Price A, Tomos A (1997) Genetic dissection of root growth in rice (*Oryza sativa* L.). II: mapping quantitative trait loci using molecular markers. Theor Appl Genet 95(1):143–152

Price A, Steele K, Moore B, Barraclough P, Clark L (2000) A combined RFLP and AFLP linkage map of upland rice (*Oryza sativa* L.) used to identify QTLs for root-penetration ability. Theor Appl Genet 100(1):49–56

Price A, Steele K, Moore B, Jones R (2002a) Upland rice grown in soil-filled chambers and exposed to contrasting water-deficit regimes: II. Mapping quantitative trait loci for root morphology and distribution. Field Crop Res 76(1):25–43

Price AH, Cairns JE, Horton P, Jones HG, Griffiths H (2002b) Linking drought-resistance mechanisms to drought avoidance in upland rice using a QTL approach: progress and new opportunities to integrate stomatal and mesophyll responses. J Exp Bot 53(371):989–1004

Price AH, Townend J, Jones MP, Audebert A, Courtois B (2002c) Mapping QTLs associated with drought avoidance in upland rice grown in the Philippines and West Africa. Plant Mol Biol 48(5–6):683–695

Puranik S, Sahu PP, Srivastava PS, Prasad M (2012) NAC proteins: regulation and role in stress tolerance. Trends Plant Sci 17(6):369–381

Qu Y, Mu P, Zhang H, Chen CY, Gao Y, Tian Y, Wen F, Li Z (2008) Mapping QTLs of root morphological traits at different growth stages in rice. Genetica 133(2):187–200

Quan R, Hu S, Zhang Z, Zhang H, Huang R (2010) Overexpression of an ERF transcription factor *TSRF1* improves rice drought tolerance. Plant Biotechnol J 8(4):476–488

Quarrie SA, Laurie DA, Zhu J, Lebreton C, Semikhodskii A, Steed A, Witsenboer H, Calestani C (1997) QTL analysis to study the association between leaf size and abscisic acid accumulation in droughted rice leaves and comparisons across cereals. Plant Mol Biol 35(1–2):155–165

Ray J, Yu L, McCouch S, Champoux M, Wang G, Nguyen H (1996) Mapping quantitative trait loci associated with root penetration ability in rice (*Oryza sativa* L.). Theor Appl Genet 92(6):627–636

Redillas MC, Jeong JS, Kim YS, Jung H, Bang SW, Choi YD, Ha SH, Reuzeau C, Kim JK (2012) The overexpression of *OsNAC9* alters the root architecture of rice plants enhancing drought resistance and grain yield under field conditions. Plant Biotechnol J 10(7):792–805

Reguera M, Peleg Z, Blumwald E (2012) Targeting metabolic pathways for genetic engineering abiotic stress-tolerance in crops. Biochim Biophys Acta 1819(2):186–194

Rivero RM, Kojima M, Gepstein A, Sakakibara H, Mittler R, Gepstein S, Blumwald E (2007) Delayed leaf senescence induces extreme drought tolerance in a flowering plant. Proc Natl Acad Sci U S A 104(49):19631–19636

Robin S, Pathan M, Courtois B, Lafitte R, Carandang S, Lanceras S, Amante M, Nguyen H, Li Z (2003) Mapping osmotic adjustment in an advanced backcross inbred population of rice. Theor Appl Genet 107(7):1288–1296

Saijo Y, Hata S, Kyozuka J, Shimamoto K, Izui K (2000) Over-expression of a single Ca^{2+}-dependent protein kinase confers both cold and salt/drought tolerance on rice plants. Plant J 23(3):319–327

Sato Y, Yokoya S (2008) Enhanced tolerance to drought stress in transgenic rice plants overexpressing a small heat-shock protein, sHSP17. 7. Plant Cell Rep 27(2):329–334

Schroeder JI, Kwak JM, Allen GJ (2001) Guard cell abscisic acid signalling and engineering drought hardiness in plants. Nature 410(6826):327–330

Seo JS, Joo J, Kim MJ, Kim YK, Nahm BH, Song SI, Cheong JJ, Lee JS, Kim JK, Choi YD (2011) OsbHLH148, a basic helix-loop-helix protein, interacts with OsJAZ proteins in a jasmonate signaling pathway leading to drought tolerance in rice. Plant J 65(6):907–921

Serraj R, Kumar A, McNally K, Slamet-Loedin I, Bruskiewich R, Mauleon R, Cairns J, Hijmans R (2009) Improvement of drought resistance in rice. Adv Agron 103:41–99

Shen L, Courtois B, McNally K, Robin S, Li Z (2001) Evaluation of near-isogenic lines of rice introgressed with QTLs for root depth through marker-aided selection. Theor Appl Genet 103(1):75–83

Shen H, Liu C, Zhang Y, Meng X, Zhou X, Chu C, Wang X (2012) OsWRKY30 is activated by MAP kinases to confer drought tolerance in rice. Plant Mol Biol 80(3):241–253

Smalle J, Vierstra RD (2004) The ubiquitin 26S proteasome proteolytic pathway. Annu Rev Plant Biol 55:555–590

Song SY, Chen Y, Chen J, Dai XY, Zhang WH (2011) Physiological mechanisms underlying OsNAC5-dependent tolerance of rice plants to abiotic stress. Planta 234(2):331–345

Steele K (2009) Novel upland rice variety bred using marker-assisted selection and client-oriented breeding released in Jharkhand, India. Bangor University

Steele K, Price A, Shashidhar H, Witcombe J (2006) Marker-assisted selection to introgress rice QTLs controlling root traits into an Indian upland rice variety. Theor Appl Genet 112(2):208–221

Steele K, Virk D, Kumar R, Prasad S, Witcombe J (2007) Field evaluation of upland rice lines selected for QTLs controlling root traits. Field Crop Res 101(2):180–186

Suji KK, Biji KR, Poornima R, Prince KS, Amudha K, Kavitha S, Mankar S, Babu RC (2012) Mapping QTLs for plant phenology and production traits using indica rice (*Oryza sativa* L.) lines adapted to rainfed environment. Mol Biotechnol 52(2):151–160

Takasaki H, Maruyama K, Kidokoro S, Ito Y, Fujita Y, Shinozaki K, Yamaguchi-Shinozaki K, Nakashima K (2010) The abiotic stress-responsive NAC-type transcription factor OsNAC5 regulates stress-inducible genes and stress tolerance in rice. Mol Genet Genomics 284(3):173–183

Tang N, Zhang H, Li X, Xiao J, Xiong L (2012) Constitutive activation of transcription factor OsbZIP46 improves drought tolerance in rice. Plant Physiol 158(4):1755–1768

Tester M, Langridge P (2010) Breeding technologies to increase crop production in a changing world. Science 327(5967):818–822

Tripathy J, Zhang J, Robin S, Nguyen TT, Nguyen H (2000) QTLs for cell-membrane stability mapped in rice (*Oryza sativa* L.) under drought stress. Theor Appl Genet 100(8):1197–1202

Umezawa T, Fujita M, Fujita Y, Yamaguchi-Shinozaki K, Shinozaki K (2006) Engineering drought tolerance in plants: discovering and tailoring genes to unlock the future. Curr Opin Biotechnol 17(2):113–122

Venuprasad R, Lafitte H, Atlin G (2007) Response to direct selection for grain yield under drought stress in rice. Crop Sci 47(1):285–293

Venuprasad R, Dalid CO, Del Valle M, Zhao D, Espiritu M, Sta Cruz MT, Amante M, Kumar A, Atlin GN (2009) Identification and characterization of large-effect quantitative trait loci for grain yield under lowland drought stress in rice using bulk-segregant analysis. Theor Appl Genet 120(1):177–190

Vikram P, Swamy BP, Dixit S, Ahmed HU, Teresa Sta Cruz M, Singh AK, Kumar A (2011) *qDTY1.1*, a major QTL for rice grain yield under reproductive-stage drought stress with a consistent effect in multiple elite genetic backgrounds. BMC Genet 12:89

Wan L, Zhang J, Zhang H, Zhang Z, Quan R, Zhou S, Huang R (2011) Transcriptional activation of OsDERF1 in *OsERF3* and *OsAP2-39* negatively modulates ethylene synthesis and drought tolerance in rice. PLoS One 6(9):e25216

Wang FZ, Wang QB, Kwon SY, Kwak SS, Su WA (2005) Enhanced drought tolerance of transgenic rice plants expressing a pea manganese superoxide dismutase. J Plant Physiol 162(4):465–472

Wang Q, Guan Y, Wu Y, Chen H, Chen F, Chu C (2008) Overexpression of a rice *OsDREB1F* gene increases salt, drought, and low temperature tolerance in both Arabidopsis and rice. Plant Mol Biol 67(6):589–602

Wang N, Xiao B, Xiong L (2011) Identification of a cluster of PR4-like genes involved in stress responses in rice. J Plant Physiol 168(18):2212–2224

Wang Y, Wan L, Zhang L, Zhang Z, Zhang H, Quan R, Zhou S, Huang R (2012) An ethylene response factor OsWR1 responsive to drought stress transcriptionally activates wax synthesis related genes and increases wax production in rice. Plant Mol Biol 78(3):275–288

Wu X, Shiroto Y, Kishitani S, Ito Y, Toriyama K (2009) Enhanced heat and drought tolerance in transgenic rice seedlings overexpressing *OsWRKY11* under the control of *HSP101* promoter. Plant Cell Rep 28(1):21–30

Xiang Y, Huang Y, Xiong L (2007) Characterization of stress-responsive *CIPK* genes in rice for stress tolerance improvement. Plant Physiol 144(3):1416–1428

Xiang Y, Tang N, Du H, Ye H, Xiong L (2008) Characterization of OsbZIP23 as a key player of the basic leucine zipper transcription factor family for conferring abscisic acid sensitivity and salinity and drought tolerance in rice. Plant Physiol 148(4):1938–1952

Xiao B, Huang Y, Tang N, Xiong L (2007) Over-expression of a *LEA* gene in rice improves drought resistance under the field conditions. Theor Appl Genet 115(1):35–46

Xiao BZ, Chen X, Xiang CB, Tang N, Zhang QF, Xiong LZ (2009) Evaluation of seven function-known candidate genes for their effects on improving drought resistance of transgenic rice under field conditions. Mol Plant 2(1):73–83

Xiong L, Yang Y (2003) Disease resistance and abiotic stress tolerance in rice are inversely modulated by an abscisic acid-inducible mitogen-activated protein kinase. Plant Cell 15(3):745–759

Xiong L, Ishitani M, Lee H, Zhu JK (2001) The Arabidopsis *LOS5/ABA3* locus encodes a molybdenum cofactor sulfurase and modulates cold stress- and osmotic stress-responsive gene expression. Plant Cell 13(9):2063–2083

Xu Y, Crouch JH (2008) Marker-assisted selection in plant breeding: from publications to practice. Crop Sci 48(2):391–407

Xu D, Duan X, Wang B, Hong B, Ho T-HD WR (1996) Expression of a late embryogenesis abundant protein gene, *HVA1*, from barley confers tolerance to water deficit and salt stress in transgenic rice. Plant Physiol 110(1):249–257

Xu C, Li X, Xue Y, Huang Y, Gao J, Xing YZ (2004) Comparison of quantitative trait loci controlling seedling characteristics at two seedling stages using rice recombinant inbred lines. Theor Appl Genet 109(3):640–647

Xu JL, Lafitte HR, Gao YM, Fu BY, Torres R, Li ZK (2005) QTLs for drought escape and tolerance identified in a set of random introgression lines of rice. Theor Appl Genet 111(8):1642–1650

Xu DQ, Huang J, Guo SQ, Yang X, Bao YM, Tang HJ, Zhang HS (2008) Overexpression of a TFIIIA-type zinc finger protein gene *ZFP252* enhances drought and salt tolerance in rice (*Oryza sativa* L.). FEBS Lett 582(7):1037–1043

Xu Y, This D, Pausch RC, Vonhof WM, Coburn JR, Comstock JP, McCouch SR (2009) Leaf-level water use efficiency determined by carbon isotope discrimination in rice seedlings: genetic variation associated with population structure and QTL mapping. Theor Appl Genet 118(6):1065–1081

Xu M, Li L, Fan Y, Wan J, Wang L (2011) *ZmCBF3* over-expression improves tolerance to abiotic stress in transgenic rice (*Oryza sativa*) without yield penalty. Plant Cell Rep 30(10):1949–1957

Yadav R, Courtois B, Huang N, McLaren G (1997) Mapping genes controlling root morphology and root distribution in a doubled-haploid population of rice. Theor Appl Genet 94(5):619–632

Yamaguchi-Shinozaki K, Shinozaki K (2006) Transcriptional regulatory networks in cellular responses and tolerance to dehydration and cold stresses. Annu Rev Plant Biol 57:781–803

Yang Z, Wu Y, Li Y, Ling HQ, Chu C (2009) OsMT1a, a type 1 metallothionein, plays the pivotal role in zinc homeostasis and drought tolerance in rice. Plant Mol Biol 70(1–2):219–229

Yang S, Vanderbeld B, Wan J, Huang Y (2010) Narrowing down the targets: towards successful genetic engineering of drought-tolerant crops. Mol Plant 3(3):469–490

Yang A, Dai X, Zhang WH (2012) A R2R3-type MYB gene, *OsMYB2*, is involved in salt, cold, and dehydration tolerance in rice. J Exp Bot 63(7):2541–2556

Ye H, Du H, Tang N, Li X, Xiong L (2009) Identification and expression profiling analysis of TIFY family genes involved in stress and phytohormone responses in rice. Plant Mol Biol 71(3):291–305

You J, Hu H, Xiong L (2012) An ornithine delta-aminotransferase gene *OsOAT* confers drought and oxidative stress tolerance in rice. Plant Sci 197:59–69

Yu SB, Xu WJ, Vijayakumar CH, Ali J, Fu BY, Xu JL, Jiang YZ, Marghirang R, Domingo J, Aquino C, Virmani SS, Li ZK (2003) Molecular diversity and multilocus organization of the parental lines used in the International Rice Molecular Breeding Program. Theor Appl Genet 108(1):131–140

Yue B, Xiong L, Xue W, Xing Y, Luo L, Xu C (2005) Genetic analysis for drought resistance of rice at reproductive stage in field with different types of soil. Theor Appl Genet 111(6):1127–1136

Yue B, Xue W, Xiong L, Yu X, Luo L, Cui K, Jin D, Xing Y, Zhang Q (2006) Genetic basis of drought resistance at reproductive stage in rice: separation of drought tolerance from drought avoidance. Genetics 172(2):1213–1228

Yue B, Xue W, Luo L, Xing Y (2008) Identification of quantitative trait loci for four morphologic traits under water stress in rice (*Oryza sativa* L.). J Genet Genomics 35(9):569–575

Zhang Q (2007) Strategies for developing Green Super Rice. Proc Natl Acad Sci U S A 104(42):16402–16409

Zhang J, Zheng H, Aarti A, Pantuwan G, Nguyen T, Tripathy J, Sarial A, Robin S, Babu R, Nguyen BD (2001a) Locating genomic regions associated with components of drought resistance in rice: comparative mapping within and across species. Theor Appl Genet 103(1):19–29

Zhang W, Shen X, Wu P, Hu B, Liao C (2001b) QTLs and epistasis for seminal root length under a different water supply in rice (*Oryza sativa* L.). Theor Appl Genet 103(1):118–123

Zhang X, Zhou S, Fu Y, Su Z, Wang X, Sun C (2006) Identification of a drought tolerant introgression line derived from Dongxiang common wild rice (*O. rufipogon* Griff.). Plant Mol Biol 62(1–2):247–259

Zhang SW, Li CH, Cao J, Zhang YC, Zhang SQ, Xia YF, Sun DY, Sun Y (2009) Altered architecture and enhanced drought tolerance in rice via the down-regulation of indole-3-acetic acid by *TLD1/OsGH3.13* activation. Plant Physiol 151(4):1889–1901

Zhang H, Liu W, Wan L, Li F, Dai L, Li D, Zhang Z, Huang R (2010a) Functional analyses of ethylene response factor JERF3 with the aim of improving tolerance to drought and osmotic stress in transgenic rice. Transgenic Res 19(5):809–818

Zhang Z, Li F, Li D, Zhang H, Huang R (2010b) Expression of ethylene response factor *JERF1* in rice improves tolerance to drought. Planta 232(3):765–774

Zhang L, Xiao S, Li W, Feng W, Li J, Wu Z, Gao X, Liu F, Shao M (2011) Overexpression of a Harpin-encoding gene *hrf1* in rice enhances drought tolerance. J Exp Bot 62(12):4229–4238

Zhang Q, Li J, Zhang W, Yan S, Wang R, Zhao J, Li Y, Qi Z, Sun Z, Zhu Z (2012a) The putative auxin efflux carrier OsPIN3t is involved in the drought stress response and drought tolerance. Plant J 72(5):805–816

Zhang S, Haider I, Kohlen W, Jiang L, Bouwmeester H, Meijer AH, Schluepmann H, Liu CM, Ouwerkerk PB (2012b) Function of the HD-Zip I gene *Oshox22* in ABA-mediated drought and salt tolerances in rice. Plant Mol Biol 80(6):571–585

Zhao L, Hu Y, Chong K, Wang T (2010) *ARAG1*, an ABA-responsive DREB gene, plays a role in seed germination and drought tolerance of rice. Ann Bot 105(3):401–409

Zheng HG, Babu RC, Pathan MS, Ali L, Huang N, Courtois B, Nguyen HT (2000) Quantitative trait loci for root-penetration ability and root thickness in rice: comparison of genetic backgrounds. Genome 43(1):53–61

Zheng B, Yang L, Zhang W, Mao C, Wu Y, Yi K, Liu F, Wu P (2003) Mapping QTLs and candidate genes for rice root traits under different water-supply conditions and comparative analysis across three populations. Theor Appl Genet 107(8):1505–1515

Zheng B, Yang L, Mao C, Huang Y, Wu P (2008) Comparison of QTLs for rice seedling morphology under different water supply conditions. J Genet Genomics 35(8):473–484

Zheng X, Chen B, Lu G, Han B (2009) Overexpression of a NAC transcription factor enhances rice drought and salt tolerance. Biochem Biophys Res Commun 379(4):985–989

Zhou W, Li Y, Zhao BC, Ge RC, Shen YZ, Wang G, Huang ZJ (2009) Overexpression of *TaSTRG* gene improves salt and drought tolerance in rice. J Plant Physiol 166(15):1660–1671

Zhu B, Su J, Chang M, Verma DPS, Fan Y-L, Wu R (1998) Overexpression of a [delta] 1-pyrroline-5-carboxylate synthetase gene and analysis of tolerance to water-and salt-stress in transgenic rice. Plant Sci 139(1):41–48

Zou GH, Mei HW, Liu HY, Liu GL, Hu SP, Yu XQ, Li MS, Wu JH, Luo LJ (2005) Grain yield responses to moisture regimes in a rice population: association among traits and genetic markers. Theor Appl Genet 112(1):106–113

Zou M, Guan Y, Ren H, Zhang F, Chen F (2008) A bZIP transcription factor, OsABI5, is involved in rice fertility and stress tolerance. Plant Mol Biol 66(6):675–683

Zou J, Liu C, Liu A, Zou D, Chen X (2012) Overexpression of *OsHsp17.0* and *OsHsp23.7* enhances drought and salt tolerance in rice. J Plant Physiol 169(6):628–635

Polyamine Biosynthesis Engineering as a Tool to Improve Plant Resistance to Abiotic Stress

5

Francisco Marco, Marta Bitrián, Pedro Carrasco, Rubén Alcázar, and Antonio F. Tiburcio

Abstract

Polyamines (PAs) are small polycationic molecules which are present in all living organisms. PAs have been involved in a wide array of metabolic plant processes, extending from development to stress protection. Most of this knowledge has been achieved through the observation of PA homeostasis and manipulation of plant PA levels mediated by different approaches. This chapter summarizes the approaches undertaken to demonstrate the relationship between PAs and the stress response and, in particular, how the genetic manipulation of polyamine levels has evolved in a useful tool for the enhancement of plant stress tolerance in many species, including crops. This chapter also includes the most recent advances in the potential mechanisms of action by which polyamines could contribute to stress protection. Apart from a protective role based of its structural properties, PAs can also play regulatory roles, either directly or indirectly by the interaction with other signalling pathways including ion channel regulation, nitric oxide, reactive oxygen species (ROS) signalling and abscisic acid (ABA).

Keywords

Abiotic stresses • Catabolism • Polyamines • Putrescine • Signalling • Spermidine • Spermine • Stress tolerance • Transgenic plants

F. Marco
Departament de Biologia Vegetal, Facultat de Farmàcia, Universitat de València,
Av Vicent Andres Estellés s/n, 46100 Burjassot, València, Spain

M. Bitrián
Institut de Biologie Moléculaire des Plantes, Centre National de la Recherche Scientifique,
12 Rue du Général Zimmer, 67084 Strasbourg, France

P. Carrasco
Departament de Bioquímica i Biologia Molecular. Facultat de Ciències Biològiques, Universitat de València, Dr. Moliner 50, 46100 Burjassot, València, Spain

R. Alcázar • A.F. Tiburcio (✉)
Unitat de Fisiologia Vegetal, Facultat de Farmàcia, Universitat de Barcelona,
Avda Joan XXIII 27-31, 08028 Barcelona, Spain
e-mail: afernandez@ub.edu

P.K. Jaiwal et al. (eds.), *Genetic Manipulation in Plants for Mitigation of Climate Change*,
DOI 10.1007/978-81-322-2662-8_5

5.1 Introduction: Plant Polyamine Metabolism

Stress is induced by adverse environmental conditions, triggering a series of changes at morphological, physiological, biochemical and molecular levels that ultimately alter plant development and yield, thus preventing crops reaching their full genetic potential. It is estimated that stress conditions are going to develop more strong and persistent with climate change, particularly global warming. Thus, it is necessary to gain a better comprehension of the molecular bases of plant stress responsiveness and the development of novel strategies to facilitate crops to survive future confronting environments while improving productivity.

Polyamines (PAs) are a group of compounds of low molecular weight and aliphatic nature, with a variable number of hydrocarbon chains and two or more primary amino groups, which at cell physiological pH appear positively charged. Their most widespread forms in living organisms are the diamine putrescine (Put), triamine spermidine (Spd) and tetramine spermine (Spm), molecules that differ in their number of positively charged amine groups, but with similar structures, since Spd and Spm are generated by the addition of aminopropyl units to the carbon frame of Put (Fig. 5.1). Previous evidences indicate that Put and Spd are required for life and growth. Put and/or Spd have been detected in all living organisms so far analysed for their PA content. Put or Spd depletion by chemical treatment or genetic manipulation is lethal in yeast, protists and plants (Hamasaki-Katagiri et al. 1998; Imai et al. 2004b; Roberts et al. 2001; Urano et al. 2005). Conversely, Spm-deficient organisms are still viable but exhibit contrasted dysfunction degrees, thus suggesting a relevant role of Spm in growth and developmental programs (Imai et al. 2004a; Minguet et al. 2008; Wang et al. 2004; Yamaguchi et al. 2007).

PAs have been involved in a wide set of key processes in plant growth and development, including hormone signalling, transcriptome regulation, cell cycle as well as abiotic and biotic stress responses (Alcazar et al. 2006b; Bagni and Tassoni 2001; Bouchereau et al. 1999; Galston and Sawhney 1990; Kumar et al. 1997; Kusano et al. 2008; Malmberg et al. 1998; Walden et al. 1997). Physiological studies show that PAs in plants are frequently regulated at intracellular level by anabolic and catabolic processes (Fig. 5.1) and could be also found conjugated to hydroxycinnamic acids or macromolecules (DNA and proteins).

The PA biosynthesis route initiates with the generation of Put that can be derived directly from ornithine (Orn) decarboxylation through ornithine decarboxylase (ODC) activity or by an alternative pathway that involves arginine (Arg) decarboxylation by arginine decarboxylase (ADC) followed by two additional consecutive reactions catalysed by agmatine iminohydrolase (AIH) and *N*-carbamoylputrescine amidohydrolase (CPA). Mammals and fungi exclusively use the ODC pathway for Put production, while plants and bacteria are provided with the alternative ADC pathway. Put is the primary substrate for subsequent Spd and Spm biosynthesis by Spd synthase (SPDS) and Spm synthase (SPMS) activities, respectively, by sequential extension of the Put carbon skeleton by union of aminopropyl units. Both enzymes use the same aminopropyl donor (decarboxylated S-adenosylmethionine (dcSAM)), generated from decarboxylation of SAM by SAM decarboxylase (SAMDC) activity. Spd and dcSAM can also be combined to form a structural isomer of spermine, thermospermine (tSpm) in a reaction catalysed by tSpm synthase (tSPMS) (Fig. 5.1).

Diamine oxidase (DAO) and PA oxidase (PAO) activities mediate PA catabolism by catalysis of oxidative deamination reactions. DAOs are enzymes with strong diamine affinity and catalyse conversion of Put into Δ^1-pyrroline, H_2O_2 and ammonia (Fig. 5.1). Δ^1-pyrroline is converted to γ-aminobutyric acid (GABA) (Fig. 5.1) that serves as a source of succinic acid, an intermediary of the citric acid cycle. Spd and Spm are oxidized by PAOs, leading to the generation of 4-aminobutanal or (3-aminopropyl)-4-aminobutanal, 1,3-diaminopropane (DAP) and H_2O_2. PAOs activity could also back convert Spm to Spd, producing 3-aminopropanal and H_2O_2. It has also been shown that some PAOs isoforms are involved in the back conversion of tSpm to

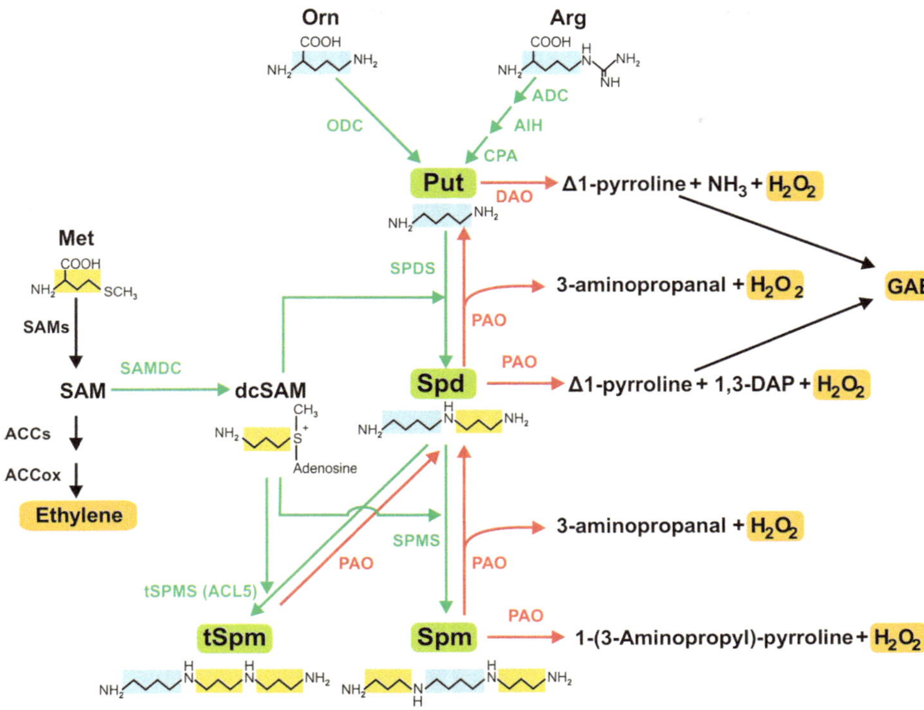

Fig. 5.1 Polyamine (PA) metabolism in plants. Biosynthesis routes are coloured in *green* and catabolism pathways in *red*. *ACCs* ACC synthase, *ACCox* ACC oxidase, *ACL5* ACAULIS5, *ADC* arginine decarboxylase, *AIH* agmatine iminohydrolase, *CPA*, N-carbamoylputrescine amidohydrolase, *DAO* diamine oxidase, *1,3-DAP* 1,3-diaminopropane, *dcSAM* decarboxylated S-adenosylmethionine, *ODC* ornithine decarboxylase, *PAO* polyamine oxidase, *SAM* S-adenosylmethionine, *SAMs* S-adenosylmethionine synthase, *SAMDC* S-adenosylmethionine decarboxylase, *SPDS* spermidine synthase, *SPMS* spermine synthase, *tSPMS* thermospermine synthase

Spm and Spm to Put (Fig. 5.1). The PA metabolic route shares common intermediaries with other cell pathways involved in the generation of relevant metabolites and signalling molecules related to stress like ethylene, GABA or H_2O_2 (Fig. 5.1) (Reviewed in Alcázar et al. 2010a and Bitrián et al. 2012).

In the model plant *Arabidopsis thaliana* (hereafter referred to as Arabidopsis), except for ODC, all the other PA biosynthesis genes are present. *Arabidopsis* carries two genes that encode for ADC (*ADC1;* Watson and Malmberg 1996 and *ADC2*; Watson et al. 1997) and one for each AIH (Janowitz et al. 2003) and CPA (Piotrowski et al. 2003). By the other side, *Arabidopsis* genome includes two genes coding for SPDS (*SPDS1* and *SPDS2*) (Hanzawa et al. 2002), single copies for genes encoding SPMS (*SPMS;* Panicot et al. 2002) and tSPMS (*ACL5;* Knott et al. 2007; Kakehi et al. 2008) and at least four copies that code for SAMDC (*SAMDC1-4)* (Urano et al. 2003). Regarding PA catabolism, the *Arabidopsis* genome also contains at least five genes coding for putative PAOs (Alcazar et al. 2006b; Takahashi et al. 2010) and 10 genes encoding putative DAOs (Planas-Portell et al. 2013).

5.2 Polyamines Are Implicated in the Plant Response to Abiotic Stress

5.2.1 Modulation of Polyamine Metabolism under Abiotic Stress Conditions

The accumulation of PAs during stress-inducing conditions has been observed in many cases since the first report by Richards and Coleman (1952) in oat plants that increased their endogenous Put levels under potassium deficiency conditions. Rise of PA levels has also been a common observation in later studies, where different plant species have been exposed to single or combined types of stresses (Alcazar et al. 2006b; Bouchereau et al. 1999; Groppa and Benavides 2008). Regarding to abiotic stress, PA accumulation correlates with enhanced tolerance to a large set of environmental challenges like salinity (Chattopadhyay et al. 1997; Liu et al. 2006; Roy and Wu 2002; Songstad et al. 1990), chilling (Shen et al. 2000; Songstad et al. 1990), osmotic and acidic stresses (Capell et al. 2004), radiation-induced oxidative stress (Deutsch et al. 2005) and others (Groppa and Benavides 2008).

A variety of genes that code for enzymes of PA biosynthesis and catabolism have been isolated in several plant species, and their expression patterns under stress conditions have also been studied (Alcazar et al. 2006b; Hussain et al. 2011; Liu et al. 2007). A common hallmark from these experiments is the stress-induced expression of some PA biosynthesis genes, with some variations in their expression kinetics. While some PA biosynthesis genes exhibit a rapid induction shortly after stress treatments followed by a steady increment or a slight variation, other genes require a more extended interval of stress to be induced. These studies suggest diverse ways of control of the expression of PA pathway genes during stress, consistent with the fact that different pathways are implicated in the regulation of PA biosynthesis (Alcázar et al. 2010a). A complementary approach has shown that 11 of the 21 genes that code for enzymes of the PA pathway are located in QTL regions associated with drought tolerance in rice (Do et al. 2013).

Kinetic studies of the PA transcriptome during stress have been performed in order to study the regulation of PA metabolism from a broader perspective (Alcazar et al. 2011, 2006a; Alet et al. 2012; Cuevas et al. 2008, 2009; Do et al. 2013; Rodríguez-Kessler et al. 2006; Urano et al. 2009).

5.2.2 Polyamine Treatment Could Modulate Plant Stress Tolerance

The changes observed in PA levels derived from stress suggest their probable involvement in the plant stress response but are not evidence enough to demonstrate that production of PAs has an intrinsic protective role against stress. Some approaches have been used to modify the endogenous levels of PAs. Early studies increased the cellular levels of PAs by exogenous application either before or during stress. Most of these works have been helpful to pinpoint correlations between PA stress accumulation and plant tolerance, showing that PAs could, in varying degrees, mitigate stress-derived cell injury and reverse growth or mitigate growth inhibition caused by stress (Alcazar et al. 2006b; Bouchereau et al. 1999; Groppa and Benavides 2008). Nevertheless, it has to be noted that this approach may have several constraints, like the damaging effects to membranes when PA are applied at high doses or the different uptake rates that could appear between replicates. Furthermore, the protective effects of individual PAs are somewhat different depending on the study considered. Some of those discrepancies may reflect differential mechanisms of PA absorption, transport and utilization among plant species.

5.2.3 Depletion of Plant Polyamine Levels Increases Their Sensitivity to Stress

Studies have reported changes in endogenous PA levels through the use of inhibitors of PA biosynthesis enzymes, such as difluoromethylarginine (DFMA), difluoromethylornithine (DFMO), D-arginine or methylglyoxal-bis (guanylhydrazone) (MGBG). Inhibitors lead to a reduction of endogenous PAs and enhancement of stress sensitivity, which is turned back by external treatment of PAs (He et al. 2002). However, in most cases, the outcome of the treatment with those molecules is variable, frequently influenced by inhibitor stability, different localizations of the inhibitor and the target enzyme or the existence of dynamic mechanisms of compensation (Kaur-Sawhney et al. 2003).

Researchers have overcome those limitations with plants that carry loss-of-function mutations in PA biosynthesis genes (Kaur-Sawhney et al. 2003; Urano et al. 2004; Watson et al. 1998). EMS mutants of *Arabidopsis thaliana spe1- 1* and *spe2-1* (which map to *ADC2*), isolated by Watson et al. (1998), display decreased ADC activity and are not able to accumulate PAs after acclimation to a high-salt environment and are less tolerant to salt stress compared to wild-type plants (Kasinathan and Wingler 2004). In addition, the knockout mutant *adc2-1*, which shows that Put levels are reduced up to 75 % of wild-type plants, shows a phenotype of increased sensitivity to salt stress, which can be partially turned back by Put treatment (Urano et al. 2004). External Put addition is also effective in the partial suppression of the freezing sensitivity phenotype described in other *ADC1* (*adc1-2*, *adc1-3*) and *ADC2* (*adc2-3*, *adc2-4*) mutant alleles (Cuevas et al. 2008). By the other side, the *Arabidopsis acl5/spms* double mutant, impaired in the production of Spm and tSpm, exhibits hypersensitivity to salt and drought stresses that can be diminished by administration of exogenous Spm (Yamaguchi et al. 2006, 2007). Single mutants *acl5-1 and spms-1* also show an accumulation of Na^+ and a lower survival's performance in long-term saline stress conditions (Alet et al. 2012). Also, hypersensitivity to heat has been described for another Arabidopsis *spms* mutant, which shows a more pronounced chlorophyll loss when subjected to heat shock (Sagor et al. 2013). Moreover, RNAi silencing of *SAMDC* genes in tobacco reduced the levels of Spd and Spm which produced a reduction in biomass under saline conditions, as well as an increase in programmed cell death (PCD) when cell suspension cultures generated from these genotypes are exposed to high-salt growth conditions (Moschou et al. 2008). Taken together, studies provide evidence to the active implication of PAs in the plant stress response. A more detailed characterization of loss-of-function mutants might be needed to unravel the precise PA mode(s) of action during stress. In this sense, the combined study of PA profiles and transcriptome studies of the entire biosynthesis pathway in *Arabidopsis* mutants defective in ADC (*adc1-3*, *adc2-3*), SPDS (*spds1-2*, *spds2-3*) or SPMS (*spms-2*) have exposed a strong metabolic canalization of Put to Spm induced by drought that can also be observed in the desiccation-tolerant plant *Craterostigma plantagineum* (Alcazar et al. 2011).

5.3 Transgenic Engineering of Polyamine Biosynthetic Pathway Improves Plant Abiotic Stress Tolerance

Another alternative strategy to produce plants with altered levels of PAs and overcome the problems raised from exogenous PAs treatment has been the generation of transgenic plants overexpressing PA biosynthesis genes, previously cloned from yeast, plant or animal genomes. Table 5.1 compiles several examples that have been obtained during the last years through the overexpression of *ODC*, *ADC*, *SAMDC*, *SPDS and SPMS* in rice, tobacco, eggplant, pear, sweet potato and *Arabidopsis*. Although the degree of alteration of the endogenous level of one or more specific PAs obtained is not the same in all plants, all transgenic lines share a common enhancement of the tolerance against one or more abiotic stresses (Table 5.1). Indeed, the improvement in tolerance observed always correlates with increased levels of Put and/or Spd and Spm.

In most cases, overexpression has been achieved by the use of constitutive promoters. Under drought stress conditions, rice plants transformed with the *Datura stramonium ADC* gene under the control of maize strong promoter ubiquitin *1* (Ubi-1) accumulate more Spd and Spm and show lower rates of leaf curling and chlorophyll loss under drought stress than the wild type (Capell et al. 2004). Heterologous overexpression of oat ADC has been reported by Prabhavathi and Rajam (2007). In this work, eggplants were transformed with *CAMV35S* promoter fused to the oat *ADC* gene, thus obtaining transgenic lines with elevated Put, Spd and Spm

Table 5.1 Transgenic approaches to obtain PA-overproducing plants and their effects on abiotic stress tolerance

Gene	Source	Transgenic plant	Overexpression	Overproduced PA	Enhanced tolerance	References
ADC	Oat	Rice	Inducible	Put	Salt	Roy and Wu (2001)
	Oat	Eggplant	Constitutive	Put, Spd and Spm	Broad sprectrum	Prabhavathi and Rajam (2007)
	Oat	*Arabidopsis*	Inducible	Put	Cold, drought	Alet et al. (2011)
	Datura stramonium	Rice	Constitutive	Spd and Spm	Drought	Capell et al. (2004)
	Arabidopsis (*ADC1*)	*Arabidopsis*	Constitutive	Put	Freezing	Tiburcio et al. (2009)
	Arabidopsis (*ADC2*)	*Arabidopsis*	Constitutive	Put	Drought	Alcázar et al. (2010b)
ODC	Mouse	Tobacco	Constitutive	Put	Salt	Kumria and Rajam (2002)
SAMDC	*Tritordeum*	Rice	Inducible	Spd and Spm	Salt	Roy and Wu (2002)
	Human	Tobacco	Constitutive	Put and Spd	Salt, osmotic	Waie and Rajam (2003)
	Human	Tomato	Constitutive	Put, Spd and Spm	Broad spectrum	Hazarika and Rajam (2011)
	Carnation	Tobacco	Constitutive	Put, Spd and Spm	Broad spectrum	Wi et al. (2006)
	Yeast	Tomato	Constitutive	Spd and Spm	Heat	Cheng et al. (2009)
	Datura stramonium	Rice	Constitutive	Spd	Drought (recovery)	Peremarti et al. (2009)
	Arabidopsis (*SAMDC1*)	*Arabidopsis*	Constitutive	Spm	Salt	Marco et al. (2011)
SPDS	*Cucurbita ficifolia*	*Arabidopsis*	Constitutive	Spd	Broad spectrum	Kasukabe et al. (2004)
	Cucurbita ficifolia	Sweet potato	Constitutive	Spd	Broad spectrum	Kasukabe et al. (2006)
	Apple	Pear	Constitutive	Spd	Broad sprectrum	Wen et al. (2009) and Wen et al. (2008)
SPMS	*Arabidopsis*	*Arabidopsis*	Constitutive	Spm	Heat	Sagor et al. (2013)
ACCs	Carnation	Tobacco	Antisense	Put and Spd	Broad spectrum	Wi and Park (2002)
ACCox	Carnation	Tobacco	Antisense	Put and Spd	Broad spectrum	Wi and Park (2002)

Adapted from Alcázar et al. (2010a) and Gill and Tuteja (2010)

levels showing a phenotype of tolerance to a wide array of abiotic stress conditions and resistance to pathogen infection. More recently, *Arabidopsis ADC* has also been overexpressed constitutively, leading to transgenic plants with elevated Put levels that show resistance to drought stress (Alcázar et al. 2010b) and freezing conditions (Tiburcio et al. 2009). Elevated levels of Put are also observed in tobacco plants overexpressing mouse *ODC*. Those transgenic plants show better germination rates under saline conditions than wild-type plants, as well as a diminished degree of wilting (Kumria and Rajam 2002).

Heterologous constitutive overexpression of *SAMDC* has been achieved in many cases, using cDNAs from different species (Table 5.1). Waie and Rajam (2003) obtained tobacco plants where *CaMV35S* promoter was used to drive expression of a human *SAMDC* gene. These plants showed an increased Spd and Put content than wild type, along with an enhanced tolerance to drought and salt stresses. Also, tomato plants overexpressing human *SAMDC* displayed high levels of PAs compared to untransformed control plants and exhibited a better tolerance to salinity, drought, cold, high temperature conditions and resistance against two critical pathogens of tomato (Hazarika and Rajam 2011). Previously, tomato plants overexpressing yeast *SAMDC* were reported, with high levels of Spm and Spd and tolerance to heat stress (Cheng et al. 2009). By the other side, it has been described that transgenic tobacco plants overexpressing carnation *SAMDC* are tolerant to a broad array of abiotic stresses (Wi et al. 2006). Moreover, elevated Spm levels and enhanced salt stress tolerance are also observed when *Arabidopsis SAMDC1* gene is overexpressed constitutively (Marco et al. 2011).

There are few reports for the overexpression of SPDS in plants, but all of them have in common the observation of increased Spd levels and tolerance to a wide array of stress conditions (Table 5.1). *Cucurbita ficifolia SPDS* has been overexpressed in *Arabidopsis*, obtaining plants with increased tolerance to salinity, drought, osmotic stress, freezing, chilling and oxidative stress by paraquat treatment (Kasukabe et al. 2004). The same gene has also been overexpressed in sweet potato (*Ipomoea batatas*), generating transgenic plants with an increased tolerance to salinity and drought (Kasukabe et al. 2006). Higher Spd titers have also been observed when apple *MDSPDS1* gene is overexpressed in pear. In this case, transgenic plants obtained display a better tolerance to saline, osmotic and heavy metal stresses compared to wild-type plants (Wen et al. 2009). Regarding to *SPMS* gene, their homologous overexpression in *Arabidopsis* leads to transgenic plants with higher level of thermotolerance than wild type (Sagor et al. 2012).

Inducible overexpression has been also used as a transgenic tool to elevate PA levels. Cold and drought tolerance has also been observed in transgenic *Arabidopsis* plants with stress-inducible promoter pRD29A (with cis-elements responsive to drought and ABA) driving the expression of oat ADC gene (Alet et al. 2011). Previously, another ABA-inducible promoter was used to control the expression of oat ADC in transgenic rice plants, which showed higher ADC activity and Put levels, as well as an increased biomass under saline conditions (Roy and Wu 2001). Later, the same authors fused this ABA-inducible promoter to *Tritordeum SAMDC* gene, and the transformed rice plants were able to accumulate three- to fourfold more Spd and Spm than wild-type plants under NaCl stress, showing also a higher biomass and stem growth than wild type under these saline stress conditions (Roy and Wu 2002).

On the other side, plants with high PA levels with better tolerance to stress have also been obtained using other transgenic approaches. Hiatt and Malmberg (1988) described a tobacco line resistant to DFMO with high PA levels and an increased tolerance to acidic exposure. Also, elevated PA levels in tobacco have been obtained by using the alternative strategy of diverting the flux of SAM from ethylene to PAs by antisense silencing of ethylene biosynthesis pathway (Fig. 5.1). This transgenic modification raised Put and Spd plant levels, increasing their tolerance to saline, acid and oxidative stresses (Wi and Park 2002). More recently, the response to oxidative stress conditions of a rice T-DNA insertion mutant line in the lysine decarboxylase-like 1 gene (*OsLDC-like 1*) has been characterized. When exposed to oxidative stress conditions, *OsLDC-like 1* mutant plants are able to produce higher levels of Put, Spd and Spm, as well as an increased tolerance compared to the wild-type rice strain (Jang et al. 2012).

5.4 Potential Mechanisms of Polyamine Action in Plant Abiotic Stress Response

In summary, research described above indicates that the elevation of PA levels is one of the responses that are induced by various stressful conditions in the plant, and this accumulation has a protective role. Despite that, the question regarding the precise way (s) of action by which PAs are able to exert this protection against external harmful conditions remains still unresolved, although some possibilities had been discussed during the last years (Alcázar et al. 2010a; Gill and Tuteja 2010).

5.4.1 Polyamines as a Protective Molecules

The idea of PAs as molecules with a protective role has come classically from their chemical structure: at physiological pH, PAs have a polycationic nature that makes them able to participate in the modulation of the cell ion balance and also interact with anionic molecules present in the cell, such as membrane lipids, proteins, DNA or RNA (Feuerstein and Marton 1989; Schuber 1989). PA binding to nucleic acids or proteins could protect these macromolecules from degradation as well as to create a safe environment where they could achieve a strong conformation, thus preventing their denaturation by extreme conditions. Stabilization of plant cell membranes and their protection from damage under stress conditions by external treatment of PAs (di- and tri- and tetra-amines) has been described in a large number of studies (He et al. 2008; Liu et al. 2007). In this sense, it has been proposed that endogenous PAs could play a role in sustaining membrane integrity (Borrell et al. 1997).

By the other side, the dual anion- and cation-binding property of PAs confers them radical-scavenging properties (Bors et al. 1989) and opens the possibility of an antioxidative role of PAs in the cell. It has been demonstrated that PAs inhibit lipid peroxidation in rat liver microsomes (Kitada et al. 1979) as well as metal-catalysed oxidative reactions (Tadolini 1988). In plants, Put, Spd and Spm reduce the level of superoxide radicals generated by senescing plant cells (Drolet et al. 1986). Additionally, one of the products of PA catabolism is H_2O_2 (Fig. 5.1), a signalling molecule that could trigger the activation of the antioxidative defence response of the plant. However, since ROS could also be generated by peroxide production, the PA-antioxidant role seems to be a matter of thresholds and subcellular localizations (Groppa and Benavides 2008).

5.4.2 Polyamines as Signalling Molecules

More recently, it has been pointed out that alteration of endogenous PA pool in the plant has a striking effect of its transcriptome, with the modification of the expression of a representative number of genes, being most of them stress related. Therefore, other mechanisms by which PAs can perform their protective action against stress could be through the activation at transcriptional level of plant defence mechanisms, which in turn would render stress tolerance. Tobacco *SAMDC* overexpressing plants have higher expression levels of several members of the enzymatic antioxidative stress machinery, such as glutathione S-transferase, superoxide dismutase and ascorbate peroxidase (Wi et al. 2006). Additionally, *Arabidopsis* plants overexpressing *SPDS* exhibit a transcriptome profile with elevated expression of some transcription factors involved in stress response pathways, such as MYB, WRKY, DREB, NAM proteins and Bbox zinc finger proteins, as well as stress-induced genes, such as rd29A or low-temperature-induced protein 78 (LTI78) (Kasukabe et al. 2004). More recently, microarray expression profiles of *Arabidopsis* plants that overexpress homologous *ADC2*, *SAMDC1* or *SPMS* have been analysed and compared, showing as common factor the induction of genes related to stress (Alcazar et al. 2005; Marco et al. 2011). Moreover, the transcriptomes of Put and Spm overproducers have in common a set of 71 up-regulated genes. This set is enriched in stress-related genes, like several putative transcription factors and Ca^{2+} signalling-related proteins, and most of them are induced in response to several abiotic stresses (Marco et al. 2011).

A possible source for some of the transcriptome variations observed in response to PA accumulation in plants could be the existence of a PA modulon expression system, similar to the system proposed in *E. coli* and yeast, where a set of several key transcription factors are stimulated at the translational level by PAs (Igarashi and Kashiwagi 2006; Uemura et al. 2009). In this sense, Hanfrey et al. (2003) have described in *Arabidopsis* the existence of a regulation of SAMDC at translation level by PAs. Also, Spd is required for the NAD-dependent formation of deoxyhypusine in the eukaryotic translation initiation factor 5A (eIF-5A) that seems necessary for its activity (Hopkins et al. 2008; Saini et al. 2009). Although not fully understood, several evidences point to the involvement of eIF5A in plant stress responses, as its overexpression in different plants produces stress-tolerant plants (Wang et al. 2012; Xu et al. 2011). Overexpression of *Tamarix androssowii eIF5A* (*TaeIF5A*) in poplar plants has revealed that TaeIF5A may mediate stress tolerance towards the increase of protein synthesis and enhancement of ROS scavenging by the induction of antioxidant enzymes, preventing chlorophyll loss and membrane damage (Wang et al. 2012). Additionally, activation of rice OSBZ8, a bZIP class of ABRE-binding transcription factor, by their specific kinase OSPDK is promoted in vitro by Spd (Gupta et al. 2012). Additional studies focused to the isolation of more transcription factors whose activity could be modulated by PAs, as well as their target genes, should be made in order to unravel the possible existence of a PA plant modulon.

Evidences indicate that some of the transcriptome profile changes observed could be the outcome of crosstalking between PAs with other signalling pathways (Alcázar et al. 2010a). Crosstalk between PAs and ABA has been described in *Arabidopsis* under drought stress, where is observed an up-regulation of PA biosynthesis genes *ADC2*, *SPDS1* and *SPMS* as well as accumulation of Put, being both processes mainly ABA-dependent responses (Alcazar et al. 2006a). By the other side, Put-accumulating *ADC2* overexpressor plants show also a diminished gibberellic acid (GA) biosynthesis, pointing to the existence of complex crosstalks between Put, ABA and GAs (Alcazar et al. 2005; Cuevas et al. 2008). Additionally, evidence of Spm-ABA crosstalk is shown in *Arabidopsis* plants overexpressing *SAMDC1*, where induction of NCED3, a key enzyme of ABA biosynthesis pathway, leads to elevated levels of this hormone (Marco et al. 2011). Several jasmonic acid (JA) biosynthesis genes as well as JA-responsive genes are also up-regulated in both Put and Spm *Arabidopsis* overproducers (Marco et al. 2011).

Furthermore, it has also been proposed a possible connection between PAs and stress responses through Ca^{2+} homeostasis (Alcázar et al. 2010a). In this sense, it has been suggested that the protective role of Spm against drought and salinity conditions could be mediated through the control of Ca^{2+} distribution by regulation of Ca^{2+} permeable channels, including vacuolar CAXs (Yamaguchi et al. 2006, 2007). Also, Spm modulation of the fluxes of cytoplasmic Ca^{2+} in the guard cells involved in stomatal movement could be one of the ways in which Spm contributes to drought tolerance. Connection between PAs and Ca^{2+} signalling is also shown in the transcriptome analysis of Put and Spm-overproducer plants, where Ca^{2+} signalling genes appear as one of the functional categories mainly up-regulated in both transgenic lines (Marco et al. 2011). Furthermore, PA accumulation is also enhanced in the *Arabidopsis* response to pathogens, and an "Spm signalling pathway" has been proposed to explain its role (Mitsuya et al. 2009; Takahashi

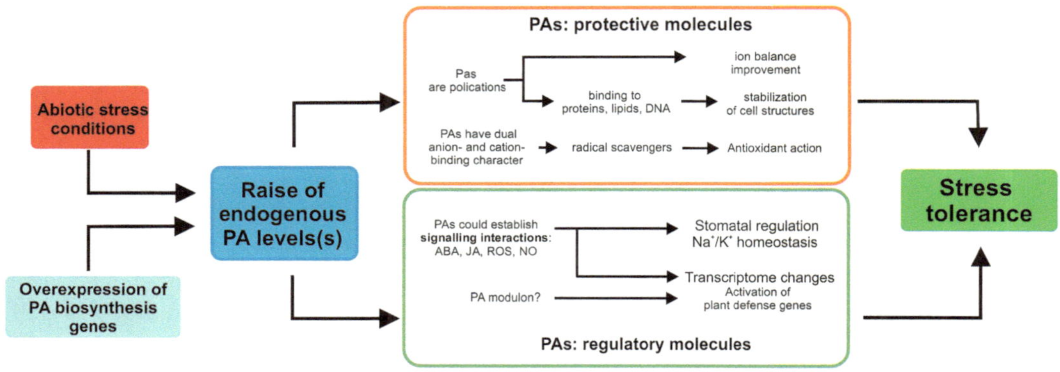

Fig. 5.2 Simplified diagram for possible mechanisms underlying roles of PAs in stress responses. Changes in PA plant levels are produced by abiotic stress conditions, as well as by overexpression of PA biosynthesis genes. Plant stress tolerance could be achieved by the combination of the dual roles of PAs: their protective role as structure stabilizers and/or antioxidant properties, as well as exerting a regulatory role by means of the activation of plant defence mechanisms by itself or by interactions with other signalling pathways, like ABA, Ca^{2+} or NO

et al. 2003). In this pathway, the combination of increased Ca^{2+} influx and H_2O_2 produced by PA oxidase activities that degrade Spm converges in a blended signal that triggers mitochondrial dysfunction and activates cell death (Takahashi et al. 2003).

It has been shown that stomatal closure is induced by PAs (An et al. 2008; Liu et al. 2000). Yamasaki and Cohen (2006) propose a model to explain the involvement of PAs on the regulation of stomatal responses where PAs participate through their interactions with nitric oxide (NO) signalling, as well as by the H_2O_2 produced by its oxidation. In this sense, ABA responses in guard cells could be promoted by the synergic action of PAs, NO and ROS (Alcázar et al. 2010a). The charged properties of PAs give them the capacity of blocking fast vacuolar cation channels, as well as affect protein kinase and/or phosphatase activities that regulate ion channel functions. Thus, stomatal closure responses could be affected by either way (Alcázar et al. 2010a).

5.5 Conclusions

In summary, PA activity in the response of the plant to stress could be the combination of the safeguard role related to their chemical properties and structure as well as by their capability to operate as members of the transduction pathways in stress responses (Fig. 5.2). Probably, the combination of both factors is the cause of the multiple stress tolerance (high salt, drought, extreme temperatures and paraquat toxicity) that has been described in plants where overproduction of one or more PAs has been obtained by transgenic modification (Table 5.1). This broad spectrum stress tolerance makes PA-content manipulation as a promising strategy to improve crop performance, since plants have often to deal with external conditions that suppose a simultaneous exposure to different sources of environmental stress. Also, harsh environments are predicted with climate change.

Acknowledgements AFT and RA acknowledge support from the Spanish Ministerio de Ciencia e Innovación (grant numbers: BIO2011-29683, CSD2007-00036) and SGR2009-1060 of the Generalitat de Catalunya. RA acknowledges support from the Ramón y Cajal Programme (RYC-2011-07847) of the Ministerio de Ciencia e Innovación (Spain) and the Marie Curie Career Integration Grant (DISEASENVIRON, PCIG10-GA-2011-303568) of the EU.

References

Alcazar R, Garcia-Martinez JL, Cuevas JC, Tiburcio AF, Altabella T (2005) Overexpression of ADC2 in Arabidopsis induces dwarfism and late-flowering through GA deficiency. Plant J 43:425–436

Alcazar R, Cuevas JC, Patron M, Altabella T, Tiburcio AF (2006a) Abscisic acid modulates polyamine metabolism under water stress in *Arabidopsis thaliana*. Physiol Plant 128:448–455

Alcazar R, Marco F, Cuevas JC, Patron M, Ferrando A, Carrasco P, Tiburcio AF, Altabella T (2006b) Involvement of polyamines in plant response to abiotic stress. Biotechnol Lett 28:1867–1876

Alcázar R, Altabella T, Marco F, Bortolotti C, Reymond M, Koncz C, Carrasco P, Tiburcio A (2010a) Polyamines: molecules with regulatory functions in plant abiotic stress tolerance. Planta 231:1237–1249

Alcázar R, Planas J, Saxena T, Zarza X, Bortolotti C, Cuevas J, Bitrian M, Tiburcio AF, Altabella T (2010b) Putrescine accumulation confers drought tolerance in transgenic Arabidopsis plants over-expressing the homologous Arginine decarboxylase 2 gene. Plant Physiol Biochem 48:547–552

Alcazar R, Bitrian M, Bartels D, Koncz C, Altabella T, Tiburcio AF (2011) Polyamine metabolic canalization in response to drought stress in Arabidopsis and the resurrection plant *Craterostigma plantagineum*. Plant Signal Behav 6:243–250

Alet AI, Sanchez DH, Cuevas JC, Del Valle S, Altabella T, Tiburcio AF, Marco F, Ferrando A, Espasandin FD, Gonzalez ME, Ruiz OA, Carrasco P (2011) Putrescine accumulation in *Arabidopsis thaliana* transgenic lines enhances tolerance to dehydration and freezing stress. Plant Signal Behav 6:278–286

Alet AI, Sánchez DH, Cuevas JC, Marina M, Carrasco P, Altabella T, Tiburcio AF, Ruiz OA (2012) New insights into the role of spermine in *Arabidopsis thaliana* under long-term salt stress. Plant Sci 182:94–100

An Z, Jing W, Liu Y, Zhang W (2008) Hydrogen peroxide generated by copper amine oxidase is involved in abscisic acid-induced stomatal closure in Vicia faba. J Exp Bot 59:815–825

Av D, Mitchell C, Williams C, Dutt K, Silvestrov N, Clement B, Abukhalaf I, Dv D (2005) Polyamines protect against radiation-induced oxidative stress. Gravit Space Biol 1005:109–110

Bagni N, Tassoni A (2001) Biosynthesis, oxidation and conjugation of aliphatic polyamines in higher plants. Amino Acids 20:301–317

Bitrián M, Zarza X, Altabella T, Tiburcio AF, Alcázar R (2012) Polyamines under abiotic stress: metabolic crossroads and hormonal crosstalks in plants. Metabolites 2:516–528

Borrell A, Carbonell L, Farràs R, Puig-Parellada P, Tiburcio AF (1997) Polyamines inhibit lipid peroxidation in senescing oat leaves. Physiol Plant 99:385–390

Bors W, Langebartels C, Michel C, Sandermann H Jr (1989) Polyamines as radical scavengers and protectants against ozone damage. Phytochemistry 28:1589–1595

Bouchereau A, Aziz A, Larher F, Martin-Tanguy J (1999) Polyamines and environmental challenges: recent development. Plant Sci 140:103–125

Capell T, Bassie L, Christou P (2004) Modulation of the polyamine biosynthetic pathway in transgenic rice confers tolerance to drought stress. Proc Natl Acad Sci U S A 101:9909–9914

Chattopadhyay MK, Gupta S, Sengupta DN, Ghosh B (1997) Expression of arginine decarboxylase in seedlings of indica rice (*Oryza sativa* L.) cultivars as affected by salinity stress. Plant Mol Biol 34:477–483

Cheng L, Zou Y, Ding S, Zhang J, Yu X, Cao J, Lu G (2009) Polyamine accumulation in transgenic tomato enhances the tolerance to high temperature stress. J Integr Plant Biol 51:489–499

Cuevas JC, Lopez-Cobollo R, Alcazar R, Zarza X, Koncz C, Altabella T, Salinas J, Tiburcio AF, Ferrando A (2008) Putrescine is involved in Arabidopsis freezing tolerance and cold acclimation by regulating abscisic acid levels in response to low temperature. Plant Physiol 148:1094–1105

Cuevas JC, Lopez-Cobollo R, Alcazar R, Zarza X, Koncz C, Altabella T, Salinas J, Tiburcio AF, Ferrando A (2009) Putrescine as a signal to modulate the indispensable ABA increase under cold stress. Plant Signal Behav 4:219–220

Do PT, Degenkolbe T, Erban A, Heyer AG, Kopka J, Kohl KI, Hincha DK, Zuther E (2013) Dissecting rice polyamine metabolism under controlled long-term drought stress. PLoS One 8:e60325

Drolet G, Dumbroff EB, Legge RL, Thompson JE (1986) Radical scavenging properties of polyamines. Phytochemistry 25:367–371

Feuerstein BG, Marton LJ (1989) Specificity and binding in polyamine/nucleic acid interactions. In: Bachrach U, Heimer YM (eds) The physiology of polyamines, vol I. CRC Press, Boca Raton, pp 109–207

Galston AW, Sawhney RK (1990) Polyamines in plant physiology. Plant Physiol 94:406–410

Gill SS, Tuteja N (2010) Polyamines and abiotic stress tolerance in plants. Plant Signal Behav 5(1):26–33

Groppa MD, Benavides MP (2008) Polyamines and abiotic stress: recent advances. Amino Acids 34:35–45

Gupta B, Gupta K, Sengupta D (2012) Spermidine-mediated in vitro phosphorylation of transcriptional regulator OSBZ8 by SNF1-type serine/threonine protein kinase SAPK4 homolog in indica rice. Acta Physiol Plant 34:1321–1336

Hamasaki-Katagiri N, Katagiri Y, Tabor CW, Tabor H (1998) Spermine is not essential for growth of Saccharomyces cerevisiae: identification of the SPE4 gene (spermine synthase) and characterization of a spe4 deletion mutant. Gene 210:195–201

Hanzawa Y, Imai A, Michael AJ, Komeda Y, Takahashi T (2002) Characterization of the spermidine synthase-related gene family in Arabidopsis thaliana. FEBS Lett 527(1–3):176–180

Hanfrey C, Franceschetti M, Mayer MJ, Illingworth C, Elliott K, Collier M, Thompson B, Perry B, Michael AJ (2003) Translational regulation of the plant S-adenosylmethionine decarboxylase. Biochem Soc Trans 31:424–427

Hazarika P, Rajam MV (2011) Biotic and abiotic stress tolerance in transgenic tomatoes by constitutive expression of S-adenosylmethionine decarboxylase gene. Physiol Mol Biol Plants 17:115–128

He L, Nada K, Kasukabe Y, Tachibana S (2002) Enhanced susceptibility of photosynthesis to low-temperature photoinhibition due to interruption of chill-induced increase of S-adenosylmethionine decarboxylase activity in leaves of Spinach (*Spinacia oleracea* L.). Plant Cell Physiol 43:196–206

He L, Ban Y, Inoue H, Matsuda N, Liu J, Moriguchi T (2008) Enhancement of spermidine content and antioxidant capacity in transgenic pear shoots overexpressing apple spermidine synthase in response to salinity and hyperosmosis. Phytochemistry 69:2133–2141

Hiatt A, Malmberg RL (1988) Utilization of putrescine in tobacco cell lines resistant to inhibitors of polyamine synthesis. Plant Physiol 86:441–446

Hopkins MT, Lampi Y, Wang T-W, Liu Z, Thompson JE (2008) Eukaryotic translation initiation factor 5A is involved in pathogen-induced cell death and development of disease symptoms in Arabidopsis. Plant Physiol 148:479–489

Hussain SS, Ali M, Ahmad M, Siddique KHM (2011) Polyamines: natural and engineered abiotic and biotic stress tolerance in plants. Biotechnol Adv 29:300–311

Igarashi K, Kashiwagi K (2006) Polyamine modulon in *Escherichia coli*: genes involved in the stimulation of cell growth by polyamines. J Biochem 139:11–16

Imai A, Akiyama T, Kato T, Sato S, Tabata S, Yamamoto KT, Takahashi T (2004a) Spermine is not essential for survival of Arabidopsis. FEBS Lett 556:148–152

Imai A, Matsuyama T, Hanzawa Y, Akiyama T, Tamaoki M, Saji H, Shirano Y, Kato T, Hayashi H, Shibata D, Tabata S, Komeda Y, Takahashi T (2004b) Spermidine synthase genes are essential for survival of Arabidopsis. Plant Physiol 135:1565–1573

Jang S, Wi S, Choi Y, An G, Park K (2012) Increased polyamine biosynthesis enhances stress tolerance by preventing the accumulation of reactive oxygen species: T-DNA mutational analysis of Oryza sativa lysine decarboxylase-like protein 1. Mol Cells 34:251–262

Janowitz T, Kneifel H, Piotrowski M (2003) Identification and characterization of plant agmatine iminohydrolase, the last missing link in polyamine biosynthesis of plants. FEBS Lett 544(1–3):258–261

Kakehi J, Kuwashiro Y, Niitsu M, Takahashi T (2008) Thermospermine is required for stem elongation in Arabidopsis thaliana. Plant Cell Physiol 49(9):1342–1349

Kasinathan V, Wingler A (2004) Effect of reduced arginine decarboxylase activity on salt tolerance and on polyamine formation during salt stress in *Arabidopsis thaliana*. Physiol Plant 121:101–107

Kasukabe Y, He L, Nada K, Misawa S, Ihara I, Tachibana S (2004) Overexpression of spermidine synthase enhances tolerance to multiple environmental stresses and up-regulates the expression of various stress-regulated genes in transgenic *Arabidopsis thaliana*. Plant Cell Physiol 45:712–722

Kasukabe Y, He L, Watakabe Y, Otani M, Shimada T, Tachibana S (2006) Improvement of environmental stress tolerance of sweet potato by introduction of genes for spermidine synthase. Plant Biotechnol 23:75–83

Kaur-Sawhney R, Tiburcio AF, Altabella T, Galston AW (2003) Polyamines in plants: an overview. J Cell Mol Biol 2:1–12

Kitada M, Igarashi K, Hirose S, Kitagawa H (1979) Inhibition by polyamines of lipid peroxide formation in rat liver microsomes. Biochem Biophys Res Commun 87:388–394

Knott JM, Römer P, Sumper M (2007) Putative spermine synthases from Thalassiosira pseudonana and Arabidopsis thaliana synthesize thermospermine rather than spermine. FEBS Lett 581(16):3081–3086

Kumar A, Taylor M, Altabella T, Tiburcio AF (1997) Recent advances in polyamine research. Trends Plant Sci 2:124–130

Kumria R, Rajam MV (2002) Ornithine decarboxylase transgene in tobacco affects polyamines, in vitro-morphogenesis and response to salt stress. J Plant Physiol 159:983–990

Kusano T, Berberich T, Tateda C, Takahashi Y (2008) Polyamines: essential factors for growth and survival. Planta 228:367–381

Liu K, Fu H, Bei Q, Luan S (2000) Inward potassium channel in guard cells as a target for polyamine regulation of stomatal movements. Plant Physiol 124:1315–1326

Liu J-H, Nada K, Honda C, Kitashiba H, Wen X-P, Pang X-M, Moriguchi T (2006) Polyamine biosynthesis of apple callus under salt stress: importance of the arginine decarboxylase pathway in stress response. J Exp Bot 57:2589–2599

Liu J-H, Kitashiba H, Wang J, Ban Y, Moriguchi T (2007) Polyamines and their ability to provide environmental stress tolerance to plants. Plant Biotechnol 24:117–126

Malmberg RL, Watson MB, Galloway GL, Yu W (1998) Molecular genetic analyses of plant polyamines. Crit Rev Plant Sci 17:199–224

Marco F, Alcazar R, Tiburcio AF, Carrasco P (2011) Interactions between polyamines and abiotic stress pathway responses unraveled by transcriptome analysis of polyamine overproducers. OMICS 15:775–781

Minguet EG, Vera-Sirera F, Marina A, Carbonell J, Blázquez MA (2008) Evolutionary diversification in polyamine biosynthesis. Mol Biol Evol 25:2119–2128

Mitsuya Y, Takahashi Y, Berberich T, Miyazaki A, Matsumura H, Takahashi H, Terauchi R, Kusano T (2009) Spermine signaling plays a significant role in the defense response of *Arabidopsis thaliana* to cucumber mosaic virus. J Plant Physiol 166:626–643

Moschou PN, Paschalidis KA, Delis ID, Andriopoulou AH, Lagiotis GD, Yakoumakis DI, Roubelakis-Angelakis KA (2008) Spermidine exodus and oxidation in the apoplast induced by abiotic stress is responsible for H_2O_2 signatures that direct tolerance responses in tobacco. Plant Cell 20:1708–1724

Panicot M, Minguet EG, Ferrando A, Alcazar R, Blazquez MA, Carbonell J, Altabella T, Koncz C, Tiburcio AF (2002) A polyamine metabolon involving aminopropyl transferase complexes in Arabidopsis. Plant Cell 14(10):2539–2551

Peremarti A, Bassie L, Christou P, Capell T (2009) Spermine facilitates recovery from drought but does not confer drought tolerance in transgenic rice plants expressing Datura stramoniumS-adenosylmethionine decarboxylase. Plant Mol Biol 70:253–264

Piotrowski M, Janowitz T, Kneifel H (2003) Plant C-N hydrolases and the identification of a plant N-carbamoylputrescine amidohydrolase involved in polyamine biosynthesis. J Biol Chem 278(3): 1708–1712

Planas-Portell J, Gallart M, Tiburcio AF, Altabella T (2013) Copper-containing amine oxidases contribute to terminal polyamine oxidation in peroxisomes and apoplast of *Arabidopsis thaliana*. BMC Plant Biol 13:109

Prabhavathi VR, Rajam MV (2007) Polyamine accumulation in transgenic eggplant enhances tolerance to multiple abiotic stresses and fungal resistance. Plant Biotechnol 24:273–282

Richards FJ, Coleman RG (1952) Occurrence of putrescine in potassium-deficient barley. Nature 170:460–460

Roberts SC, Jiang Y, Jardim A, Carter NS, Heby O, Ullman B (2001) Genetic analysis of spermidine synthase from *Leishmania donovani*. Mol Biochem Parasitol 115:217–226

Rodríguez-Kessler M, Alpuche-Solís A, Ruiz OA, Jiménez-Bremont JF (2006) Effect of salt stress on the regulation of maize (*Zea mays* L.) genes involved in polyamine biosynthesis. Plant Growth Regul 48:175–185

Roy M, Wu R (2001) Arginine decarboxylase transgene expression and analysis of environmental stress tolerance in transgenic rice. Plant Sci 160:869–875

Roy M, Wu R (2002) Overexpression of S-adenosylmethionine decarboxylase gene in rice increases polyamine level and enhances sodium chloride-stress tolerance. Plant Sci 163:987–992

Sagor GH, Takahashi H, Niitsu M, Takahashi Y, Berberich T, Kusano T (2012) Exogenous thermospermine has an activity to induce a subset of the defense genes and restrict cucumber mosaic virus multiplication in *Arabidopsis thaliana*. Plant Cell Rep 31:1227–1232

Sagor GHM, Berberich T, Takahashi Y, Niitsu M, Kusano T (2013) The polyamine spermine protects Arabidopsis from heat stress-induced damage by increasing expression of heat shock-related genes. Transgenic Res 22:595–605

Saini P, Eyler DE, Green R, Dever TE (2009) Hypusine-containing protein eIF5A promotes translation elongation. Nature 459:118–121

Schuber F (1989) Influence of polyamines on membrane functions. Biochem J 260:1–10

Shen W, Nada K, Tachibana S (2000) Involvement of polyamines in the chilling tolerance of cucumber cultivars. Plant Physiol 124:431–440

Songstad DD, Duncan DR, Widholm JM (1990) Proline and polyamine involvement in chilling tolerance of maize suspension cultures. J Exp Bot 41:289–294

Tadolini B (1988) Polyamine inhibition of lipoperoxidation. The influence of polyamines on iron oxidation in the presence of compounds mimicking phospholipid polar heads. Biochem J 249:33–36

Takahashi Y, Berberich T, Miyazaki A, Seo S, Ohashi Y, Kusano T (2003) Spermine signalling in tobacco: activation of mitogen-activated protein kinases by spermine is mediated through mitochondrial dysfunction. Plant J 36:820–829

Takahashi Y, Cong R, Sagor GH, Niitsu M, Berberich T, Kusano T (2010) Characterization of five polyamine oxidase isoforms in *Arabidopsis thaliana*. Plant Cell Rep 29:955–965

Tiburcio AF, Altabella T, Ferrando A (2009) Plant with resistance to low temperature and method of production thereof. Spanish patent application WO2010/004070

Uemura T, Higashi K, Takigawa M, Toida T, Kashiwagi K, Igarashi K (2009) Polyamine modulon in yeast – stimulation of COX4 synthesis by spermidine at the level of translation. Int J Biochem Cell Biol 41:2538–2545

Urano K, Yoshiba Y, Nanjo T, Igarashi Y, Seki M, Sekiguchi F, Yamaguchi-Shinozaki K, Shinozaki K (2003) Characterization of Arabidopsis genes involved in biosynthesis of polyamines in abiotic stress responses and developmental stages. Plant Cell Environ 26(11):1917–1926

Urano K, Yoshiba Y, Nanjo T, Ito T, Yamaguchi-Shinozaki K, Shinozaki K (2004) Arabidopsis stress-inducible gene for arginine decarboxylase AtADC2 is required for accumulation of putrescine in salt tolerance. Biochem Biophys Res Commun 313:369–375

Urano K, Hobo T, Shinozaki K (2005) Arabidopsis ADC genes involved in polyamine biosynthesis are essential for seed development. FEBS Lett 579:1557–1564

Urano K, Maruyama K, Ogata Y, Morishita Y, Takeda M, Sakurai N, Suzuki H, Saito K, Shibata D, Kobayashi M, Yamaguchi-Shinozaki K, Shinozaki K (2009) Characterization of the ABA-regulated global responses to dehydration in Arabidopsis by metabolomics. Plant J 57:1065–1078

Waie B, Rajam MV (2003) Effect of increased polyamine biosynthesis on stress responses in transgenic tobacco by introduction of human S-adenosylmethionine gene. Plant Sci 164:727–734

Walden R, Cordeiro A, Tiburcio AF (1997) Polyamines: small molecules triggering pathways in plant growth and development. Plant Physiol 113:1009–1013

Wang X, Ikeguchi Y, McCloskey DE, Nelson P, Pegg AE (2004) Spermine synthesis is required for normal viability, growth, and fertility in the mouse. J Biol Chem 279:51370–51375

Wang L, Xu C, Wang C, Wang Y (2012) Characterization of a eukaryotic translation initiation factor 5A homolog from *Tamarix androssowii* involved in plant abiotic stress tolerance. BMC Plant Biol 12:118

Watson MB, Malmberg RL (1996) Regulation of Arabidopsis thaliana (L.) Heynh arginine decarboxylase by potassium deficiency stress. Plant Physiol 111(4):1077–1083

Watson MW, Yu W, Galloway GL, Malmberg RL (1997) Isolation and characterization of a second arginine decarboxylase cDNA from Arabidopsis (Accession No AF009647). Plant Physiol 114(4):1569

Watson MB, Emory KK, Piatak RM, Malmberg RL (1998) Arginine decarboxylase (polyamine synthesis) mutants of *Arabidopsis thaliana* exhibit altered root growth. Plant J 13:231–239

Wen XP, Pang XM, Matsuda N, Kita M, Inoue H, Hao YJ, Honda C, Moriguchi T (2008) Over-expression of the apple spermidine synthase gene in pear confers multiple abiotic stress tolerance by altering polyamine titers. Transgenic Res 17:251–263

Wen X-P, Ban Y, Inoue H, Matsuda N, Moriguchi T (2009) Aluminum tolerance in a spermidine synthase-overexpressing transgenic European pear is correlated with the enhanced level of spermidine via alleviating oxidative status. Environ Exp Bot 66:471–478

Wi S, Park K (2002) Antisense expression of carnation cDNA encoding ACC synthase or ACC oxidase enhances polyamine content and abiotic stress tolerance in transgenic tobacco plants. Mol Cells 13:209–220

Wi S, Kim W, Park K (2006) Overexpression of carnation S-adenosylmethionine decarboxylase gene generates a broad-spectrum tolerance to abiotic stresses in transgenic tobacco plants. Plant Cell Rep 25:1111–1121

Xu J, Zhang B, Jiang C, Ming F (2011) RceIF5A, encoding an eukaryotic translation initiation factor 5A in Rosa chinensis, can enhance thermotolerance, oxidative and osmotic stress resistance of *Arabidopsis thaliana*. Plant Mol Biol 75:167–178

Yamaguchi K, Takahashi Y, Berberich T, Imai A, Miyazaki A, Takahashi T, Michael AJ, Kusano T (2006) The polyamine spermine protects against high salt stress in *Arabidopsis thaliana*. FEBS Lett 580:6783–6788

Yamaguchi K, Takahashi Y, Berberich T, Imai A, Takahashi T, Michael AJ, Kusano T (2007) A protective role for the polyamine spermine against drought stress in Arabidopsis. Biochem Biophys Res Commun 352:486–490

Yamasaki H, Cohen MF (2006) NO signal at the crossroads: polyamine-induced nitric oxide synthesis in plants? Trends Plant Sci 11:522–524

Enhancing Nutrient Starvation Tolerance in Rice

6

K.K. Vinod

Abstract

Nutrient starvation occurs in plants either by the insufficiency of nutrients in the soil or by their unavailability in plant absorbable form. Nutrient malnutrition is an age-old problem, aggravated by the human demand for more food which had led to the development of nutrient-hungry crop varieties. Ironically, what once lauded as a boon to mankind, the intensive agriculture, is turning to be a multifaceted bane in the form of depletion of natural reserves of inorganic fertilisers, price rise of farm inputs, environmental degradation due to nutrient residues and socio-economic and political divide among farming communities and nations. With the low availability of nutrients, plants are subjected to tremendous stress that jeopardises their normal physiology and survival itself. Rice, the major staple crop on earth is set to suffer any or all of the above problems in the near future. Immediate reduction of fertiliser input is the only viable solution to this problem, but it is going to trigger low production from farmlands. Therefore, nutrient input reduction should be done in conjunction with the development of low nutrient happy rice varieties. There is enough variability for nutrient response within the rice gene pool including low nutrient tolerance, which is to be tapped for the development of new varieties. In addition, low nutrient tolerant varieties can help in producing the best out of marginal lands that are rendered unsuitable for high-yielding varieties due to low nutrient status. This chapter overviews the developments in breeding towards nutrient deficiency tolerant rice varieties as a sustainable solution for future agriculture.

Keywords

Rice • Low nutrient stress • Nutrient deficiency tolerance • Breeding

K.K. Vinod (✉)
Division of Genetics, ICAR-Indian Agricultural Research Institute, Rice Breeding & Genetics Research Centre, 612101 Aduthurai, Tamil Nadu, India
e-mail: kkvinod@iari.res.in

P.K. Jaiwal et al. (eds.), *Genetic Manipulation in Plants for Mitigation of Climate Change*,
DOI 10.1007/978-81-322-2662-8_6

6.1 Introduction

Feeding more than half of the world population, rice cultivation requires significant quantum input of mineral nutrients than any other agricultural crop. Rice occupies nearly one third of the total cereal cultivated area of the world (Guerra et al. 1998), largely confined to tropics where soil nutrient status dwindles and plant nutrient availability is dependent on vagaries of climate and water availability (Takijima and Gunawardena 1969). Although there are different systems of rice cultivation, globally, irrigated lowland (wetland) system accounts for 55 % of the rice area producing 75 % of the rice production. Together with upland rice, wetland rice accounts for 80 % of the global area harvesting 90 % of the global production (Dobermann et al. 2004). When compared to other major grain crops, cropping intensity under irrigated rice is very high owing to relatively shorter duration of modern varieties, producing 10–15 t ha^{-1} $year^{-1}$ of grains (Dobermann and Fairhurst 2000). High-intensity cultivation depletes soil nutrient reserves far quicker, requiring continuous repletion in large quantities through addition of fertilisers. However, continuous fertilisation poses a looming threat of extinction of natural fertiliser reserves, which may imperil future agriculture. Ensuing threat of climate change adds to this uncertainty, especially in the demographically dense tropical crop production zones, where rice cultivation is a century-old tradition. Further, the swelling world population asserts tremendous pressure of higher grain demand, requiring rice production to go up. Consequent increase in fertiliser demand is making them expensive by every day exacerbating the unaffordability to poor and marginal rice farmers who practise subsistence farming.

Over a century now, tremendous advances in the science of genetics, has driven crop improvement worldwide through giant leaps, taking the food production to new levels of self-sufficiency. In rice, the introduction of semidwarf varieties in the 1960s has expanded crop production to staggering new levels, as these varieties realised excellent yield under best crop management practices including sufficient fertilisation. Concerted breeding efforts to push yield levels to newer heights have resulted in selection of rice varieties that responded to high levels of fertilisation. However, the cultivation of high-yielding varieties has removed significant quantum of soil nutrients through harvested grains and biomass enforcing continuous fertilisation to sustain the rice farming. This has generated overdependence on fertilisers resulting in their indiscriminate use. Serious soil limitations for major and few secondary nutrients have been reported in rice cultivation that drives the crop into serious stress of nutrient starvation. Nutrient starvation is becoming a contemporary problem that requires immediate attention of breeders and agronomists for sustaining the rice production in the future. This chapter deals with those nutrients for which considerable genetic research is ongoing to address nutrient starvation stress in rice.

6.2 Sustenance of Rice Mineral Nutrition

Rice requires sixteen elemental nutrients categorised into primary nutrients, secondary nutrients and micronutrients. Except for the three major non-mineral nutrients, carbon (C), hydrogen (H) and oxygen (O) that are naturally supplied by water and air, all other mineral nutrients are supplied through soil. Among the mineral nutrients, nitrogen (N), phosphorus (P) and potassium (K) are required in large quantities (major nutrients) and are soil added as fertilisers. Of the remaining, calcium (Ca), magnesium (Mg) and sulphur (S) that are required in smaller quantities (secondary nutrients) are supplied as supplementary fertilisers. The micronutrients, boron (B), copper (Cu), chlorine (Cl), iron (Fe), manganese (Mn), molybdenum (Mo) and zinc(Zn) that are required in very little quantities are generally available in soil in sufficient quantities. The micronutrients require only contextual application in case of any deficiency (De Datta 1981). Although quantity requirement varies, all the nutrients are essentially required for the rice plant to carry out various metabolic functions to complete its life cycle (Fageria 2013; Dobermann and Fairhurst 2000; Yoshida 1981; De Datta 1981). The limitation of any or more of these nutrients causes deficiency symptoms, which are the manifestation of various metabolic impairments. Furthermore, mineral deficiency can predispose plants vulnerable to attack by various biotic and abiotic agents, leading to mortality, morbidity and total yield loss (Dordas 2008).

Nutrient cycling in food grain crops including rice is a partial non-cyclic affair, because every crop harvest removes a significant quantum of nutrients without returning them back to the soil. A major portion of the loss is through grains entering into food chain and channelised elsewhere, especially into the sewers through human and animal waste and irrecoverably lost (Cordell et al. 2009). Together with the benefits of increased food production and poverty alleviation, the green revolution had counteracting effects of increased nutrient input demand, soil nutrient drainage and loss of nutrients through large volume of grains harvested. In India, rice tops the list of crops with 31.8 % share of fertiliser consumption of which 22.2 % is consumed by irrigated lowland rice. Estimates show that irrigated rice is applied with 2.48 Mt of N, 0.79 Mt of P_2O_5 and 0.45 Mt of K_2O annually, while the corresponding share of upland rice is 1.17, 0.30 and 0.13 Mt, respectively (FAO 2005).

Rice crop removes significant quantities of N and K from soil followed by P, Ca, Mg and S (Table 6.1). Net nutrient removal through harvested grains is 60 % for N, 67 % for P and 15 % for K, and the balance is retained in the straw (Dobermann and Fairhurst 2000). Compared to earlier period, when a significant quantity of rice straw was recycled back into the soil through cattle feeding and soil incorporation of either straw or cattle waste or both, nutrient recycling in modern systems is dropping due to alternate straw disposal methods such as open field burning (Jain et al. 2014; Kumar and Joshi 2013); usage in biofuel generation such as aerobic digestion, pyrolysis and gasification (Silalertruksa and Gheewala 2013); and usage in paper board industries (El-Kassasa and Mourad 2013).

To bring nutritional sustenance to rice crop production, there are several methods particularly related to agro-management and genetic interventions. The most common practice of nutrient repletion through fertilisation, however, is not going to be an enduring solution as the fertiliser resources are fast waning from the face of earth at the current consumption levels (Magdoff 2013). In order to prolong the soil nutrient availability, nutrient drainage needs to be curtailed while making the bio-unavailable nutrients into plant available form through solubilisation. One of the methods to contain nutrient loss through harvested rice straw is to develop strategies that aid in hay recycling such as composting (Saha et al. 2012). To minimise the grain-related loss through food chains, recently, efforts are on to recycle solids from sewers to produce struvite, which can be substituted as N and P fertiliser, thereby bringing the lost nutrients back to the crop nutrient cycle (Morales et al. 2013; Liu et al. 2011). However there is a respite in the most intensively cultivated parts of Asia where almost all of the organic wastes including animal wastes produced in confinement and all crop residues not used for fuel or in manufacturing were recycled (Smil 2000). Notwithstanding, management strategies do not reduce nutrient input demand.

Genetic improvement of rice is a particularly tenable strategy to address the reduction of nutrient input demand because it is possible to bring in genotype improvement in nutrient use traits such as growth under reduced fertilisation. This is possible by improving nutrient foraging behaviour, faster uptake, nutrient-solubilising ability, microbial symbiosis, terminal mobilisation of nutrients and yield conversion efficiency. Further, developing varieties that export less nutrients to grains into anti-nutritional forms such as phytate can aid in preventing nutrient loss through unsolicited means (Lott et al. 2000; Rose et al. 2013b).

Table 6.1 Nutrient requirement and removal in rice

Nutrient	Major plant available form	Nutrient removal (Kg.ha^{-1})	Critical deficiency levels (mg.kg^{-1})		Method of extraction[a]
			Plant	Soil	
Nitrogen (N)	NH_4^+, NO_3^-	123.00	<25,000.0	<186.0	Alkaline $KMnO_4$
Phosphorus (P)	$H_2PO_4^-$, HPO_4^-	21.40	<1000.0	<6.7	Olsen P
Potassium (K)	K^+	120.00	<10,000.0	<72.0	NH_4OAc
Calcium (Ca)	Ca^{2+}	32.00	<1500.0	<200.0	NH_4OAc
Magnesium (Mg)	Mg^{2+}	27.00	<1000.0	<120.0	NH_4OAc
Sulphur (S)	SO_4^{2-}	12.80	<1000.0	<10.0	Hot water
Iron (Fe)	Fe^{3+}, Fe^{2+}	1.70		2.5	DTPA+$CaCl_2$
Manganese (Mn)	Mn^{2+}	4.40	<20.0	1.0	DTPA+$CaCl_2$
Zinc (Zn)	Zn^{2+}	0.29	<15.0	0.5	DTPA+$CaCl_2$
Copper (Cu)	Cu^{2+}	0.06	<6.0	0.2	DTPA+$CaCl_2$
Boron (B)	BO_3^{3-}	0.09	<20.0	0.1	Hot water

[a]*DTPA* diethylene triamine pentaacetic acid, *NH_4OAc* ammonium acetate, *$CaCl_2$* calcium chloride, *$KMnO_4$* potassium permanganate

6.3 Nutrient Deficiency in Rice Soils

Soil nutrient deficiency is becoming a compelling problem in rice-growing areas that requires serious intervention in the days to come. In India, recent soil nutrient deficiency status indicates that 62 % of the soils are low in N, followed by 42 % low in P and 13 % low in K. Among the secondary nutrients, S deficiency occurs in 40 % of the soils, and among micronutrients, Zn deficiency is common in 49 % of soils, followed by 12 % low Fe soils, 5 % low Mn soils and 3 % low Cu soils (FAO 2005).

Predominant soil N forms such as nitrate and ammonium are either applied or native and are highly water soluble rendering N highly mobile in rice ecosystem (Haynes and Goh 1978) expediting quick lose through leaching. Additionally, the denitrification processes in soil can release N in volatile form, either as oxides of N or as atmospheric N. Further, escape of N as gaseous ammonia (NH_3) can also favour N loss from soil (Galloway et al. 2003). Without regular N repletion through fertilisation, N deficiency is a constant problem in rice soils worldwide. Soils that are prone to N deficiency are those with very low soil organic matter content (<0.5 %) such as coarse-textured acid soils, as well as low organic carbon containing alkaline and calcareous soils. Soils that are constrained with natural N supply such as acid sulphate soils, saline soils, P-deficient soils and poorly drained wetland soils where the amount of N mineralisation or biological N_2 fixation is meagre also suffer from N deficiency. Alkaline soils have potential risk of high NH_3 volatilisation.

P is available to plants in the form of orthophosphate (Pi) such as $H_2PO_4^-$ and HPO_4^{2-} (Raghothama 1999), which is assimilated into plant system as Pi monoesters and compounds containing phosphoanhydride bonds such as adenosine di- and triphosphates. However, Pi does not remain in soil, as it gets quickly fixed into insoluble form in the presence of polyvalent cations such as Al^{3+}, Fe^{3+} and Ca^{2+}. P deficiency is a widespread problem in all major rice ecosystems (Fageria 2013), especially in lowland soils that possess high native P-fixing capacity such as acid soils. In soils with low P supplying capacity, P deficiency occurs as long as Pi supplementation is insufficient. Notwithstanding, high erosion loss can also incite P deficiency in soils such as upland soils. P deficiency occurs in coarse-textured soils containing small amounts of organic matter and minimal P reserves such as sandy soils; highly weathered, clayey, acid upland soils with high P-fixation capacity such as ultisols and oxisols; degraded lowland soils; calcareous, saline, sodic soils; volcanic soils with high P-sorption capacity; peat soils (histosols); and acid sulphate soils in which large amounts of active Al and Fe result in the formation of insoluble P compounds at low pH. P deficiency is also found associated with Fe toxicity at low pH, Zn deficiency, Fe deficiency and salinity in alkaline soils (Dobermann and Fairhurst 2000). Furthermore, P immobilisation occurs due to excessive liming and excessive N fertilisation with insufficient P application.

K is a rare limiting factor in major rice-growing soils (Dobermann et al. 1996; Reichardt et al. 1998); however, K deficiency occurs either due to native limitation of K availability or due to inhibition of K due to fixation. Recent reports, however indicate that K deficiency is on the increase throughout rice-growing regions of the world (Jia et al. 2008; Hasan 2002; Naidu et al. 2011; Datta 2011). K deficiency typically occurs in coarse-textured soils such as sandy soils, highly weathered soils such as acid upland soils and degraded lowland clay soils and well-drained organic soils due to poor K reserves. These soils remain K deficient if not supplemented through fertilisation. In contrast, in clayey soils with large K content, presence of significant amount of 2:1 layer clay minerals causes K fixation, making them practically K inhibited. Similarly soils with very wide (Ca+Mg)/K ratio release little K to soil solution due to stronger K adsorption to cation exchange sites. K deficiency also occurs in highly leached acid sulphate soils and poorly drained and strongly reducing soils in the presence of Fe^{2+} ions, organic acids and hydrogen sulphide.

Deficiency of secondary and micronutrients is not a widespread problem in rice soils around the world. However, there are regions which suffer from acute deficiency of one more of these mineral nutrients that needs serious attention. S deficiency is reported as a problem in South Asian countries, especially in Indonesia (Ismunadji et al. 1991) which in general is addressed through addition of S-containing fertilisers. A shift to non-S fertilisers may aggravate the deficiency problem in such areas. S deficiency is not particularly a common problem in irrigated lowland soils; however it can occur in coarse soils that are well drained such as sandy soils, soils that are rich in aluminium silicate clay, soils with low organic matter and in highly weathered soils rich in Fe oxides (Dobermann and Fairhurst 2000).

Recent reports suggest that Zn deficiency, the major micronutrient deficiency in rice, is ranked after N and P deficiency in lowland rice soils (Wissuwa et al. 2006) that calls for immediate attention in terms of deficiency alleviation by addition of Zn fertilisers (Abilay and De Datta 1978) and/or by improving genotypic efficiency to acquire more Zn when soil availability is limited (Rose et al. 2013a). Further, micronutrient malnutrition due to Zn deficiency together with vitamin A deficiency is receiving serious attention internationally, due to serious health risks it incites such as mental retardation, poor immunity and overall poor health (Cakmak 2009). Zn deficiency occurs in high-intensity cropped areas and in poorly drained soils. In Indian soils, 49 % of 250,000 soil samples collected from 20 states were tested Zn deficient having DTPA-extractable Zn content of less than 0.6 mg kg^{-1} (Singh 2009) indicating the enormity of Zn deficiency problem. Among the crops and soils, lowland rice is said to suffer more from Zn deficiency (Rattan et al. 2009). Zn deficiency occurs along with S deficiency, especially in neutral and calcareous soils rich in bicarbonate, which can affect rice crop in all phenological stages. Other type of soils that can have Zn deficiency are saline and sodic soils, peat soils, sandy soils and highly weathered and leached acid soils. Zn deficiency occurs either due to Zn precipitation as sparingly soluble zinc hydroxide or zinc sulphide or due to the formation of various complexes in soil either in combination with inorganic and organic ions.

Deficiencies of other micronutrients such as Ca, Mg, Mn and Cu can occur in highly drained coarse sandy soils and highly leached weathered acid soils either on lowlands or uplands. Deficiencies of these elements occur due to limited soil availability caused by excessive drainage either by physical means such as leaching or through crop removal due to intense cultivation. Ca uptake from soil is affected under excessive fertilisation, due to proportional imbalance of Ca with other nutrient elements causing Ca deficiency. Mg deficiency occurs due to the suppressed Mg absorption caused by decreased Mg uptake due to a wide exchangeable K-Mg ratio. Mn deficiency occurs either due to interference of mineral ions such as Fe, Ca, Mg, Zn and ammonium in soil solution or by the formation of complexes by organic matter or adsorbed and occluded by Fe and Al hydroxides and oxides. Cu deficiency can occur in soil by adsorption or by competition with Zn, while excessive NPK fertilisation can exhaust Cu from soil through removal by rapid vegetation growth.

6.4 Need for Nutrient Deficiency Tolerance

In modern-day agriculture, due to continuous crop removal, soil nutrient deficiency has become coexistent with crop production systems and amelioration of soil nutrient status though amendment of fertilisers has become a standard practice. Compelling demand for more food for the growing world population is exerting tremendous pressure on fertiliser use and cultivation of modern-day fertiliser-responsive high-yielding cultivars. Since the fertiliser production is dependent on the finite resources on earth, production and availability of fertilisers are starting to decline gradually. Its after-effects are now appearing, fertiliser scarcity is on the rise along with escalation in fertiliser costs, which is particularly making fertilisers unaffordable to poor and marginal rice farmers, thereby bringing more areas under nutrient deficiency map rapidly. Under high-intensity production systems, however, indiscriminate fertiliser use is causing environmental hazards (Vinod and Heuer 2012). Arising out of the aforementioned issues, there are three major reasons that compel reduction of fertiliser input into agricultural systems. These are (a) low nutrient use efficiency, (b) environmental degradation due to residual nutrients and (c) depletion of natural fertiliser reserves.

6.4.1 Low Genotypic Nutrient Use Efficiency

Nutrient use efficiency is a generic term that integrates different mechanisms of nutrient homeostasis in plants. Although definitions differ contextually, nutrient use efficiency, defined as the ratio between realised yields to applied fertiliser, represents only genotype response under added nutrition. In broader sense, use efficiency can be divided into uptake and utilisation efficiencies, although the mechanisms may differ between nutrients. Earlier breeding efforts for nutrient use efficiency were focussed on the genotypes that yielded well under high fertilisation, thus accumulating genes that helped to realise high biomass and grain yield. Since these genotypes were not subjected to nutrient-deficient situations, the selection was working against the mechanisms that enabled them to tolerate nutrient-deficient conditions. It should be noted here that since the selection was done under surplus fertilisation, uptake efficiencies of the genotypes hardly mattered in the selected genotypes, because irrespective of their uptake efficiency, there were always enough nutrients to meet the internal demands to support growth and yield. Remarkable breeding success has already been achieved in selection of varieties with high internal efficiency under high-input systems (Vinod and Heuer 2012). Therefore, modern varieties are low nutrient use efficient and need surplus nutrients to yield better. The net result of low nutrient use efficiency of genotypes under excessive fertilisation is accumulation of nutrient elements in soil resulting in imbalances with other elements that are not artificially supplied. This causes induced deficiency of other elements requiring further amendment of chemicals to soil. To increase nutrient use efficiency from the current level, we need to reduce the fertiliser input without compromising the quantum of nutrients taken up by the plants. This can be achieved by improving of uptake efficiency without compromising the internal efficiency to produce the potential yield.

To sustain the agricultural production in the future, the efficiency of rice varieties to harvest more nutrients from soil needs to be improved from the current level (Singh et al. 2000). This should however be achieved with lesser external nutrient input. In fact, soil is rich in almost all of the nutrients majority of which are not plant available or are at depths well beyond the crop root zones. However, there are innate mechanisms in rice crop to utilise unavailable nutrient fractions by solubilising them into available forms, such as production of nutrient-solubilising root exudates and encouraging the colonisation of nutrient-solubilising microflora through a symbiotic relationship (Hayat et al. 2010; Khan 2006). These traits are genetically controlled and hence can be targeted in crop improvement programmes. The development of nutrient-efficient rice varieties under marginal sufficiency will help in realising the yield gain even at the regulated low-level fertiliser input. Therefore, under reduced fertiliser input, the nutrient requirement for rice is to be considered vis-à-vis soil net nutrient supply, nutrient use efficiency and nutrient deficiency tolerance.

6.4.2 Environmental Contamination Due to Surplus Nutrients

Intensive agriculture in the contemporary world demands high input of fertilisers as most of the modern high-yielding varieties are input responsive (Ramesh et al. 2005). Before their introduction, even the poor and marginal soils could supply sufficient nutrients to support rice cultivation as the net nutrient removal of the traditional varieties was low (Dobermann et al. 1998). In practice, several farmers apply excess quantity of fertilisers to modern varieties expecting to realise higher yield since they are known respond to high dose of fertilisers. In reality, not all of the applied nutrients are utilised, either due to poor nutrient use efficiency or due to excessive nutrients over and above the threshold uptake capacity of varieties, leading to serious environmental hazards such as accumulation into water bodies polluting them causing eutrophication (Smith and Crews 2014).

Environmental contamination due to surplus nutrients is a serious contemporary problem. Most of the residual nutrients, especially excess N and P, get leached into groundwater or accumulate in surface water causing eutrophication (Raven and Taylor 2003). This causes harmful algal blooms (HABs), in marine, estuarine and fresh waters threatening the health of the environment, plants, animals and people. Cyanobacterial and red tide (*Karenia brevis*) blooms can block sunlight penetration in water, deplete oxygen and produce toxins that seriously threaten aquatic life (Heisler et al. 2008). However, excess K is not known to create any environmental problem.

Atmospheric pollution due to excess nutrients occurs particularly in case of N, due to volatilisation. Major volatile forms of N are NH_3, nitrous oxide (N_2O) and nitric oxide (NO). N_2O is a greenhouse gas that contributes to about 6 % of the anthropogenic greenhouse effect and causes the depletion of stratospheric ozone. NO is readily converted into nitrogen dioxide (NO_2) in atmosphere, and at high concentration levels, NO_2 is potentially injurious to plants that reduce growth and yield. Even at lower concentrations, NO_2 can be potentially injurious in combination with either ozone (O_3) or sulphur dioxide (SO_2). Nitrogen oxides (NO_X) play a critical role in soil acidification, inflicted through acid rains and dry deposition. NH_3 is a major air pollutant which neutralises acidic vapours of SO_2 and NO_x. Neutralised acids form a major constituent of atmospheric aerosols and deposition of which brings back NH_3 to surface which causes eutrophication and acidification (FAO 2001).

6.4.3 Waning Natural Fertiliser Resources

Most of the fertilisers are inorganic chemicals that release one or more major nutrients to soil, and almost all of them are obtained or chemically synthesised from natural sources. These natural sources are fast depleting due to continuous mining. As most of the natural fertiliser reserves are confined to particular geographical regions, geopolitical issues and monopolisation for fertiliser dependence are growing concerns for the future. Additionally, chemical synthesis of fertiliser is a high energy-consuming process that requires the usage of non-renewable energy sources such as petroleum and natural gas (Giampietro and Pimental 1993).

The production of nitrogenous fertilisers such as urea and ammonium salts from commercial NH_3 generated through Haber-Bosch process uses natural gas as the major feedstock (Erisman et al. 2008). In the Haber process, H is obtained from natural gas by reacting it with steam at high temperatures. This H is combined with N supplied from air in the presence of Fe as catalyst to form NH_3. Urea is produced by reacting NH_3 with CO_2 in the presence of an intermediate product ammonium carbamate. The world's estimated reserve of natural gas stood at 185.7 trillion cubic metres by the end of 2013, which is estimated to last only for the next 55.1 years (BP 2014).

Phosphate rock or phosphorite that largely contains phosphates of calcium is the only natural source for phosphatic fertilisers. 82 % of the mined rock phosphate is used for fertiliser production (Cordell and White 2013), as well as used directly as P fertiliser in the form of rock phosphate. Rock phosphate is insoluble in water and hence is a slow-release fertiliser. Reacting rock phosphate with sulphuric acid produces phosphoric acid which is used for the production of soluble phosphatic fertilisers such as single super phosphate, triple super phosphate and ammonium phosphate (Bolland and Gilkes 1998). Phosphate rocks are non-renewable, and the recent estimates suggest that phosphate rock reserves of the world stand at 290 billion metric tonnes which may last only for next 300–400 years at the current rate of production. Further, 68 % of the global geological phosphate rock reserves are confined to countries such as Morocco, USA and China (Van Kauwenbergh 2010).

The major source of natural K fertilisers is from the deep salt pans in earth's crust that may either contain salt deposits or rock-forming silicate minerals (Manning 2010). Almost all of the commercial potash deposits come from marine sources, either ancient seas that are now buried or from salt water brines (Fixen 2009). Unlike that of N and P, natural sources of K are in many mineral forms predominantly made of chlorides (sylvinite, sylvite, carnallite, kainite), sulphates (polyhalite, langbeinite, schoenite) and nitrates (nitre) of K. Fertilisers are produced from the salt deposits by different separation and purification processes. The world has an estimated 250 billion metric tonnes of K_2O resources that may last longer than 600 years as per the current estimates (USGS 2011). Natural sources of K are also non-renewable, and hence, a future threat of mineral depletion cannot be ruled out. Furthermore, K mines are also confined to certain geographical regions of the world, wherein Canada and Russia share almost 80 % of the global reserves (Fixen 2009; Mohnot et al. 2005; Roberts and Stewart 2002).

Earth's crust contains S as one of its common constituents, which is available either as elemental S or as several compounded forms. However, most of the S used today is predominantly extracted from natural gas and crude oil, and some S is recovered from coal, by metallurgical processing and from pyrites mining (Fixen 2009). Of this, 80–85 % is used to manufacture sulphuric acid. Although S-containing fertilisers are not primarily produced, 50 % of the world's sulphuric acid production is used mainly to convert phosphates to water-soluble forms of phosphatic fertilisers. About one ton of S is required to produce about 2 tons of diammonium phosphate (DAP). The USA, Canada, China and Russia are the leading countries in S production accounting for about 50 % of the global S production (Fixen 2009).

Around the world, the overall average soil concentration of Zn is estimated to be 50 mg.kg^{-1}, but the levels vary widely from 4 mg.kg^{-1} to 3500 mg.kg^{-1} depending on geographical occurrence and concentrations of ore minerals such as sphalerite (ZnS), smithsonite ($ZnCO_3$), zincite (ZnO), zinkosite ($ZnSO_4$), franklinite ($ZnFe_2O_4$) and hopeite ($Zn_3(PO_4)_2.4H_2O$) (Alloway 2008).

Unlike other fertilisers, Zn fertilisers are not directly produced from natural sources, but are obtained from products or by-products of processes using Zn metal as primary component. Although there are several sources of Zn used as fertilisers, the most common source is $ZnSO_4$ used either in granular or crystalline forms. In soils, Zn fertilisation has a very high residual effect, sometimes prolonging for many years, and hence, continuous Zn fertilisation may not be required if Zn level raises to adequate levels (Mortvedt and Gilkes 1993).

6.5 Mechanisms for Nutrient Starvation Tolerance

Reducing the application of nutrients exacerbates problem of nutrient starvation in agriculture, and hence, one of the most sustainable solution is to breed for nutrient-efficient genotypes that require low nutrient input. Although it is known that nutrient starvation tolerance in rice is a function of nutrient efficiency, despite several studies on nutrient uptake and utilisation, information accumulated are insufficient to assess the potential for breeding rice cultivars with improved use efficiencies. Notwithstanding, breeding efforts for low nutrient tolerance are mainly targeted on major nutrients N, P and K and other essential nutrients such as S and Zn.

6.5.1 Nitrogen

Presently N use efficiency of rice varieties remains at 30–50 %, indicating that 50–70 % of the applied N is lost from the soil (Peoples et al. 1995) either through volatilisation, leaching or by other means. Increasing the efficiency can therefore counterbalance the loss and can help in reducing N input. Genotypic responses to lowered N in rice, as reported from several studies, predominantly include root variations such as enhancement of surface area, density, volume, distribution and root-shoot ratio (Marschner et al. 1986; Fan et al. 2010; Ogawa et al. 2014), accelerated uptake of N at early growth stages (Peng et al. 1994) and efficient internal recycling at terminal stages (Mae 1997; Mae and Ohira 1981; Tabuchi et al. 2007). Further, low N response in rice genotypes varies at different growth stages (Sheehy et al. 1998) as well as to varying N supply levels during growth phases. When N accumulation becomes sufficient enough for internal N homeostasis, internal mechanisms take over the role of N assimilation from uptake machinery, leading to optimal growth and yield. Nevertheless, the uptake remains the key factor determining yield particularly under low N conditions (Singh et al. 1998; Witcombe et al. 2008). For ultimate realisation of yield, N uptake prior to panicle initiation is critical in building up the internal N reservoir (Vinod and Heuer 2012), which is accomplished through a complex network of internal and external signalling under low N conditions.

The primary N acquisition from soil takes place in biphasic fashion (Britto and Kronzucker 2005) involving low- and high-affinity transporter systems in the plasmalemma, irrespective of the form of N being taken up. Acquired N subsequently undergoes primary and secondary assimilation process before being incorporated into amino acids. N metabolism in rice is a complex process that involves several gene families, including nitrate and ammonium transporters and primary and secondary N assimilation genes, urea transporters, amino acid transporters, transcription factors and other regulatory elements (Lea and Miflin 2011; Nischal et al. 2012). However, several of these gene systems are constitutively expressed and self-regulated. It is therefore pertinent to identify those genes that give a 'push' on the entire machinery to work under N-deficient conditions, triggering mechanisms particularly targeting extensive and intensive foraging under limitations of N supply (Pathak et al. 2008). This may involve those genes that provide an early signal to the N deficiency in the seedling stage itself, accelerating mechanisms for the development of extended and deeper root system while provisioning enough energy to support the signalling and early development mechanisms. Therefore, early seedling vigour under low N may be a crucial factor in early identification of low N-efficient genotypes. Other traits that may be of interest in realising yield under N deficiency are prolonged photosynthetic efficiency and translocation of photosynthates (Hawkesford and Howarth 2011) that is vital for maintaining canopy longevity to support continuous remobilisation of N and starch accumulation. In this context, functional stay green (Fu and Lee 2008) is a trait to look for as it can support prolonging of canopy greenness by delaying leaf senescence that can also sustain photosynthesis in maturing plants (Vinod and Heuer 2012). To identify such genotypes, it is better to screen the germplasm under reduced N input rather than no input, because no input cannot be a sustained practice in agriculture and signals to low N may fail below a critical threshold, because acute N deficiency may jeopardise the survival of the plant itself.

The development of molecular breeding and identification of several genes involved in N uptake, transport and assimilation (Li et al. 2009a; Kant et al. 2011) and quantitative trait loci (QTLs) for N use efficiency under reduced N conditions (Vinod and Heuer 2012) provide new hope in repeating the success of introgression of QTLs such as *Sub1* for submergence tolerance (Singh et al. 2010; Septiningsih et al. 2009) and *Saltol* for salt tolerance (Thomson et al. 2010) in rice.

6.5.2 Phosphorus

Modern rice genotypes typically have soil P recovery efficiency less than 20 % (Fageria 2013) much less than other major nutrients, N and K. This is in contrast to the internal P use efficiency, which is much higher than that of N and K. P content in rice grain is much higher than in straw; therefore, the P harvest index in rice is as high as 72 %. This data indicates that to improve overall P use efficiency of rice, it is necessary to improve P uptake rather than internal P utilisation.

The paradox of high Pi mobility within the rice plant system while growing in a pool of immobile soil P warrants the need for the solubilisation of soil P before being mobilised into plants. Under Pi deprivation, rice plants elicit a variety of mechanisms at genotype levels that are combinations of morphological, physiological and biochemical/metabolic adaptations collectively known as P starvation response (PSR) (Plaxton and Tran 2011). PSR aids in foraging for more P under deficiency, such as dramatic changes in root system and production of several kinds of root exudates besides encouraging microbial symbiosis. Rice genotypes show varying levels of expression for these adaptive traits under P starvation gaining access to more P when the bioavailable P in the soil is low (Heuer et al. 2013). When confronted with P limitation, plants grow more roots, increase root uptake, retranslocate P from older leaves and deplete the vacuolar P reserves (Schachtman et al. 1998). In rice, root morphological adaptations are reported by several workers in response to limited P availability that includes increase of root hair growth, increase in root volume and root morphological and architectural modifications (Heuer et al. 2013). Recent reports on the functionality of the QTL for P deficiency tolerance, *phosphorus uptake 1* (*Pup1*), suggest that the major gene responsible for the P starvation tolerance, *PSTOL1*, enhances root growth under P-deficient conditions (Heuer et al. 2009; Chin et al. 2011; Gamuyao et al. 2012).

Root efflux of variety of compounds such as organic acids, membrane proteins, acid phosphatases and ribonucleases (RNAses) is reported under P deficiency in rice. Root exudation of low molecular weight organic acids (LMOAs) such as citrates, malates and oxalates is reported to be a prominent root-based mechanism that aids in enhanced solubilisation of P from the soil (Kirk et al. 1999; Hocking et al. 2000; Hocking 2001; Rose et al. 2011). LMOAs also have significant chelation capacity, lowering the activities P-fixing cations, Fe^{3+}, Al^{3+} and Ca^{2+}, thus preventing Pi loss (Singh et al. 1991). Zhang et al. (2011a) reported P deficiency enhanced activities of both H^+-ATPase and proton pump, which contributed to rhizosphere acidification in rice roots. Rhizosphere acidification occurs by the proton release from roots, as a primary process of H^+-ATPase activity in the plasmalemma of root cells

(Schubert 1995). The mechanism of H^+ release balances excess intake of cations over anions by acting as a primary proton pump, thereby creating pH and electric potential differences across the plasmalemma. Since the Pi uptake may involve proton cotransport, the H^+-ATPase activity may improve plant P nutrition by enhancing the electrochemical proton gradient that drives ion transport across the cell membrane. Further, H^+-ATPase activity may couple with LMOA efflux to improve the P mobilisation as an adaptation to P deficiency.

Root secretion of acid phosphatases (APases) is entailed as a response to P deficiency in rice (Lim et al. 2003; Li et al. 2009b). APases play a role in P nutrition by release and mobilisation of Pi from organic complexes in soil through hydrolysis. Under P-starved conditions, plants increase the level of secreted APases, along with intercellular phosphatases for efficient uptake and mobilisation of Pi from the soil (Tian et al. 2012). In response to P starvation, along with the increased production of APases, significant production of RNAses are also induced in rice (Plaxton and Tran 2011) that are involved in the acquisition of exogenous P (Nürnberger et al. 1990). The secreted RNAses degrade large quantum of nucleic acids from decaying soil organic matter to release Pi mobilised for root uptake (Fang et al. 2009).

Yet another mechanism for P acquisition is through arbuscular mycorrhizal (AM) symbiosis, which helps plants to source more P channelised through symbiotic associations (Vallino et al. 2009). Mycorrhizal fungi are aerobic symbionts, and therefore, AM associations in rice were a matter of debate due to increased anaerobic conditions of lowland soils. However, evidences indicate that such associations do exist in rice under submerged conditions significantly contributing to nutrient uptake including P (Solaiman and Hirata 1997; Hajiboland et al. 2009; Watanarojanaporn et al. 2013). It is likely that the AM association to rice roots in wetlands is maintained by oxygen supply via aerenchyma (Ipsilantis and Sylvia 2007).

Pi uptake from soils is regulated by a large gene family of P transporters, falling under low-affinity and high-affinity transport systems. Some of these P transporters are also involved in AM symbiosis indicating their complex role in P uptake under P-deprived conditions (Vinod and Heuer 2012; Heuer et al. 2013). In addition, several other genes such as MYB-type transcription factors, stress response genes such as peroxidases and metallothioneins, glycolytic enzymes, transmembrane proteins, DNA and protein degradation enzymes are also unregulated under P deprivation in rice roots (Li et al. 2009b).

6.5.3 Potassium

As the most abundant cation in plant system, K constitutes about 10 % of the plant dry weight (Véry and Sentenac 2003). Use efficiency of K in rice is high, unlike that of N and P, because K is not subjected to loss either due to volatilisation or soil fixation. K is highly recycled between crop rotations and is relatively immobile in most of the soils. K recovery from soil can range from 20 to 60 % (Roberts 2008). Since K is harmless to the environment, reduction of K fertiliser arises only in the context of bringing down the cost of K fertiliser input and thereby cost of cultivation, because major rice-growing countries import K fertilisers. Besides, K depletion is occurring in high intensity in rice systems in Asia (Dobermann and Cassman 2002) which is threatening sustained rice production. Breeding K-efficient varieties will help in the endurable reduction of K input in the future. Although significant genotype variations for K use efficiency in rice have been reported (Liu et al. 1987; Yang et al. 2003, 2004), data still remains insufficient to assess the potential for breeding for low K-tolerant rice varieties.

Rice uptakes K in its monovalent cationic form K^+, driven by biphasic transport systems at the plasmalemma. Charged metallic nutrient cations like K^+ and other micronutrients do not freely diffuse across lipid bilayer membranes, which require transporter proteins to ferry them in and out of cells and intracellular compartments (Eide 2005). Under K deprivation, high-affinity transport plays a significant role in K^+ loading in plants (Szczerbab et al. 2009), mediated either by proton-pumping ATPases (Briskin and Gawienowski 1996) or K^+/Na^+ uniporters/symporters of the HKT (high-affinity K^+ transporter) gene family that regulate differential uptake of K^+/Na^+ ions depending upon the external ionic conditions (Hauser and Horie 2010). The presence of Ca^{2+} ions increases K^+ uptake, as it mediates K^+ transport by activating tandem pore K (TPK) channels (Lebaudy et al. 2007). Induced K deficiency under saline conditions is particularly important in rice, because rice is a predominant crop grown under salinised environments of the tropics (Vinod et al. 2013). Under saline conditions, Na^+ ions compete and interfere with K^+ uptake, resulting in Na toxicity and K deficiency in the plants. High-affinity K^+ transporters are sensitive to salinity (Fuchs et al. 2005) as well as to NH_4^+, the predominant form of N nutrition in rice. However in the presence of NH_4^+, the K uptake in rice is not found to be affected by salinity because high-affinity K^+ transporters are not sensitive to NH_4^+ (Szczerba et al. 2008). Although the presence of Ca^{2+} ions in the soil helps in K^+ uptake, because Ca^{2+} regulates K^+ channel, under high saline conditions, K^+ channel fails in the presence of excess Na^+ ions. Nonetheless, in rice, uptake under K-deprived conditions is regulated by several other factors such as shaker channel proteins (Obata et al. 2007), aquaporins (Maurel et al. 2008), ion-sensing receptor proteins (Szczerbab et al. 2009), guanine nucleotide-binding proteins (G proteins, Urano et al. 2013), vacuolar ion channels (Isayenkov et al. 2011) and tonoplast proteins (Bañuelos et al. 2002).

6.5.4 Sulphur

Sulphur is an essential secondary nutrient that is involved in several metabolic processes and constituent of many biomolecules especially of S-containing amino acids, cysteine and methionine (Saito 2000). Rice plants uptake S in the form of SO_4^{2-} ions and convert them into organic sulphides in the biomolecules (Smith et al. 1995). S deficiency is an ensuing problem in many parts of the intensely cultivated regions where availability of S-tolerant genotypes can be an added advantage (Blair et al. 1978). S uptake in rice is mediated by proton-SO_4^{2-} symporters (Takahashi et al. 2000) that load SO_4^{2-} ions across plasma membrane, from which it is transported symplastically through roots (Godwin et al. 2003). SO_4^{2-} ions are then loaded into xylem vessels and delivered into chloroplasts wherein the ions gets adenylated and reduced to sulphite and then to sulphide by sulphite reductase enzyme, before being incorporated into various biomolecules in the plant system (Davidian and Kopriva 2010). S assimilation is a highly regulated process, expressed by internal signals for S demand and repressed by the accumulation of reduced S. As in the case of any other nutrients, root S uptake is under biphasic control, in which high-affinity transporters play a major role under S starvation (El Kassis et al. 2007). Under low S conditions, these transporters are upregulated by SO_4^- limitation stimuli (Yoshimoto et al. 2007). Notwithstanding, synergetic effects of high- and low-affinity systems are believed to contribute improved retrieval of SO_4^- in response to S starvation (Kataoka et al. 2004a, b). Apart from the S transporters, S uptake and translocation is regulated by several other genes, transcriptional regulators such as *S-responsive element* (*SURE*) and *S limitation 1*(*SLIM1*), post-transcriptional regulators and micro-RNAs (Davidian and Kopriva 2010). This complicated regulatory systems of S metabolism works in a coordinated fashion to achieve overall S use efficiency in plants that also drives the responses towards S deficiency tolerance (Davidian and Kopriva 2010).

6.5.5 Zinc

Cationic uptake of Zn by rice plants as Zn^{2+} ions can be seriously impaired in wetland soils due to precipitation particularly by the fusion of Zn with free sulphides (Hafeez et al. 2013). Under the submerged conditions, Zn precipitation is predisposed in the presence of Mn and Fe oxides, commonly forming an amorphous sesquioxide precipitate or franklinite ($ZnFe_2O_4$) (Sajwan and Lindsay 1988).

Mechanism of Zn uptake in rice, particularly under wetland conditions are subject of recent investigations (Weiss et al. 2005; Arnold et al. 2010; Impa et al. 2013) and reviews (Bashir et al. 2012; Yu et al. 2012; Rose et al. 2013a) focussing on the improvement of Zn uptake and accumulation in grains. Intricate mechanisms of Zn metabolism in rice (Yu et al. 2012) include those involved in the uptake processes such as Zn sufficiency sensing by roots that control expression of metal-binding transporters and root excretions of ligand molecules such as phytosiderophores (PS), shoot translocation and final retranslocation to grains. Zn use in rice, like any other nutrient, is under strict regulatory control of several genes that control uptake, translocation, accumulation, mobilisation and remobilisation. Several of them belong to large gene families such as Zn-regulated transporters (ZRT) and Fe-regulated transporter (IRT) like proteins (ZIP) (Eide 2005), nicotianamine synthases (NAS) and yellow stripe-like (YSL) protein transporters. Further, the involvement of genes for PS synthesis (Arnold et al. 2010), PS transporters and metal ion-PS complex transporters (Nozoye et al. 2011; Murata et al. 2006) are also indicated in the uptake process. The root uptake of Zn from soil is suggested to be under control of two ZIP genes, *OsZIP1* and *OsZIP3* (Ramesh et al. 2003), while *OsZIP4*, *OsZIP5* and *OsZIP8* are involved in shoot translocation (Lee et al. 2010a, b).

PS are low molecular weight, non-protein amino acids released by graminaceous plants that form soluble metal ion-PS complexes with cations such as Fe^{3+} and several micronutrient ions such as Zn, Mn and Cu that aid in cationic mobilisation into the plants (Römheld 1991; Marschner 1995). In plants, together with nicotianamine (NA), mugineic acid (MA) family of phytosiderophores are well recognised in transport of Fe^{3+} (Bashir et al. 2010, 2012). NA is a metal cation chelator for Fe^{2+} and Zn^{2+}, and it is biosynthesised from S-adenosyl methionine via NA synthase (Higuchi et al. 1999). NA is produced by all plants, but MA secretion is characteristic of graminaceous plants (Marschner 1995). Deoxymugineic acid (DMA) is one of the early species of MA identified in graminaceous plants, which is synthesised from NA by NA aminotransferase (NAAT) and DMA synthase (DMAS) in a conserved pathway (Bashir et al. 2012). There are several modelling studies that implicate a significant role of DMA in Zn uptake and mobilisation in rice (Arnold et al. 2010; Ptashnyk et al. 2011), although preference for Zn^{2+} over Zn-DMA is reported in uptake process while Zn-DMA is involved in translocation process (Suzuki et al. 2008). Latest evidences such as predominance of Zn-NA in the phloem sap (Nishiyama et al. 2012) indicate that, in rice, Zn transport may be occurring as Zn-NA complex, rather than Zn-DMA complex. However, an effective transporter for Zn-NA or Zn-DMA is yet to be identified in rice (Bashir et al. 2012).

6.6 Breeding for Nutrient Starvation Tolerance

Nutrient starvation tolerance is a complex trait, manifested through a maze of morpho-physiological responses depending on the environment under which nutrient deprivation is manifested. Therefore, nutrient starvation itself does not have a common definition across nutrients and environments (Rose and Wissuwa 2012). This implies that breeding targets and component traits may differ under various circumstances and across nutrients and therefore conventional breeding may likely to bring little progress. Rice genotypes display varying levels of adaptive responses and variation in nutrient use under different nutrient levels leveraging the potential for breeding for nutrient starvation tolerance. However, as in the case of N use efficiency, drawing a distinction between uptake- and utilisation-efficient genotypes remains difficult. This may also have a negative bearing on using conventional breeding methods, because the exact traits to be selected for uptake and utilisation efficiencies are difficult to distinguish and therefore would be difficult to screen. However, molecular breeding has shown tremendous success under complex situations in rice, such as submergence tolerance (Septiningsih et al. 2009), salinity (Thomson et al. 2010), yield under drought (Kumar et al. 2013) and P starvation (Heuer et al. 2013). Sourcing a sustainable level of tolerance itself is critical because most of the modern high-yielding varieties are highly responsive under nutrient sufficiency and may fail totally under deficient conditions. They may contribute little towards uptake efficiency as they were selected for ultimate yield under sufficiently fertile environemnts that may or may not include uptake efficiency. These varieties possess high yield potential together with internal utilisation efficiency that can translate into higher yield under ideal conditions. On the other hand, many of the low-yielding traditional and niche-adapted landraces that survive under harsher nutrient-starved environments would be ideal sources for the tolerance to nutrient starvation, since they have better adaptive mechanisms under nutrient deprivation as well as better uptake mechanisms. Hence, it is logical to look for uptake efficiency in landraces and under-explored germplasm from areas of natural nutrient limitations to combine the target traits into utilisation-efficient modern varieties. Exploring naturally adapted multi-stress-tolerant genetic group such as *aus* varieties (Lafitte et al. 2004) is one such option. Under P-deficient situations, Wissuwa et al. (2009) found that most of the identified tolerant genotypes belonged to landraces or traditional plant types. Therefore, ideally, breeding for nutrient starvation tolerance should target combining uptake efficiency from the older varieties with high internal efficiency of the modern varieties. Although conclusive data is still lacking, it may be advantageous to have poor or nutrient-deficient soils as the target environment for selection for nutrient deficiency tolerance.

Breeding for tolerance to nutrient starvation, irrespective of the underlying factors, requires thorough understanding of the target traits that drives nutrient uptake and assimilation under stress. Under N-limiting conditions, genetic variation in nitrogen use efficiency (NUE) is particularly important in improving uptake efficiency (Gallais and Coque 2005), with several component adaptive mechanisms such as root development and architecture, delayed leaf senescence, early uptake of N especially at vegetative stages, increased symbiotic relations with arbuscular mycorrhiza and other N-fixing symbionts and increased activity of enzymes involved in N mobilisation and translocation (Vinod and Heuer 2012; Fess et al. 2011). Although several rice QTLs for N deficiency tolerance has been mapped from many biparental populations (Wei et al. 2012; Tong et al. 2011; Lian et al. 2005; Feng et al. 2010; Senthilvel et al. 2008; Wang et al. 2009b; Shan et al. 2005; Cho et al. 2007), most of them are found to have either little phenotypic contribution or large QTL-by-environment interactions or both, rendering them practically unusable in breeding (Senthilvel et al. 2008).

Heuer et al. (2013) outlined several factors to be considered while breeding for P deficiency tolerant rice cultivars, such as better solubilisation potential of fixed P, root modifications, high internal P use efficiency and reduced grain P concentration. However, efforts so far to improve P starvation tolerance in rice have gone primarily into improving P acquisition efficiency – the most remarkable achievement in this direction is the identification, characterisation and use of *Pup1*, a major QTL conferring P deficiency tolerance (Wissuwa et al. 1998; Heuer et al. 2009; Chin et al. 2011).

Under K-starved conditions, improving root traits that expedite uptake process is recognised as a major breeding target in rice. Major traits on focus are root architecture, high surface uptake capacity and the mobilisation capacity of the non-exchangeable K by root exudates (Rengel and Damon 2008). Substantiating this view, Jia et al. (2008) reported that K deficiency tolerant genotypes had maintained a better root architecture than intolerant ones. However, no specific target trait other than quantification of biomass or yield per unit K taken up has been defined for K utilisation efficiency in rice (Yang et al. 2003, 2004). Unlike that of other nutrients, there are not many QTLs reported for K starvation tolerance in rice. For K efficiency, the report on QTLs underlying K deficiency in rice by Wu et al. (1998) perhaps remains as the only report in crop plants (Rengel and Damon 2008). However, there are few other QTLs for shoot and root K concentrations in rice (Lin et al. 2004; Ren et al. 2005), reported particularly in association with salt tolerance.

Although S limitation is reported to cause reduction in dry matter production (Tsujimto et al. 2013), information on the nature of S deficiency tolerance in rice remains scanty. Further, as far as our knowledge goes, no QTL has been reported to confer S deficiency tolerance in rice. While improving tolerance to Zn deficiency, shoot Zn concentration at early vegetative stage below 15–20 mg.kg^{-1} is considered critical in lowland rice (Dobermann and Fairhurst 2000). Further, Zn uptake and root-to-shoot Zn translocation are two important parameters that determine tolerance under moderate and severe Zn deficiency and continued uptake of Zn at terminal stages is found critical to grain loading (Impa et al. 2013).

6.7 Engineering Nutrient Starvation Tolerance

There are several reports on candidate genes imparting varying levels of influence on nutrient use parameters in rice and other crop plants (Abrol et al. 1999; Hirel et al. 2007; Ramaekers et al. 2010; Rose et al. 2011; McAllister et al. 2012; Vinod and Heuer 2012; Veneklaas et al. 2012; Rose et al. 2013a, b; Zhang et al. 2014). However, genetic manipulations for nutrient use in rice have so far been limited to experimental validation of few candidate genes.

In an attempt to develop N use efficient plants, antisense technology was used to develop transgenic rice plants with reduced Rubisco content by transformation with the Rubisco small subunit *OsRbcS* antisense gene under the control of the *OsRbcS* promoter, resulting in plants with better N use efficiency under conditions of saturating CO_2 and high irradiance (Makino et al. 1997). However, overexpression (OX) of *OsrbcS1* in transgenic lines showed no significant change in photosynthesis and in fact showed reduction in tiller number (Morita et al. 2014). Screening of transgenic rice lines carrying cytosolic glutamine synthetase genes from rice (*OsGS1;1* and *OsGS1:2*) and *Escherichia coli* (*glnA*) showed overall increase of total leaf GS activity, elevated soluble protein concentration, higher total amino acid content, and increased total N content in the whole plant. However, transgenic lines failed to give higher grain yield but had lower grain amino acid concentrations (Cai et al. 2009). Furthermore, transgenic *OsGS1;2*OX rice lines constitutionally driven by maize ubiquitin promoter were reported to show better utilisation efficiency leading to better harvest index, spikelet fertility and grain number than the wild types. Contrary to the expectations, there was no significant advantage shown by the transgenics under N-limiting conditions than N-sufficient conditions (Brauer et al. 2011). Recently, rice ammonium transporter gene, *OsAMT1;1* was overexpressed in transgenic rice lines, to demonstrate high NH_4^+ permeability and high NH_4^+ accumulation, resulting in greater N assimilates; increased chlorophyll, starch, and sugars; and increased grain yield when grown under suboptimal and optimal N conditions (Ranathunge et al. 2014). Shrawat et al. (2008) introduced a barley *AlaAT* (alanine aminotransferase) cDNA driven by a rice tissue-specific promoter (*OsAnt1*) to produce transgenic rice lines with increased biomass and grain yield under sufficient N supply. These transgenic plants had significant increase in key metabolites and total N content, indicating increased N uptake efficiency. Further testing of these transgenic lines under different N levels revealed improved NUE at medium and high N supply (Beatty et al. 2013). Positive effects on NUE were reported in transgenic plants expressing an early nodulin gene *OsENOD93-1*, which increased shoot dry biomass and seed yield, which also accumulated higher concentration of amino acids in xylem sap, especially under N stress (Bi et al. 2009). Further, *OsENOD93-1*OX transgenic rice plants driven by a constitutive ubiquitin promoter achieved 23 % and 16 % more yield and biomass, respectively, compared to wild-type plants when grown under N-limited conditions. These lines further expressed enhanced accumulation of total amino acids in the roots and xylem sap (Guevara et al. 2014). Arginine is considered as an important amino acid for N transport and storage, playing a crucial role during seedling development. Transgenic introduction of arginine hydrolysis enzyme (*OsARG*), a key enzyme in the arginine catabolism, was reported to improve grain number per plant under N-limited conditions in *OsARG*OX rice lines, with strong *OsARG* expression in developing panicles (Ma et al. 2013). A high-affinity urea transporter *OsDUR3* that was found upregulated in rice roots under N deficiency and urea resupply after N starvation, when introduced into *Arabidopsis* transgenic lines, exhibited marked improvement in growth on low urea and increased root urea uptake (Wang et al. 2012). Transgenic introduction of a calcium-dependent protein kinase, *OsCDPK12*, was reported to improve N content and dry weight of the plants under N-limited conditions, indicating its role in signal transduction under N-deprived conditions (Asano et al. 2010). Under N-limited conditions, transgenic rice lines expressing the maize transcription factor (*ZmDof1*) were demonstrated to have improved growth, enhanced accumulation of carbon and N accompanied by predominant distribution of N to roots, significant root biomass increase and modification of the shoot-to-root ratio (Kurai et al. 2011). Similarly, rice transcription factor *OsDof25*, when introduced into *Arabidopsis*, was found to increase expression levels of high- and low-affinity ammonium transporters (*AtAMT1.1* and *AtAMT2.1*, respectively) and repressed the high-affinity nitrate transporter (*AtNRT2.1*). Together with an increase in amino acid content and elevated levels of enzymes involved in carbon metabolism the transgenic lines suggested that *OsDof25* was involved in NH_4^+ uptake and organic acid metabolism in plants (Santos et al. 2012).

For P starvation tolerance, functional analysis of *Pup1* QTL was carried out by cloning, and transgenic expression of the serine-threonine protein kinase gene, *OsPSTOL1*, constitutively driven by 35S promoter, showed that *OsPSTOL1* enhanced crown root growth at an early developmental stage in rice plants, concomitantly increasing root surface area and thereby enabling the plants to forage a larger soil area and to take up more P and other nutrients (Gamuyao et al. 2012). Several transcription factors are implicated in imparting P starvation tolerance in rice. Transgenic OX of a rice transcription factor (*OsPTF1*) involved in the phosphate starvation response in phloem cells of the primary root, leaves and lateral roots in rice lines manifested enhanced tolerance to P starvation. Microarray data on this *OsPTF*OX transgenic rice plants showed an enhanced expression of rice proton-translocating pyrophosphatases (H^+-PPases) (Yi et al. 2005). H^+-PPases are highly conserved sequences in plant genomes that are known to respond to various abiotic stresses such as salinity, drought (Gaxiola et al. 2001) and P starvation (Yi et al. 2005). Rice lines overexpressing the *Arabidopsis* vacuolar H^+-pyrophosphatase gene (*AtAVP1*), a type 1 H^+-PPase, exhibited sustained shoot growth under Pi-deficient conditions as against the controls which showed poor growth (Yang et al. 2007). Moreover, these lines developed enhanced rhizosphere acidification capacity, enhanced size and density of root hairs and more robust root systems than controls in both Pi-sufficient and Pi-deficient conditions (Gaxiola et al. 2011). *Arabidopsis* phosphate starvation response regulator 1 (*AtPHR1*) is an MYB transcription factor known as *PHR1*-binding sequences (*P1BS*) that play a key role in P starvation signalling by binding to a cis-element motif GnATATnC (Rubio et al. 2001). *OsPHR1* and *OsPHR2* are rice orthologs of *AtPHR1*. Transgenic rice lines overexpressing *OsPHR2* were characterised by excess Pi accumulation and Pi toxicity under P-replete conditions (Zhou et al. 2008; Wang et al. 2009a). In rice roots, the low-affinity Pi transporter gene, *OsPT2* is positively regulated by *OsPHR2* through physical interaction and upstream regulation of rice phosphate over accumulator 2, *OsPHO2*. *OsPHO2* is an E2 enzyme having an ubiquitin-conjugating (UBC) domain, which regulates Pi uptake, allocation and remobilization (Dong et al., 1998). *OsPT2* is responsible for most of the *OsPHR2*-mediated accumulation of excess shoot Pi (Liu et al. 2010). Under exposure to Pi starvation, transgenic *OsPHR2*OX rice lines mimicked Pi starvation response together with the induction of P starvation-induced (PSI) genes (Wu et al. 2013) along with the upregulation of purple acid phosphatase (PAP) genes as evidenced by an increase in both acid phosphatase and surface secretory acid phosphatase activities in rice roots (Zhang et al. 2011b). A negative regulator of *OsPHR2* in rice, *OsSPX1*, an SPX (SYG/PHO81/XPR1) domain gene (Wang et al. 2009a) supresses the accumulation of excess shoot Pi, in transgenic plants by interfering with the *OsPT2* expression by *OsPHR2* (Liu et al. 2010). An R2R3 MYB transcription factor, *OsMYB2P-1*, was found to improve tolerance to P starvation as well as root system architecture, in *OsMYBP-1*OX lines exposed to Pi deficiency (Dai et al. 2012). Another group of transcription factors in rice, auxin response factors 16 and 19 (*OsARF16* and *OsARF19*), regulates auxin-responsive genes and thereby increases the lateral root growth as a response to P starvation (Shen et al. 2012; Wang et al. 2014). Another gene, leaf tip necrosis (*OsLTN1*), was demonstrated to act as a negative regulator of P uptake and plays a major role in P signalling in rice (Hu et al. 2011). There are also reports of micro-RNAs associated with P starvation response in rice (Kuo and Chiou 2011). Transgenic analysis of miR827, which is highly upregulated under Pi deficiency, showed complex regulation with respect to its target genes, *OsSPX-MFS1* and *OsSPX-MFS2*, which encode SPX-major facilitator superfamily (SPX-MFS) proteins predicted to be implicated in phosphate (Pi) sensing or transport (Lin et al. 2010).

In contrast, most common genes in the P assimilation pathway such as P transporters were seldom found to make significant P starvation response. Jia et al. (2011) analysed the effect of the rice P transporter *OsPht1;8* by OX and repression through RNAi and found that P uptake in the transgenics was altered as expected, resulting in a significant reduction in the number and size of panicles, as well as >80 % spikelet sterility. In another study, transgenic rice plants overexpressing *OsPht1;1*, showed high Pi accumulation in leaves under P-replete conditions (Sun et al. 2012). In another investigation, only few of the transgenic rice plants overexpressing the tobacco transporter *NtPT1* were found to outperform the controls, only to yield less than the controls on an average (Park et al. 2010). Expression of a P transporter *OsPT11* that is specifically induced during arbuscular mycorrhizal (AM) symbiosis, in yeast knocked down mutants for high-affinity Pi transporter *PhO1* was found to complement the defect in phosphate uptake in the mutants (Paszkowski et al. 2002).

To prevent loss of P as anti-nutritional factor through grains, reduction of grain phytate content in rice was attempted through RNAi-mediated silencing of myo-inositol-3-phosphate synthase gene (*OsMIPS*) that catalyses the first step of phytic acid biosynthesis in the developing rice seed (Feng and Yoshida 2004). Transgenic plants carrying silenced *OsMIPS* gene driven by seed-specific *Oleosin18* (*Ole18*) promoter had reduced phytate content expressed in the aleurone layer and embryo, but had undesired effects on seed myo-inositol metabolisms (Qu and Takaiwa 2004). In a recent attempt, Ali et al. (2013) employed a modified approach in which inositol 1,3,4,5,6-pentakisphosphate 2-kinase gene (*OsIPK1*) that catalyses the last step of phytic acid biosynthesis was silenced using RNAi techniques without affecting the initial steps. Transgenic rice lines expressing silenced *OsIPK1* showed reduced phytate content in grains, with a concomitant increase in the amount of Pi and iron and no adverse effect on seed germination or in any of the agronomic traits.

6.8 Conclusions and Perspectives

In rice-growing regions around the world, nutrient starvation occurs in agricultural soils either naturally in poor and marginal lands or by nutrient depletion due to continuous farming. Nutrient limitation together with the need for increased food production enforces demand for artificial nutrient supplementation to soil. However, natural reserves of the fertiliser sources are rapidly depleting around the world, triggering price rise and sociopolitical divides. Additionally, in areas where intensive rice cultivation is practised, excessive fertilisation is causing environmental degradation. All these issues of contemporary agriculture converge on a single sustainable solution of reduction of fertiliser input in agriculture, feasibility of which necessitates the development of low nutrient-friendly rice varieties.

Rice gene pool harbours enough variability to nutrient use that has already been successfully exploited in developing high-yielding varieties using conventional approaches. The selection of these varieties occurred under conditions of nutrient sufficiency and therefore could have favoured genes that are responsible to internal nutrient use and against those genes that triggered responses for nutrient deficiency. Nutrient starvation responses are mainly driven by factors associated with increased uptake under nutrient-deprived conditions in order to maintain the internal nutrient homeostasis. Experiences show that favourable genes for deficiency response are to be mined from older varieties rather than the modern ones. Combining the uptake efficiencies of older varieties with internal utilisation efficiency and high-yielding ability of the modern varieties should therefore be the ideal approach towards breeding varieties for nutrient starvation tolerance.

Responses to nutrient starvation are very complex, and the breeding objective itself is a reversal to what has been done towards the development of modern HYVs. Molecular approaches may be better suited to tailor nutrient deficiency responses into elite genetic backgrounds that combine quantity and quality of yield. Successful introgression of QTLs related to various biotic and abiotic stress tolerances has already been demonstrated in rice. The success of the development of P deficiency tolerant rice varieties in the elite backgrounds such as IR64 and IR74 by marker-assisted introgression of *Pup1* QTL (Chin et al. 2011) is a major impetus towards replicating similar success in case of other nutrients. De facto pyramiding of several loci conferring biotic stress tolerance especially to various diseases has been achieved in rice using marker-assisted selection strategy (Singh et al. 2011, 2013). Efforts are ongoing to pyramid various abiotic stress tolerance QTLs in rice, which also include *Pup1* (Wissuwa 2011). However, not many QTLs/genes for nutrient deficiency tolerance have been identified as of today.

Concerted efforts are to be bolstered to identify phenotypic components of nutrient assimilation in rice as well as to identify related novel genomic regions that can confer tolerance to nutrient-limited situations, by looking beyond the horizons of nutrient uptake (Heuer et al. 2013). Currently several state-of-the-art phenotyping platforms are getting ready in major laboratories around the world that can monitor plants very closely throughout its lifespan for phonological behaviour (Tung et al. 2010). Furthermore, present-day rice improvement programmes are augmented with molecular-based technologies to identify and introgress novel genomic regions targeting specific traits of long-term interest. Marching ahead from structured population-based QTL mapping strategies, genomic-assisted tools such as the development of association mapping (AM) panels are becoming common platforms for the identification of novel genomic regions today. Genome-wide and nested association mapping panels are to be explored through single-nucleotide polymorphism (SNP) and sequence-based polymorphisms to identify novel variations. A close variant of nested AM population called multi-parent advanced generation inter-cross (MAGIC) population is being tested in rice (Zhao et al. 2011). Further, genomic selection is also in the pipeline that can accelerate the development of targeted trait-based crop improvement (Xu 2013). With efficient technologies in hand, formidable challenge for breeders is to identify relevant traits and ideal phenotyping tools for improving nutrient starvation tolerance in rice.

References

Abilay WP, De Datta SK (1978) Management practices for correcting zinc deficiency in transplanted and direct-seeded wetland rice. Philipp J Crop Sci 3:190–194

Abrol YP, Chatterjee SR, Kumar PA, Jain V (1999) Improvement in nitrogen use efficiency: physiological and molecular approaches. Curr Sci 76:1357–1364

Ali N, Paul S, Gayen D, Sarkar SN, Datta K, Datta SK (2013) Development of low phytate rice by RNAi mediated seed-specific silencing of inositol 1,3,4,5,6-pentakisphosphate 2-kinase gene (*IPK1*). PLoS One 8:e68161

Alloway BJ (2008) Zinc in soils and crop nutrition, 2nd edn. IZA/IFA, Brussels/Paris, 135p

Arnold T, Kirk GJD, Wissuwa M, Frei M, Zhao FJ, Mason TFD, Weiss DJ (2010) Evidence for the mechanisms of zinc uptake by rice using isotope fractionation. Plant Cell Environ 33:370–381

Asano T, Wakayama M, Aoki N, Komatsu S, Ichikawa H, Hirochika H, Ohsugi R (2010) Overexpression of a calcium-dependent protein kinase gene enhances growth of rice under low-nitrogen conditions. Plant Biotechnol 27:369–373

Bañuelos MA, Garciadeblas B, Cubero B, Rodríguez-Navarro A (2002) Inventory and functional characterization of the HAK potassium transporters of rice. Plant Physiol 130:784–795

Bashir K, Ishimaru Y, Nishizawa NK (2010) Iron uptake and loading into rice grains. Rice 3:122–130

Bashir K, Ishimaru Y, Nishizawa NK (2012) Molecular mechanisms of zinc uptake and translocation in rice. Plant Soil 361:189–201

Beatty PH, Carroll RT, Shrawat AK, Guevara D, Good AG (2013) Physiological analysis of nitrogen-efficient rice overexpressing alanine aminotransferase under different N regimes. Botany 91:866–883

Bi YM, Kant S, Clarke J, Gidda S, Ming F, Xu J, Rochon A, Shelp BJ, Hao L, Zhao R, Mullen RT, Zhu T, Rothstein SJ (2009) Increased nitrogen-use efficiency in transgenic rice plants over-expressing a nitrogen-responsive early nodulin gene identified from rice expression profiling. Plant Cell Environ 32:1749–1760

Blair GJ, Mamaril CP, Momuat E (1978) Sulfur nutrition of wetland rice, IRRI research paper series 21. International Rice Research Institute, Manila, 29p

Bolland MDA, Gilkes RJ (1998) The chemistry and agronomic effectiveness of phosphate fertilizers. J Crop Prod 1:139–163

BP (2014) BP statistical review of world energy 2014. 45p. Accessed online on 29 June 2014 from http://www.bp.com

Brauer EK, Rochon A, Bi YM, Bozzo GG, Rothstein SJ, Shelp BJ (2011) Reappraisal of nitrogen use efficiency in rice overexpressing glutamine synthetase 1. Physiol Plant 141:361–372

Briskin DP, Gawienowski MC (1996) Role of the plasma membrane H^+-ATPase in K^+ transport. Plant Physiol 111:1199–1207

Britto DT, Kronzucker HJ (2005) Plant nitrogen transport and its regulation in changing soil environments. J Crop Improv 15:1–23

Cai H, Zhou Y, Xiao J, Li X, Zhang Q, Lian X (2009) Overexpressed glutamine synthetase gene modifies nitrogen metabolism and abiotic stress responses in rice. Plant Cell Rep 28:527–537

Cakmak I (2009) Enrichment of fertilizers with zinc: an excellent investment for humanity and crop production in India. J Trace Elem Med Bio 23:281–289

Chin JH, Gamuyao R, Dalid C, Bustamam M, Prasetiyono J, Moeljopawiro S, Wissuwa M, Heuer S (2011) Developing rice with high yield under phosphorus deficiency: *Pup1* sequence to application. Plant Physiol 156:1202–1216

Cho Y, Jiang WZ, Chin JH, Piao ZP, Cho YG, McCouch SR, Koh HJ (2007) Identified QTLs associated with physiological nitrogen use efficiency in rice. Mol Cells 23:72–79

Cordell D, White S (2013) Sustainable phosphorus measures: strategies and technologies for achieving phosphorus security. Agronomy 3:86–116

Cordell D, Drangert JO, White S (2009) The story of phosphorus: global food security and food for thought. Glob Environ Change 19:292–305

Dai X, Wang Y, Yang A, Zhang WH (2012) *OsMYB2P-1*: an R2R3 MYB transcription factor, is involved in the regulation of phosphate starvation responses and root architecture in rice. Plant Physiol 159:169–183

Datta SC (2011) Potassium dynamics and status in Indian soils. Karnataka J Agric Sci 24:7–11

Davidian JC, Kopriva S (2010) Regulation of sulfate uptake and assimilation – the same or not the same? Mol Plant 3:314–325

De Datta SK (1981) Principles and practice of rice production. Wiley, New York, 618p

Dobermann A, Cassman KG (2002) Plant nutrient management for enhanced productivity in intensive grain production systems of the United States and Asia. Plant Soil 247:153–175

Dobermann A, Fairhurst TH (2000) Rice: nutrient disorders and nutrient management. Potash & Phosphate Institute, Potash & Phosphate Institute of Canada, Singapore and International Rice Research Institute, Los Baños, 191p

Dobermann A, Cruz PCS, Cassman KG (1996) Fertilizer inputs, nutrient balance, and soil nutrient-supplying power in intensive, irrigated rice systems. I. Potassium uptake and K balance. Nutr Cycl Agroecosyst 46:1–10

Dobermann A, Cassman KG, Mamaril CP, Sheehy JE (1998) Management of phosphorus, potassium, and sulfur in intensive, irrigated lowland rice. Field Crop Res 56:113–138

Dobermann A, Witt C, Dawe D (2004) Increasing productivity of intensive rice systems through site-specific nutrient management. Science Publishers, Enfield and International Rice Research Institute, Los Baños, 410 p

Dong B, Rengel Z, Delhaize E (1998) Uptake and translocation of phosphate by pho2 mutant and wild-type seedlings of *Arabidopsis thaliana*. Plant Physiol 205:251–256

Dordas C (2008) Role of nutrients in controlling plant diseases in sustainable agriculture. Agron Sustain Dev 28:33–46

Eide DJ (2005) The ZIP family of zinc transporters. In: Iuchi S, Kuldell N (eds) Zinc finger proteins: from atomic contact to cellular function. Kluwer Academic/Plenum Publishers, New York, pp 261–264

El Kassis E, Cathala N, Rouached H, Fourcroy P, Berthomieu P, Terry N, Davidian J-C (2007)

Characterization of a selenate-resistant *Arabidopsis thaliana* mutant. Root growth as a potential target for selenate toxicity. Plant Physiol 143:1231–1241

El-Kassasa AM, Mourad AHI (2013) Novel fibers preparation technique for manufacturing of rice straw based fiber boards and their characterization. Mater Des 50:757–765

Erisman JW, Sutton MA, Galloway J, Klimont Z, Winiwarter W (2008) How a century of ammonia synthesis changed the world. Nat Geosci 1:636–639

Fageria NK (2013) Mineral nutrition of rice. CRC Press, Boca Raton, 586p

Fan JB, Zhang YL, Turner D, Duan YH, Wang DS, Shen QR (2010) Root physiological and morphological characteristics of two rice cultivars with different nitrogen-use efficiency. Pedosphere 20:446–455

Fang ZY, Shao C, Meng YJ, Wu P, Chen M (2009) Phosphate signaling in *Arabidopsis* and *Oryza sativa*. Plant Sci 176:170–180

FAO (2001) Global estimates of gaseous emissions of NH_3, NO and N_2O from agricultural land. International Fertilizer Industry Association, Paris and Food and Agriculture Organization of the United Nations, Rome, 106p

FAO (2005) Fertiliser use by crops in India. Food and Agriculture Organization of the United Nations, Rome, 45p

Feng X, Yoshida KT (2004) Molecular approaches for producing low-phytic-acid grains in rice. Plant Biotechnol 21:183–189

Feng Y, Cao LY, Wu WM, Shen XH, Zhan XD, Zhai RR, Wang RC, Chen DB, Cheng SH (2010) Mapping QTLs for nitrogen-deficiency tolerance at seedling stage in rice (*Oryza sativa* L.). Plant Breed 129:652–656

Fess TL, Kotcon JB, Benedito VA (2011) Crop breeding for low input agriculture: a sustainable response to feed a growing world population. Sustainability 3:1742–1772

Fixen PE (2009) World fertilizer nutrient reserves – a view to the future. Better Crops 93:8–11

Fu JD, Lee BW (2008) Changes in photosynthetic characteristics during grain filling of functional stay-green rice SNUSG1 and its F_1 hybrids. J Crop Sci Biotechnol 11:75–82

Fuchs I, Stölzle S, Ivashikina N, Hedrich R (2005) Rice K^+ uptake channel *OsAKT1* is sensitive to salt stress. Planta 221:212–221

Gallais A, Coque M (2005) Genetic variation and selection for nitrogen use efficiency in maize: a synthesis. Maydica 50:531–547

Galloway JN, Aber JD, Erisman JW, Seitzinger SP, Howarth RW, Cowling EB, Cosby BJ (2003) The nitrogen cascade. BioScience 53:341–356

Gamuyao R, Chin JH, Pariasca-Tanaka J, Pesaresi P, Dalid C, Slamet-Loedin I, Tecson-Mendoza EM, Wissuwa M, Heuer S (2012) The protein kinase *OsPSTOL1* from traditional rice confers tolerance of phosphorus deficiency. Nature 488:535–539

Gaxiola R, Li J, Undurraga S, Dang L, Allen G, Alper S, Fink G (2001) Drought and salt-tolerant plants result from overexpression of the AVP1 H^+-pump. Proc Natl Acad Sci U S A 98:11444–11449

Gaxiola RA, Edwards M, Elser JJ (2011) A transgenic approach to enhance phosphorus use efficiency in crops as part of a comprehensive strategy for sustainable agriculture. Chemosphere 84:840–845

Giampietro M, Pimentel D (1993). The tightening conflict: population, energy use, and the ecology of agriculture. NPG Forum, 8 p. Negative Population Growth Inc., Teaneck. http://www.npg.org/forum_series/TheTighteningConflict.pdf

Godwin RM, Rae AL, Carroll BJ, Smith FW (2003) Cloning and characterization of two genes encoding sulfate transporters from rice (*Oryza sativa* L.). Plant Soil 257:113–123

Guerra LC, Bhuiyan SI, Tuong TP, Baker R (1998) Producing more rice with less water from irrigated systems. International Rice Research Institute, Manila, Discussion Paper Series No. 29, 18 p

Guevara D, Bi YM, Rothstein S (2014) Identification of regulatory genes to improve nitrogen use efficiency. Can J Plant Sci 94:1009–1012. doi:10.4141/CJPS2013-154

Hafeez B, Khanif YM, Samsuri AW, Radziah O, Zakaria W, Saleem M (2013) Direct and residual effect of zinc on zinc efficient and inefficient rice genotypes grown under less zinc content submerged acidic condition. Commun Soil Sci Plant Anal 44:2233–2252

Hajiboland R, Aliasgharzad N, Barzeghar R (2009) Phosphorus mobilization and uptake in mycorrhizal rice (*Oryza sativa* L.) plants under flooded and non-flooded conditions. Acta Agriculturae Slovenica 93:153–161

Hasan R (2002) Potassium status of soils of India. Better Crop Int 16:3–5

Hauser F, Horie T (2010) A conserved primary salt tolerance mechanism mediated by HKT transporters: a mechanism for sodium exclusion and maintenance of high K^+/Na^+ ratio in leaves during salinity stress. Plant Cell Environ 33:552–565

Hawkesford MJ, Howarth JR (2011) Transcriptional profiling approaches for studying nitrogen use efficiency. In: Foyer C, Zhang H (eds) Nitrogen metabolism in plants in the post-genomic era, vol 42, Annual plant reviews. Blackwell Publishing Ltd, West Sussex, pp 41–62

Hayat R, Ali S, Amara U, Khalid R, Ahmed I (2010) Soil beneficial bacteria and their role in plant growth promotion: a review. Ann Microbiol 60:579–598

Haynes RJ, Goh KM (1978) Ammonium and nitrate nutrition of plants. Biol Rev 53:465–510

Heisler J, Glibert PM, Burkholder JM, Anderson DM, Cochlan W, Dennison WC, Dortch Q, Gobler CJ, Heil CA, Humphries E, Lewitus A, Magnien R, Marshall HG, Sellner K, Stockwell DA, Stoecker DK, Suddleson M (2008) Eutrophication and harmful algal blooms: a scientific consensus. Harmful Algae 8:3–13

Heuer S, Lu X, Chin JH, Tanaka JP, Kanamori H, Matsumoto T, De Leon T, Ulat VJ, Ismail AM, Yano M, Wissuwa M (2009) Comparative sequence analyses of the major quantitative trait locus *phosphorus uptake 1* (*Pup1*) reveal a complex genetic structure. Plant Biotechnol J 7:456–457

Heuer S, Chin JH, Gamuyao R, Haefele SM, Wissuwa M (2013) Molecular breeding for phosphorus-efficient rice. In: Varshney RK, Tuberosa R (eds) Translational genomics for crop breeding, vol II, Abiotic stress, yield and quality. John Wiley & Sons, Ames, pp 65–82

Higuchi K, Suzuki K, Nakanishi H, Yamaguchi H, Nishizawa NK, Mori S (1999) Cloning of nicotianamine synthase genes, novel genes involved in the biosynthesis of phytosiderophores. Plant Physiol 119:471–480

Hirel B, Chardon F, Durand J (2007) The contribution of molecular physiology to the improvement of nitrogen use efficiency in crops. J Crop Sci Biotechnol 10:123–132

Hocking PJ (2001) Organic acids exuded from roots in phosphorus uptake and aluminum tolerance of plants in acid soils. Adv Agron 74:63–97

Hocking PJ, Randall PJ, Delhaize E, Keerthisinghe G (2000) The role of organic acids exuded from roots in phosphorus nutrition and aluminium tolerance acidic soils. In: Management and conservation of tropical acid soils for sustainable crop production. International Atomic Energy Agency, Vienna, pp 61–70

Hu B, Zhu C, Li F, Tang J, Wang Y, Lin A, Liu L, Che R, Chu C (2011) *LEAF TIP NECROSIS1* plays a pivotal role in the regulation of multiple phosphate starvation responses in rice. Plant Physiol 156:1101–1115

Impa SM, Morete MJ, Ismail AM, Schulin R, Johnson-Beebout SE (2013) Zn uptake, translocation and grain Zn loading in rice (*Oryza sativa* L.) genotypes selected for Zn deficiency tolerance and high grain Zn. J Exp Bot 64:2739–2751

Ipsilantis I, Sylvia DM (2007) Interactions of assemblages of mycorrhizal fungi with two Florida wetland plants. Appl Soil Ecol 35:261–271

Isayenkov S, Isner JC, Maathuis FJM (2011) Rice two-pore K^+ channels are expressed in different types of vacuoles. Plant Cell 23:756–768

Ismunadji M, Blair G, Lefroy R (1991) S research on rice in of grain analysis to assess nutrient status for yield Indonesia. In: Sulfur fertilizer policy for lowland and upland rice cropping systems in Indonesia. Australian Centre for International Agricultural Research, Canberra, pp 87–90

Jain N, Bhatia A, Pathak H (2014) Emission of air pollutants from crop residue burning in India. Aerosol Air Qual Res 14:422–430

Jia Y, Yang X, Feng Y, Jilani G (2008) Differential response of root morphology to potassium deficient stress among rice genotypes varying in potassium efficiency. J Zhejiang Univ Sci B 9:427–434

Jia H, Ren H, Gu M, Zhao J, Sun S, Zhang X, Chen J, Wu P, Xu G (2011) The phosphate transporter gene *OsPht1;8* is involved in phosphate homeostasis in rice. Plant Physiol 156:1164–1175

Kant S, Bi YM, Rothstein SJ (2011) Understanding plant response to nitrogen limitation for the improvement of crop nitrogen use efficiency. J Exp Bot 62:1499–1509

Kataoka T, Hayashi N, Yamaya T, Takahashi H (2004a) Root-to-shoot transport of sulfate in Arabidopsis. Evidence for the role of *SULTR3;5* as a component of low-affinity sulfate transport system in the root vasculature. Plant Physiol 136:4198–4204

Kataoka T, Watanabe-Takahashi A, Hayashi N, Ohnishi M, Mimura T, Buchner P, Hawkesford MJ, Yamaya T, Takahashi H (2004b) Vacuolar sulfate transporters are essential determinants controlling internal distribution of sulfate in Arabidopsis. Plant Cell 16:2693–2704

Khan AG (2006) Mycorrhizoremediation – an enhanced form of phytoremediation. J Zhejiang Univ Sci B 7:503–514

Kirk GJD, Santos EE, Santos MB (1999) Phosphate solubilization by organic anion excretion from rice growing in aerobic soil: rates of excretion and decomposition, effects on rhizosphere pH and effects on phosphate solubility and uptake. New Phytol 142:185–200

Kumar P, Joshi L (2013) Pollution caused by agricultural waste burning and possible alternate uses of crop stubble: a case study of Punjab. In: Nautiyal S, Rao KS, Kaechele H, Raju KV, Schaldach R (eds) Knowledge systems of societies for adaptation and mitigation of impacts of climate change. Springer, Berlin/Heidelberg, pp 367–385

Kumar A, Dixit S, Henry A (2013) Marker-assisted introgression of major QTLs for grain yield under drought in rice. In: Varshney RK, Tuberosa R (eds) Translational genomics for crop breeding: abiotic stress, yield and quality, vol 2. John Wiley & Sons, Ames, pp 47–64

Kuo HF, Chiou TJ (2011) The role of microRNAs in phosphorus deficiency signaling. Plant Physiol 156:1016–1024

Kurai T, Wakayama M, Abiko T, Yanagisawa S, Aoki N, Ohsugi R (2011) Introduction of the *ZmDof1* gene into rice enhances carbon and nitrogen assimilation under low-nitrogen conditions. Plant Biotechnol J 9:826–837

Lafitte HR, Ismail A, Bennett J (2004) Abiotic stress tolerance in rice for Asia: progress and the future. New directions for a diverse planet. In: 4th International Crop Science Congress, Brisbane, Australia, pp 1–17

Lea PJ, Miflin BJ (2011) Nitrogen assimilation and its relevance to crop improvement. Annu Plant Rev 42:1–40

Lebaudy A, Véry AA, Sentenac H (2007) K^+ channel activity in plants: genes, regulations and functions. FEBS Lett 581:2357–2366

Lee S, Jeong H, Kim S, Lee J, Guerinot M, An G (2010a) *OsZIP5* is a plasma membrane zinc transporter in rice. Plant Mol Biol 73:507–517

Lee S, Kim SA, Lee J, Guerinot ML, An G (2010b) Zinc deficiency-inducible *OsZIP8* encodes a plasma membrane-localized zinc transporter in rice. Mol Cells 29:551–558

Li BZ, Merrick M, Li SM, Li HY, Zhu SW, Shi WM, Su YH (2009a) Molecular basis and regulation of ammonium transporter in rice. Rice Sci 16:314–322

Li LH, Qiu XH, Li XH, Wang SP, Lian XM (2009b) The expression profile of genes in rice roots under low phosphorus stress. Sci China Ser C Life Sci 52:1055–1064

Lian X, Xing Y, Yan H, Xu C, Li X, Zhang Q (2005) QTLs for low nitrogen tolerance at seedling stage identified using a recombinant inbred line population derived from an elite rice hybrid. Theor Appl Genet 112:85–96

Lim JH, Chung IM, Ryu SS, Park MR, Yun SJ (2003) Differential responses of rice acid phosphatase activities and isoforms to phosphorus deprivation. J Biochem Mol Biol 36:597–602

Lin HX, Zhu MZ, Yano M, Gao JP, Liang ZW, Su WA, Hu XH, Ren ZH, Chao DY (2004) QTLs for Na^+ and K^+ uptake of the shoots and roots controlling rice salt tolerance. Theor Appl Genet 108:253–260

Lin SI, Santi C, Jobet E, Lacut E, El Kholti N, Karlowski WM, Verdeil JL, Breitler JC, Périn C, Ko SS, Guiderdoni E, Chiou TJ, Echeverria M (2010) Complex regulation of two target genes encoding SPX-MFS proteins by rice miR827 in response to phosphate starvation. Plant Cell Physiol 51:2119–2131

Liu XG, Liu ZX, Liu FX (1987) Screening of rice genotypes tolerant to low K and their K uptake characteristics. J Fujian Agric Acad 2:10–17

Liu F, Wang Z, Ren H, Shen C, Li Y, Ling HQ, Wu C, Lian X, Wu P (2010) *OsSPX1* suppresses the function of *OsPHR2* in the regulation of expression of *OsPT2* and phosphate homeostasis in shoots of rice. Plant J 62:508–517

Liu Y, Kumar S, Kwag J, Kim J, Kim J, Ra C (2011) Recycle of electrolytically dissolved struvite as an alternative to enhance phosphate and nitrogen recovery from swine wastewater. J Hazard Mater 195:175–181

Lott JNA, Ockendena I, Raboya V, Battena GD (2000) Phytic acid and phosphorus in crop seeds and fruits: a global estimate. Seed Sci Res 10:11–33

Ma X, Cheng Z, Qin R, Qiu Y, Heng Y, Yang H, Ren Y, Wang X, Bi J, Ma X, Zhang X, Wang J, Lei C, Guo X, Wang J, Wu F, Jiang L, Wang H, Wan J (2013) *OsARG* encodes an arginase that plays critical roles in panicle development and grain production in rice. Plant J 73:190–200

Mae T (1997) Physiological nitrogen efficiency in rice: nitrogen utilization, photosynthesis, and yield potential. Plant and Soil 196:201–210

Mae T, Ohira K (1981) The remobilization of nitrogen related to leaf growth and senescence in rice plants (*Oryza sativa* L.). Plant Cell Physiol 22:1067–1074

Magdoff F (2013) Global resource depletion – is population the problem? Mon Rev 64:13–28. Accessed online http://monthlyreview.org/2013/01/01/global-resource-depletion)

Makino A, Shimada T, Takumi S, Kaneko K, Matsuoka M, Shimamoto K, Nakano H, Miyao-Tokutomi M, Mae T, Yamamoto N (1997) Does decrease in Ribulose-1,5-bisphosphate carboxylase by antisense *RbcS* lead to a higher N-use efficiency of photosynthesis under conditions of saturating CO_2 and light in rice plants? Plant Physiol 114:483–491

Manning DAC (2010) Mineral sources of potassium for plant nutrition. A review. Agron Sustain Dev 30:281–294

Marschner H (1995) Mineral nutrition of higher plants. Academic, Boston

Marschner H, Romheld V, Horst WJ, Martin P (1986) Root induced changes in the rhizosphere: importance for mineral nutrition of plants. Z Pflanzenernachr Bodenkd 149:441–456

Maurel C, Verdoucq L, Luu DT, Santoni V (2008) Plant aquaporins: membrane channels with multiple integrated functions. Annu Rev Plant Biol 59:595–624

McAllister CH, Beatty PH, Good AG (2012) Engineering nitrogen uses efficient crop plants: the current status. Plant Biotechnol J 10:1011–1025

Mohnot JK, Prasad VVR, Verma HK (2005) Investment opportunities for potash mining in India as an import substitute mineral. In: Proceedings of the 1st Indian Mineral Congress, Dhanbad

Morales N, Boehler MA, Buettner S, Liebi C, Siegrist H (2013) Recovery of N and P from urine by struvite precipitation followed by combined stripping with digester sludge liquid at full scale. Water 5:1262–1278

Morita K, Hatanaka T, Misoo S, Fukayama H (2014) Unusual small subunit that is not expressed in photosynthetic cells alters the catalytic properties of Rubisco in rice. Plant Physiol 164:69–79

Mortvedt JJ, Gilkes RJ (1993) Zinc fertilizers. In: Robson AD (ed) Zinc in soils and plants. Kluwer Academic Publishers, Dordrecht, pp 33–44

Murata Y, Ma JF, Yamaji N, Ueno D, Nomoto K, Iwashita T (2006) A specific transporter for iron(III)–phytosiderophore in barley roots. Plant J 46:563–572

Naidu LGK, Ramamurthy V, Sidhu GS, Sarkar D (2011) Emerging deficiency of potassium in soils and crops of India. Karnataka J Agric Sci 24:12–19

Nischal L, Mohsin M, Khan I, Kardam H, Wadhwa A, Abrol YP, Iqbal M, Ahmad A (2012) Identification and comparative analysis of microRNAs associated with low-N tolerance in rice genotypes. PLoS One 7:e50261. doi:10.1371/journal.pone.0050261

Nishiyama R, Kato M, Nagata S, Yanagisawa S, Yoneyama T (2012) Identification of Zn–nicotianamine and Fe–2'-deoxymugineic acid in the phloem sap from rice plants (*Oryza sativa* L). Plant Cell Physiol 53:381–390

Nozoye T, Nagasaka S, Kobayashi T, Takahashi M, Sato Y, Sato Y, Uozumi N, Nakanishi H, Nishizawa NK (2011) Phytosiderophore efflux transporters are crucial for iron acquisition in graminaceous plants. J Biol Chem 286:5446–5454

Nürnberger T, Abel S, Jost W, Glund K (1990) Induction of an extracellular ribonuclease in cultured tomato cells upon phosphate starvation. Plant Physiol 92:970–976

Obata T, Kitamoto HK, Nakamura A, Fukuda A, Tanaka Y (2007) Rice shaker potassium channel *OsKAT1* confers tolerance to salinity stress on yeast and rice cells. Plant Physiol 144:1978–1985

Ogawa S, Selvaraj MG, Fernando AJ, Lorieux M, Ishitani M, McCouch S, Arbelaez FD (2014) N and P mediated seminal root elongation response in rice seedlings. Plant Soil 375:305–315

Park MR, Tyagi K, Baek SH, Kim YJ, Rehman S, Yun SJ (2010) Agronomic characteristics of transgenic rice with enhanced phosphate uptake ability by overexpressed tobacco high affinity phosphate transporter. Pak J Bot 42:3265–3273

Paszkowski U, Kroken S, Roux C, Briggs SP (2002) Rice phosphate transporters include an evolutionarily divergent gene specifically activated in arbuscular mycorrhizal symbiosis. Proc Natl Acad Sci U S A 99:13324–13329

Pathak RR, Ahmad A, Lochab S, Raghuram N (2008) Molecular physiology of plant NUE and biotechnological options for its enhancement. Curr Sci 94:1395–1403

Peng S, Khush GS, Cassman KG (1994) Evolution of the new plant ideotype for increased yield potential. In: Cassman KG (ed) Breaking the yield barrier. International Rice Research Institute, Manila, pp 57–60

Peoples MB, Herridge DF, Ladha JK (1995) Biological nitrogen fixation: an efficient source of nitrogen for sustainable agricultural production. Plant Soil 174:3–28

Plaxton WC, Tran HT (2011) Metabolic adaptations of phosphate-starved plants. Plant Physiol 156:1006–1015

Ptashnyk M, Roose T, Jones DL, Kirk GJD (2011) Enhanced zinc uptake by rice through phytosiderophore secretion: a modelling study. Plant Cell Environ 34:2038–2046

Qu L, Takaiwa F (2004) Evaluation of tissue specificity and expression strength of rice seed component gene promoters in transgenic rice. Plant Biotechnol J 2:113–125

Raghothama KG (1999) Phosphate acquisition. Annu Rev Plant Physiol Plant Mol Biol 50:665–693

Ramaekers L, Remans R, Rao IM, Blair MW, Vanderleyden J (2010) Strategies for improving phosphorus acquisition efficiency of crop plants. Field Crop Res 117:169–176

Ramesh SA, Shin R, Eide DJ, Schachtman DP (2003) Differential metal selectivity and gene expression of two zinc transporters from rice. Plant Physiol 133:126–134

Ramesh P, Singh M, Rao AS (2005) Organic farming: its relevance to the Indian context. Curr Sci 88:561–568

Ranathunge K, El-kereamy A, Gidda S, Bi YM, Rothstein SJ (2014) AMT1;1 transgenic rice plants with enhanced NH_4^+ permeability show superior growth and higher yield under optimal and suboptimal NH_4^+ conditions. J Exp Bot 65:965–979

Rattan RK, Kumar M, Narwal RP, Singh AP (2009) Soil health and nutritional security – micronutrients. In: Proceedings of the platinum jubilee symposium. Indian Society of Soil Science, New Delhi, pp 249–265

Raven JA, Taylor R (2003) Macroalgal growth in nutrient enriched estuaries: a biogeochemical and evolutionary perspective. Water Air Soil Pollut 3:7–26

Reichardt W, Dobermann A, George T (1998) Intensification of rice production systems: opportunities and limits. In: Dowling NG, Greenfield SM, Fischer KS (eds) Sustainability of rice in the global food system. Pacific Basin Study Center/International Rice Research Institute, Davis/Manila, pp 127–144

Ren ZH, Gao JP, Li LG, Cai XL, Huang W, Chao DY, Zhu MZ, Wang ZY, Luan S, Lin HX (2005) A rice quantitative trait locus for salt tolerance encodes a sodium transporter. Nat Genet 37:1141–1146

Rengel Z, Damon PM (2008) Crops and genotypes differ in efficiency of potassium uptake and use. Physiol Plant 133:624–636

Roberts TL (2008) Improving nutrient use efficiency. Turk J Agric For 32:177–182

Roberts TL, Stewart WM (2002) Inorganic phosphorus and potassium production and reserves. Better Crop 86:6–7

Römheld V (1991) The role of phytosiderophores in acquisition of iron and other micronutrients in graminaceous species: an ecological approach. Plant Soil 130:127–134

Rose TJ, Wissuwa M (2012) Rethinking internal phosphorus utilization efficiency: a new approach is needed to improve PUE in grain crops. Adv Agron 116:185–217

Rose TJ, Rose MT, Pariasca-Tanaka J, Heuer S, Wissuwa M (2011) The frustration with utilization: why have improvements in internal phosphorus utilization efficiency in crops remained so elusive? Front Plant Nutr 2:73. doi:10.3389/fpls.2011.00073

Rose TJ, Impa SM, Rose MT, Pariasca-Tanaka J, Mori A, Heuer S, Johnson-Beebout SE, Wissuwa M (2013a) Enhancing phosphorus and zinc acquisition efficiency in rice: a critical review of root traits and their potential utility in rice breeding. Ann Bot 112:331–345

Rose TJ, Liu L, Wissuwa M (2013b) Improving phosphorus efficiency in cereal crops: is breeding for reduced grain phosphorus concentration part of the solution? Front Plant Sci 4:444. doi:10.3389/fpls.2013.00444

Rubio V, Linhares F, Solano R, Martin AC, Iglesias J, Leyva A, Paz-Ares J (2001) A conserved MYB transcription factor involved in phosphate starvation signaling both in vascular plants and in unicellular algae. Genes Dev 15:2122–2133

Saha N, Mukherjee D, Sen S, Sarkar A, Bhattacharaya KK, Mukhopadyay N, Patra PK (2012) Application of highly efficient lignocellulolytic fungi in co-composting of paddy straw amended poultry droppings for the production of humus rich compost. Compost Sci Util 20:239–244

Saito K (2000) Regulation of sulphate transport and synthesis of sulphur containing amino acids. Curr Opin Plant Biol 3:188–195

Sajwan KS, Lindsay WL (1988) Effect of redox, zinc fertilization and incubation time on DTPA-extractable zinc, iron and manganese. Commun Soil Sci Plant Anal 583(19):1–11

Santos LA, de Souza SR, Fernandes MS (2012) *OsDof25* expression alters carbon and nitrogen metabolism in Arabidopsis under high N-supply. Plant Biotechnol Rep 6:327–337

Schachtman DP, Reid RJ, Ayling SM (1998) Phosphorus uptake by plants: from soil to cell. Plant Physiol 116:447–453

Schubert S (1995) Proton release by plant roots. In: Singh BB, Mengel K (eds) Plant physiology and biochemistry. Panina Publishing Corporation, New Delhi, pp 97–119

Senthilvel S, Vinod KK, Malarvizhi P, Maheswaran M (2008) QTL and QTL× environment effects on agronomic and nitrogen acquisition traits in rice. J Integr Plant Biol 50:1108–1117

Septiningsih EM, Pamplona AM, Sanchez DL, Neeraja CN, Vergara GV, Heuer S, Ismail AM, Mackill DJ (2009) Development of submergence-tolerant rice cultivars: the *Sub1* locus and beyond. Ann Bot 103:151–160

Shan YH, Wang YL, Pan XB (2005) Mapping of QTLs for nitrogen use efficiency and related traits in rice (*Oryza sativa* L). Acta Agron Sin 4:721–727

Sheehy JE, Dionora MJA, Mitchell PL, Peng S, Cassman KG, Lemaire G, Williams RL (1998) Critical nitrogen concentrations: implications for high-yielding rice (*Oryza sativa* L.) cultivars in the tropics. Field Crop Res 59:31–41

Shen C, Wang S, Zhang S, Xu Y, Qian Q, Qi Y, Jiang DA (2012) OsARF16: a transcription factor, is required for auxin and phosphate starvation response in rice (*Oryza sativa* L.). Plant Cell Environ 36:607–620

Shrawat AK, Carroll RT, DePauw M, Taylor GJ, Good AG (2008) Genetic engineering of improved nitrogen use efficiency in rice by the tissue-specific expression of alanine aminotransferase. Plant Biotechnol J 6:722–732

Silalertruksa T, Gheewala SH (2013) A comparative LCA of rice straw utilization for fuels and fertilizer in Thailand. Bioresour Technol 150:412–419

Singh MV (2009) Micronutrient nutritional problems in soils of India and improvement for human and animal health. Indian J Fert 5:11–16

Singh Y, Khind CS, Singh B (1991) Efficient management of leguminous green manures in wetland rice. Adv Agron 45:135–189

Singh U, Ladha JK, Castillo EG, Punzalan G, Tirol-Padre A, Duqueza M (1998) Genotypic variation in nitrogen use efficiency in medium- and long-duration rice. Field Crop Res 58:35–53

Singh RB, Woodhead T, Papademetriou MK (2000) Strategies to sustain and enhance Asia-Pacific rice production. In: Papademetriou MK, Dent FJ, Herath EM (eds) Bridging the rice yield gap in the Asia-Pacific region. Food and Agriculture Organisation of United nations, Rome, p 222

Singh N, Dang TT, Vergara GV, Pandey DM, Sanchez D, Neeraja CN, Septiningsih EM, Mendioro M, Tecson-Mendoza EM, Ismail AM, Mackill DJ, Heuer S (2010) Molecular marker survey and expression analyses of the rice submergence-tolerance gene *SUB1A*. Theor Appl Genet 121:1441–1453

Singh AK, Gopalakrishnan S, Singh VP, Prabhu KV, Mohapatra T, Singh NK, Sharma TR, Nagarajan M, Vinod KK, Singh D, Singh UD, Chander S, Atwal SS, Seth R, Singh VK, Ellur RK, Singh A, Anand D, Khanna A, Yadav S, Goel N, Singh A, Shikari AB, Singh A, Marathi B (2011) Marker assisted selection: a paradigm shift in Basmati breeding. Indian J Genet Plant Breed 71:120–128

Singh VK, Singh A, Singh SP, Ellur RK, Singh D, Krishnan SG, Bhowmick PK, Nagarajan M, Vinod KK, Singh UD, Mohapatra T, Prabhu KV, Singh AK (2013) Marker assisted simultaneous but stepwise backcross breeding for pyramiding blast resistance genes *Piz5* and *Pi54* into an elite Basmati rice restorer line "PRR78". Plant Breed 132:486–495

Smil V (2000) Phosphorus in the environment: natural flows and human interferences. Annu Rev Energy Environ 25:53–88

Smith VH, Crews T (2014) Applying ecological principles of crop cultivation in large-scale algal biomass production. Algal Res 4:23–34

Smith FW, Ealing PM, Hawkesford MJ, Clarkson DT (1995) Plant members of a family of sulfate transporters reveals functional subtypes. Proc Natl Acad Sci U S A 92:9373–9377

Solaiman MZ, Hirata H (1997) Effect of arbuscular mycorrhizal fungi inoculation of rice seedlings at the nursery stage upon performance in the paddy field and greenhouse. Plant Soil 191:1–12

Sun S, Gu M, Cao Y, Huang X, Zhang X, Ai P, Zhao J, Fan X, Xu G (2012) A constitutive expressed phosphate transporter, *OsPht1;1*, modulates phosphate uptake and translocation in phosphate-replete rice. Plant Physiol 159:1571–1581

Suzuki M, Tsukamoto T, Inoue H, Watanabe S, Matsuhashi S, Takahashi M, Nakanishi H, Mori S, Nishizawa NK (2008) Deoxymugineic acid increases Zn translocation in Zn-deficient rice plants. Plant Mol Biol 66:609–617

Szczerba MW, Britto DT, Ali SA, Balkos KD, Kronzucker HJ (2008) NH_4^+-stimulated and -inhibited components of K^+ transport in rice (*Oryza sativa* L.). J Exp Bot 59:3415–3423

Szczerbab MW, Britto DT, Kronzucker HJ (2009) K^+ transport in plants: physiology and molecular biology. J Plant Physiol 166:447–466

Tabuchi M, Abiko T, Yamaya T (2007) Assimilation of ammonium ions and reutilization of nitrogen in rice (*Oryza sativa* L). J Exp Bot 58:2319–2327

Takahashi H, Watanabe-Takahashi A, Smith FW, Blake-Kalff M, Hawkesford MJ, Saito K (2000) The roles of three functional sulphate transporters involved in uptake and translocation of sulphate in *Arabidopsis thaliana*. Plant J 23:171–182

Takijima Y, Gunawardena SDIE (1969) Nutrient deficiency and physiological disease of lowland rice in Ceylon. Soil Sci Plant Nutr 15:259–266

Thomson MJ, Ocampo M, Egdane J, Rahman MA, Sajise AG, Adorada DL, Tumimbang-Raiz E, Blumwald E, Seraj ZI, Singh RK, Gregorio GB, Ismail AM (2010) Characterizing the *Saltol* quantitative trait locus for salinity tolerance in rice. Rice 3:148–160

Tian J, Wang C, Zhang Q, He X, Whelan J, Shou H (2012) Overexpression of *OsPAP10a*, a root-associated acid phosphatase, increased extracellular organic phosphorus utilization in rice. J Integr Plant Biol 54:631–639

Tong H, Chen L, Li W, Mei H, Xing Y, Yu X, Xu X, Zhang S, Luo L (2011) Identification and characterization of quantitative trait loci for grain yield and its components under different nitrogen fertilization levels in rice (*Oryza sativa* L.). Mol Breed 28:495–509

Tsujimto Y, Yamamoto Y, Hayashi K, Zakari A, Inusah Y, Hatta T, Fosu M, Sakagami JI (2013) Topographic distribution of the soil total carbon content and sulfur deficiency for rice cultivation in a floodplain ecosystem of the Northern region of Ghana. Field Crop Res 152:74–82

Tung CW, Zhao K, Wright MH, Ali ML, Jung J, Kimball J, Tyagi W, Thomson MJ, McNally K, Leung H, Kim H, Ahn SN, Reynolds A, Scheffler B, Eizenga G, McClung A, Bustamante C, McCouch SR (2010) Development of a research platform for dissecting phenotype-genotype associations in rice (*Oryza* spp.). Rice 3:205–217

Urano D, Chen J-G, Botella JR, Jones AM (2013) Heterotrimeric G protein signalling in the plant kingdom. Open Biol 3:120186. http://dx.doi.org/10.1098/rsob.120186

USGS (2011) Mineral commodity summaries 2011. US Geological Survey, Reston, pp 197

Vallino M, Greppi D, Novero M, Bonfante P, Lupotto E (2009) Rice root colonisation by mycorrhizal and endophytic fungi in aerobic soil. Ann Appl Biol 154:195–204

Van Kauwenbergh SJ (2010) World phosphate rock reserves and resources. International Fertilizer Development Centre, Muscle Shoals, 48p

Veneklaas EJ, Lambers H, Bragg J, Finnegan PM, Lovelock CE, Plaxton WC, Price CA, Scheible WR, Shane MW, White PJ, Raven JA (2012) Opportunities for improving phosphorus-use efficiency in crop plants. New Phytol 195:306–320

Véry AA, Sentenac H (2003) Molecular mechanisms and regulation of K^+ transport in higher plants. Annu Rev Plant Biol 54:575–603

Vinod KK, Heuer S (2012) Approaches towards nitrogen- and phosphorus-efficient rice. AoB Plant 2012:pls028. doi:10.1093/aobpla/pls028

Vinod KK, Krishnan SG, Babu NM, Nagarajan M, Singh AK (2013) Improving salt tolerance in rice: looking beyond the conventional. In: Ahmad P et al (eds) Salt stress in plants: signalling, omics and adaptations. Springer, New York. doi:10.1007/978-1-4614-6108-1_10

Wang C, Ying S, Huang H, Li K, Wu P, Shou H (2009a) Involvement of *OsSPX1* in phosphate homeostasis in rice. Plant J 57:895–904

Wang Y, Sun YJ, Chen DY, Yu SB (2009b) Analysis of quantitative trait loci in response to nitrogen and phosphorus deficiency in rice using chromosomal segment substitution lines. Acta Agron Sin 35:580–587

Wang WH, Köhler B, Cao FQ, Liu GW, Gong YY, Sheng S, Song QC, Cheng XY, Garnett T, Okamoto M, Qin R, Mueller-Roeber B, Tester M, Liu LH (2012) Rice *DUR3* mediates high-affinity urea transport and plays an effective role in improvement of urea acquisition and utilization when expressed in Arabidopsis. New Phytol 193:432–444

Wang S, Zhang S, Sun C, Xu Y, Chen Y, Yu C, Qian Q, Jiang DA, Qi Y (2014) Auxin response factor (*OsARF12*), a novel regulator for phosphate homeostasis in rice (*Oryza sativa*). New Phytol 201:91–103

Watanarojanaporn N, Boonkerd N, Tittabutr P, Longtonglang A, Young JPW, Teaumroong N (2013) Effect of rice cultivation systems on indigenous arbuscular mycorrhizal fungal community structure. Microbes Environ 28:316–324

Wei D, Cui K, Ye G, Pan J, Xiang J, Huang J, Nie L (2012) QTL mapping for nitrogen-use efficiency and nitrogen-deficiency tolerance traits in rice. Plant Soil 359:281–295

Weiss DJ, Mason TFD, Zhao FJ, Kirk GJD, Coles BJ, Horstwood MSA (2005) Isotopic discrimination of zinc in higher plants. New Phytol 165:703–710

Wissuwa M (2011) Utilization of abiotic stress tolerance genes. In: Trends of international rice research and Japanese scientific contribution – support to GRiSP and CARD. JIRCAS international symposium 2011, Tsukuba

Wissuwa M, Yano M, Ae N (1998) Mapping of QTLs for phosphorus-deficiency tolerance in rice (*Oryza sativa* L.). Theor Appl Genet 97:777–783

Wissuwa M, Ismail AM, Yanagihara S (2006) Effects of zinc deficiency on rice growth and genetic factors contributing to tolerance. Plant Physiol 142:731–741

Wissuwa M, Mazzola M, Picard C (2009) Novel approaches in plant breeding for rhizosphere-related traits. Plant Soil 321:409–430

Witcombe JR, Hollington PA, Howarth CJ, Reader S, Steele KA (2008) Breeding for abiotic stresses for sustainable agriculture. Philos Trans R Soc Lond B Biol Sci 363:703–716

Wu P, Ni JJ, Luo AC (1998) QTLs underlying rice tolerance to low-potassium stress in rice seedlings. Crop Sci 38:1458–1462

Wu P, Shou H, Xu G, Lian X (2013) Improvement of phosphorus efficiency in rice on the basis of understanding phosphate signaling and homeostasis. Curr Opin Plant Biol 16:205–212

Xu S (2013) Genetic mapping and genomic selection using recombination breakpoint data. Genetics 195:1103–1115

Yang XE, Liu JX, Wang WM, Li H, Luo AC, Ye ZQ, Yang YA (2003) Genotypic differences and associated plant traits in potassium internal use efficiency of lowland rice (*Oryza sativa* L.). Nutr Cycl Agroecosyst 67:273–282

Yang XE, Liu JX, Wang WM, Ye ZQ, Luo AC (2004) Potassium internal use efficiency relative to growth vigor, potassium distribution, and carbohydrate allocation in rice genotypes. J Plant Nutr 27:837–852

Yang H, Knapp J, Koirala P, Rajagopal D, Peer WA, Silbart L, Murphy A, Gaxiola R (2007) Enhanced phosphorus nutrition in monocots and dicots overexpressing a phosphorus-responsive type I H^+-pyrophosphatase. Plant Biotechnol J 5:735–745

Yi K, Wu Z, Zhou J, Du L, Guo L, Wu Y, Wu P (2005) *OsPTF1*, a novel transcription factor involved in tolerance to phosphate starvation in rice. Plant Physiol 138:2087–2096

Yoshida S (1981) Fundamentals of rice crop science. International Rice Research Institute, Manila, 269p

Yoshimoto N, Inoue E, Watanabe-Takahashi A, Saito K, Takahashi H (2007) Post-transcriptional regulation of high-affinity sulfate transporters in Arabidopsis by sulfur nutrition. Plant Physiol 145:378–388

Yu YJ, Liao HB, Chen WR, Tian SK, Yang XE (2012) Mechanism of Zn uptake, translocation in rice plant and Zn-enrichment in rice grain. Chin J Rice Sci 26:365–372

Zhang R, Liu G, Wu N, Gu M, Zeng H, Zhu Y, Xu G (2011a) Adaptation of plasma membrane H^+ ATPase and H^+ pump to P deficiency in rice roots. Plant Soil 349:3–11

Zhang Q, Wang C, Tian J, Li K, Shou H (2011b) Identification of rice purple acid phosphatases related to phosphate starvation signalling. Plant Biol 13:7–15

Zhang Z, Liao H, Lucas WJ (2014) Molecular mechanisms underlying phosphate sensing, signaling, and adaptation in plants. J Integr Plant Biol 56:192–220. doi:10.1111/jipb.12163

Zhao K, Tung CW, Eizenga GC, Wright MH, Ali ML, Price AH, Norton GJ, Islam MR, Reynolds A, Mezey J, McClung AM, Bustamante CD, McCouch SR (2011) Genome-wide association mapping reveals a rich genetic architecture of complex traits in *Oryza sativa*. Nat Commun 2:467

Zhou J, Jiao F, Wu Z, Li Y, Wang X, He X, Zhong W, Wu P (2008) OsPHR2 is involved in phosphate-starvation signaling and excessive phosphate accumulation in shoots of plants. Plant Physiol 146:1673–1686

Engineered Plants for Heavy Metals and Metalloids Tolerance

7

Parul Rana Tomar, Anirudha R. Dixit,
Pawan Kumar Jaiwal, and Om Parkash Dhankher

Abstract

Agricultural soils around the world are slightly to moderately polluted with heavy metals such as As, Cd, Hg, Cr, Cu, Pb, Ni, and Zn as a result of industrialization, widespread application of pesticides, fertilizers, and anthropogenic activities. Plants experience toxic effects of heavy metals in the form of oxidative stress, reduction in overall growth, and productivity. To minimize these toxic effects, plants have evolved an arsenal of mechanisms such as preventing uptake via roots or blocking transport to aboveground parts. If everything fails, the toxic metal inside the cell is dealt with using a range of detoxification and storage strategies including chelation with thiols and amino acids and subsequent sequestration into subcellular compartments. In this chapter, we have reviewed general strategies for heavy metal tolerance and detoxification by plants. Also plants engineered for heavy metal transport, oxidative stress tolerance pathways, and other mechanisms such as stress-associated protein have substantially advanced our understanding of heavy metal tolerance by plants. In future, as a result of ongoing climate change, frequent floods, storms, and more use of underground and recycled water from industrial and municipal wastes for crop irrigation can further increase the heavy metals in the agricultural soils. Therefore, to minimize the impact of heavy metals on global agricultural production, it will be of utmost importance to further our knowledge of heavy metal tolerance and detoxification by plants.

P.R. Tomar • O.P. Dhankher (✉)
Stockbridge School of Agriculture, University of Massachusetts Amherst, Amherst, MA 01002, USA
e-mail: parkash@umass.edu

A.R. Dixit
John F. Kennedy Space Center, Merritt Island, FL 32899, USA

P.K. Jaiwal
Centre for Biotechnology, Maharishi Dayanand University, Rohtak 124001, Haryana, India

P.K. Jaiwal et al. (eds.), *Genetic Manipulation in Plants for Mitigation of Climate Change*,
DOI 10.1007/978-81-322-2662-8_7

Keywords
Heavy metals • Metalloids • Metal chelation • GSH • Phytochelatins • Metals transporters • Detaoxifications • Transgenic plants

7.1 Introduction

Heavy metals and metalloids are of significant environmental concerns. Exposure to heavy metals and metalloids not only adversely affects human and environmental health but also inhibits growth and productivity of crop plants and thus poses serious threats to global agricultural production and food security (Bray et al. 2000). Toxicity to plants and their growth inhibition can occur due to elevated concentrations of both essential and nonessential heavy metals present in the soil. Heavy metals cause cellular damage by either inducing oxidative stress such as lipid peroxidation through production of reactive oxygen species (ROS) or free radicals or by binding to the sulphydryl groups of various enzymes and proteins such as the transcription factors (Assche and Clijsters 1990; Dietz et al. 1999). Because of the global climate change, the burden of environmental stresses including heavy metals on crop plants is likely to increase in future. With decrease in groundwater use for irrigation, metals that naturally occur in soil and water are increasing in concentration and thus enter the food chain affecting human and environmental health. Water deficiency is also induced by other abiotic stresses such as salinity, heat, and cold; thus, all these stresses go hand in hand, making it absolutely essential to understand the responses of plant to water deficiency (Bohnert et al. 1995). Apart from drought and salinity, heavy metals and metalloids such as mercury (Hg), cadmium (Cd), lead (Pb), chromium (Cr), nickel (Ni), copper (Cu), arsenic (As), and selenium (Se) constitute of inorganic pollutants and are present as positive or negatively charged ions in the soil (Paulose et al. 2008).

7.1.1 Heavy Metal Toxicity in Plants and Effects on Crop Productivity

7.1.1.1 Arsenic

Arsenic (As) contamination is widespread in the environment. Natural processes such as volcanic eruptions and hot water springs along with human intervention with activities such as mining, smelting, use of pesticides, herbicides etc. contribute to arsenic pollution in the environment. It is a carcinogen and has been shown to cause major health problems around the world including but not limiting to cancer of the liver, kidney, and lungs (Kaiser 2001). Arsenic-contaminated soils, sediments, and water supplies are major sources of food chain contamination. High levels of As have been reported in the underground drinking water and food crops such as rice (*Oryza sativa*) grown in many parts of Southeast Asia especially India and Bangladesh where this staple food is widely irrigated using arsenic-contaminated groundwater subjecting millions of people to arsenic poisoning risk (Clark et al. 2000). There have been several reports of unacceptable levels of arsenic being present in edible crops, which were grown on contaminated lands (Larsen et al. 1992; Das et al. 2004; Williams et al. 2005). Arsenic is present in soil and water in organic forms – monomethylarsenate (MMA) and dimethylarsenate (DMA) as well as inorganic forms – arsenate (AsO_4^{-3}, referred to as AsV) and arsenite (AsO_3^{-3}, referred to as AsIII), but it is the latter that is a more toxic form and needs remediation (Bentley and Chasteen 2002; Chen et al. 2005). The metalloid arsenic (As) and other heavy metals are phytotoxic and the elevated concentrations of As in soil causes a significant loss of crop yield (Xiong et al. 1987;

Marin et al. 1993; Zhu et al. 2008). In addition, arsenate (AsV), being a phosphate analogue, competes with phosphate uptake, causes the inhibition of phosphate and other nutrient uptake, and thus further decreases crop production (Meharg and Macnair 1992; Abedin et al. 2002; Dhankher et al. 2006; LeBlanc et al. 2013). It has been reported that in 2003, 1/5th of the total cultivable land of China had already been accounted toward heavy metal-contaminated land. These widespread heavy metal contamination has led to a loss of about 1000 million tons of grain production directly impacting the economy (Gu et al. 2003). Arsenic gets accumulated in the soil due to the use of sewage sludge and recycled water on the agricultural land due to shortage of surface water. Therefore, it is essential to develop strategies to resist crops from uptaking arsenic or if absorbed then need to be converted to a nontoxic form.

7.1.1.2 Mercury

Mercury is among the most hazardous of the heavy metals (Keating et al. 1997), primarily because its charged species have great affinity for the thiol group on cysteine residues of proteins and other important biological molecules (Liu et al. 1992). Early studies demonstrated that mercury species inactivate metabolic enzymes and structural proteins (Boyer 1954; Falchuk et al. 1977). The strong interaction of mercury species with cellular ligands may also account for its tendency to accumulate in organisms. Organomercurial species are more toxic in some eukaryotes and are more likely to biomagnify across trophic levels than ionic mercury [Hg (II)] (Bizily et al. 1999). Fish contaminated with monomethylmercury (CH_3-Hg^+) at Minamata bay in Japan was a widespread ecological disaster (Minamata Disease Research Group 1968). Consumption of the fish contaminated with CH_3-Hg^+ leads to severe neurodegeneration in birds, cats, and humans. Methylmercury has also been found in lakes and estuaries into which only inorganic forms of mercury have been released (Balogh et al. 2006; Hammerschmidt et al. 2006). For plants, Hg is highly toxic, affecting major cellular metabolic pathways, and thus causes severe reduction in plant growth and crop yields (Boening 2000; Patra and Sharma 2000; Patra et al. 2004).

7.1.1.3 Cadmium

The regulatory limit of Cd in agricultural soil is 100 mg/kg (Salt et al. 1995). Cadmium pollution has increased in the environment due to mining, industrial usage, and anthropogenic activities. Cd is abundantly used in surface coatings, pigment formulation, manufacture of batteries, stabilization of polyvinylchloride (plastics), manufacture of automobiles, and aerospace and military applications. Cd has application where high stability and resistance to heat, cold, and light are required. Cd released to the environment tends to concentrate in soils and sediments, where it is potentially available to rooted plants (Prasad 1995). In bean plants (*Phaseolus vulgaris*), leaf cell expansion growth and relative water content of primary leaves decreased by about 10 % compared to control after 48-h exposure to 3 μM Cd (Barceló and Poschenreider 1990). Cd is an effective inhibitor of photosynthesis (Krupa 1988; Greger and Ogren 1991; Krupa et al. 1993). The linear relationship between net photosynthesis and inhibition of transpiration observed in clover, lucerne, and soybean suggests the closure of stomata by Cd (Huang et al. 1987). In experiments with bean plants (*P. vulgaris*), Cd inhibited net photosynthesis by increasing stomatal and mesophyll resistance to CO_2 uptake (Krupa et al. 1993). Therefore, exposure to Cd causes reduction in photosynthesis, water, and nutrient uptake.

7.1.1.4 Zinc

Zinc is an essential nutrient for plants as well as human beings. It plays important roles as component of enzymes for protein synthesis and energy production and maintains structural integrity of biomolecules (Hänsch and Mendel 2009). Zn^{2+} is an integral part of a large number of zinc finger containing proteins, transcription factors, oxidoreductases, and hydrolytic enzymes such as metalloproteases (Krämer and Clemens 2006). Concentrations of Zn in contaminated soils frequently exceed 150–300 mg/kg of soil (de Vries et al. 2007). Similar to Cd, exposure to excess levels of Zn results in chlorosis, reduced growth of both roots and shoots, and early senescence in plants (Choi et al. 1996; Ebbs and Kochian 1997; Fontes and Cox 1998). Excess Zn can also cause Mn, Cu, and phosphate deficiencies in plants (Lee et al. 1996).

7.1.1.5 Copper

Copper (Cu) is also an essential micronutrient for plant growth and plays important roles in processes such as CO_2 assimilation, ATP synthesis, mitochondrial respiration, oxidative stress protection, and cell wall synthesis. Under physiological conditions, copper exists in the two-oxidation states Cu^{1+} and Cu^{2+} and can interchange between these forms (monovalent copper is unstable). This allows copper to function as a reducing or oxidizing agent in biochemical reactions. But at the same time, this property also makes copper potentially toxic as copper ions can catalyze the production of free radicals, in particular through Fenton chemistry, thus leading to the damage of proteins, DNA, and other biomolecules (Hänsch and Mendel 2009). Exposure of plants to excess Cu generates oxidative stress and ROS (Stadtman and Oliver 1991). Oxidative stress causes disturbance of metabolic pathways and damage to macromolecules and affects crop yields (Hegedüs et al. 2001).

7.1.1.6 Lead

Lead naturally occurs in the earth's crust in small amounts as a bluish-gray metalloid (Gupta et al. 2010). Contamination of Pb occurs from its use in pesticides and fertilizers, combustion of Pb containing fossil fuels, and Pb additive containing municipal sewage dumping, from its use in industrial processes such as mining and smelting (Gupta et al. 2010). Many commercially available products such as batteries, medical equipment (i.e., x-ray shields, fetal monitors), paints, ceramic glazes, television glass, ammunition, etc. also contain small concentrations of Pb (Gupta et al. 2010). Pb has the tendency to accumulate in the body organs (i.e., brain), which may lead to poisoning or even death and has also shown to affect the gastrointestinal tract, kidneys, and central nervous system. Pb exposure has detrimental effect on small children and they can have impaired development, lower IQ, shortened attention span, hyperactivity, and mental deterioration.

Basic forms of Pb released into the soil and water including ground and surface are ionic lead (Pb^{2+}), Pb oxides and hydroxides, and Pb-metal oxyanion complexes with ionic Pb and Pb hydroxyl complexes being the most stable forms (Evanko and Dzombak 1997). Pb is very difficult to remove once it is introduced in the soil matrix as it is a very sticky metal because of its ability to form a precipitate within the soil matrix along with anions such as phosphate ions which decreases its solubility and, in many cases, is not readily bioavailable (Gupta et al. 2010). The top few inches of soil are where Pb is mainly found, bound to the organic matter through adsorption, ion exchange, precipitation, and/or complexation (Hart et al. 1999; Gill and Tuteja 2011). Many plants transport a very small concentration of Pb^{2+} in the aboveground plant tissues and retain maximum concentration in their roots via sorption and precipitation. Pb exposure causes stunted root growth and accelerated cell death in rice (Huang and Huang 2008). As with any other heavy metals, one of the major consequences of Pb toxicity is the enhanced production of reactive oxygen species (ROS) including superoxide radicals, hydroxyl radicals, and hydrogen peroxide, H_2O_2 (Shu et al. 2012).

7.1.1.7 Chromium

Chromium (Cr) is a heavy metal that causes serious environmental contamination in soil, sediments, and groundwater (Shanker et al. 2005). The tanning industry is one of the major consumers of water and most of it is discharged as wastewater, which contains high amount of Cr (1.07–7.80 mg/l). Inorganic Cr exists in two forms, trivalent Cr (III) and hexavalent Cr (VI), the latter being more toxic, powerful epithelial irritant and a proven human carcinogen established by the International Agency for Research on Cancer (IARC), the Environmental Protection Agency (EPA), and the World Health Organization (WHO). Toxicity of Cr has been studied in many plants. Excess of Cr causes inhibition of plant growth, chlorosis in young leaves, nutrient imbalance, wilting of tops, and root injury (Chatterjee

2000; Dixit et al. 2002; Sharma et al. 2003; Scoccianti et al. 2006).

7.1.1.8 Selenium

Selenium (Se) is a metalloid that naturally occurs in the shale rocks and is also produced by some anthropogenic sources such as oil refineries, power plants, etc. and thus is present in excess, potentially in toxic levels in the environment (Wilber 1983; Fordyce 2005). At low levels, Se serves as an essential nutrient with normal dietary intake in the range of 50–400 μg Se/day and has several health benefits like prevention of heart diseases, muscle disorders, and cancer and is also involved in viral suppression and functioning of the immune system. Higher levels are highly toxic and cause stomach cancer, defects in the development, and failure in the reproductive system (Terry et al. 2000; Ellis and Salt 2003). Excessive levels of Se can also accumulate in food chain, which can cause serious health defects in human and wildlife population. In soil and water, Se is present in several forms such as selenide (Se^{2-}), elemental selenium (Se^0), selenite (Se^{4+}), and selenate (Se^{6+}) and many organic forms like dimethyl selenide (DMSe) and dimethylselenenylsulfide (DMSeS), which have different bioavailabilities. Selenate (Se^{6+}) form of Se is formed in alkaline, well-aerated soils, making it highly soluble, and does not form stable adsorption complexes with other components present in the soil. As a consequence, they are more bioavailable for plant uptake, whereas acidic soils favor the formation of elemental selenium (Se^0) and selenides (Se^{2-}) that are quite insoluble and therefore are less bioavailable for plant uptake (Zayed et al. 1998; Terry et al. 2000).

7.2 Strategies for Heavy Metal Detoxification and Enhanced Tolerance in Plants

Some plants have the natural ability to detoxify, accumulate, and tolerate high levels of heavy metal stresses using processes at the cellular levels, and other plants could be engineered to carry out such processes. The major challenge to engineer food crops for heavy metal tolerance is that the genes encoding metal-binding peptides such as metallothioneins (MTs), glutathione (GSH), and phytochelatins (PCs) caused increased metal uptake along with metal tolerance. The phytotoxic effects suffered by crops grown on soil with heavy metal and metalloid residues could be overcome by developing crops resistant to these metals. However, progress toward developing such genetics-based strategies has been hindered by the lack of understanding of the basic molecular and biochemical mechanisms of heavy metal uptake and detoxification in plants. Despite these limitations, a significant progress has been made to engineer plants with genes, either individually or in combination, for increasing tolerance and detoxification of heavy metals such as As, Hg, Cd, Pb, Se, Ni, etc. Further plants have been engineered for remediation of several metals and metalloids (Pilon-Smits et al. 1999; Rugh 2001; Dhankher et al. 2002, 2006; van Huysen et al. 2003; Li et al. 2005; Dixit and Dhankher 2011; Paulose et al. 2013).

Plants utilize several strategies/mechanisms to achieve tolerance toward heavy metals and metalloids. These include thiol-mediated chelation followed by sequestration (Schmoger 2000; Song et al. 2010), uptake or exclusion through transporters, and complexation with phytochelatins followed by vacuolar storage (Dhankher et al. 2002) and chelation by metallothioneins and anthocyanins or by binding with carboxylic acids such as citrate, malate, or amino acids such as histidine and nicotianamine. A list of genes used for engineering plants for enhanced tolerance to heavy metals and metalloids is presented in Table 7.1.

7.2.1 Chelation with Metal-Binding Peptides

In non-hypertolerant plants, binding of metals by strong ligands is the main detoxification strategy. The best-known types of ligands for this purpose are thiols, including GSH and its precursor gamma-glutamylcysteine (γEC dipeptide), phytochelatins (PCs), and metallothioneins (MTs)

Table 7.1 Heavy metal and metalloid tolerant genes used for engineering plants for enhanced tolerance to various toxic metals and metalloids

Plant species	Gene source	Gene	Metal/Metalloid tolerance	Phenotypes	Reference
Arabidopsis thaliana	*Escherichia coli*	*merB*	Hg	Tolerance to Hg	Bizily et al. (1999)
A. thaliana	*Nicotiana tabacum*	*CBP4*	Pb, Ni	Higher Pb accumulation and reduced uptake of Ni in shoots	Bizily et al. (1999)
A. thaliana	*Escherichia coli*	*ArsC, γ-ECS*	As	Enhanced tolerance and accumulation of As	Dhankher et al. (2002)
A. thaliana	*Arabidopsis thaliana*	*ACR2*	As	Enhanced accumulation of As in shoots	Dhankher et al. (2006)
Nicotiana glauca	*Triticum aestivum*	*PCS1*	Pb, Cd	Enhanced tolerance to Pb and Cd	Gisbert et al. (2003)
Nicotiana tabacum	*Arabidopsis thaliana*	*MT2b*	As	Enhanced As root-to-shoot transport	Grispen et al. (2009)
A. thaliana	*Allium sativum, Saccharomyces cerevisiae*	*PCS1, GSH1*	Cd, As	Enhanced tolerance and accumulation of As and Cd	Guo et al. (2008)
A. thaliana	*Allium sativum, Saccharomyces cerevisiae*	*PCS1, YCF1*	As, Cd	Enhanced tolerance to and accumulation of As and Cd	Guo et al. (2012)
Brassica oleracea,	*Saccharomyces cerevisiae*	*CUP1*	Cd	Enhanced tolerance to Cd	Hasegawa et al. (1997)
A. thaliana	*Arabidopsis thaliana*	*ZIF*	Zn	Enhanced tolerance and accumulation of Zn	Haydon and Cobbett (2007a, b)
A. thaliana	*Schizosaccharomyces pombe*	*HMT1*	Cd	Enhanced tolerance to and accumulation of Cd, Cu, As and Zn	Huang et al. (2012)
Pteris vittata	*Pteris vittata*	*ACR3*	As	Tolerance and hyperaccumulation of As	Indriolo et al. (2010)
A. thaliana	*Arabidopsis thaliana*	*ATM3*	Cd	Enhanced resistance to Cd	Kim et al. (2006)
Brassica juncea	*Astragalus bisculcatus*	*SMT*	Se	Enhanced tolerance to Se	LeDuc et al. (2004)
Brassica juncea	*Arabidopsis thaliana, Astragalus bisculcatus*	*SMT, APS1*	Se	Enhanced accumulation of Se	LeDuc et al. (2006)
Vicia faba	*Arabidopsis thaliana*	*MT2a, MT3*	Cd	Tolerance to Cd by reducing ROS production	Lee et al. (2004)
A. thaliana	*Arabidopsis thaliana*	*HMA3*	Cd, Co, Pb, Zn	Increased tolerance to and accumulation of Cd, Co, Pb and Zn	Morel et al. (2009)
A. thaliana	*Arabidopsis thaliana*	*ABCC1*	Cd	Increased resistance to Cd	Park et al. (2012)
Brassica juncea	*Arabidopsis thaliana*	*APS1*	Se	Increased uptake, reduction and tolerance of Se	Pilon-Smits et al. (1999)
A. thaliana	*Escherichia coli*	*merA*	Hg	Resistance to toxic levels of Hg	Rugh et al. (1996)
Nicotiana tabacum, A. thaliana	*Escherichia coli*	*merA, merB*	Hg	Increased tolerance and accumulation of Hg and phenyl-Hg	Ruiz et al. (2003)
Oryza sativa	*Triticum aestivum*	*HsfA4a*	Cd	Enhanced tolerance to Cd	Shim et al. (2009)
A. thaliana	*Saccharomyces cerevisiae*	*YCF1*	Cd, Pb	Increased tolerance to and accumulation of Cd and Pb	Song et al. (2003)
A. thaliana	*Arabidopsis thaliana*	*ABCC1ABCC2*	As	Increased tolerance to As	Song et al. (2010)
Nicotiana tabacum	*Saccharomyces cerevisiae*	*CUP1*	Cu	Increased accumulation of Cu	Thomas et al. (2003)
A. thaliana	*Arabidopsis thaliana*	*ZAT*	Zn	Increased tolerance to and accumulation of Zn	van der Zaal et al. (1999)
Brassica juncea	*Escherichia coli*	*GSH1*	Cd	Enhanced tolerance and accumulation of Cd	Zhu et al. (1999)

Note: This table does not represent all the genes in the published literature but includes only a subset of genes used for engineering plants for heavy metal tolerance

(Leitenmaier and Kupper 2013). Most of the divalent cations (Cd, Hg, Pb, Cu, Zn) and oxyanions (AsV and AsIII) are highly reactive and has a strong affinity toward thiol groups such as those in γEC, GSH, PCs, and MTs (Schmoger 2000; Dhankher et al. 2002; Zimeri et al. 2005).

7.2.1.1 Chelation with Glutathione

Glutathione (GSH) is the key redox molecule in plant cells. It plays important roles in protecting cells from oxidative stress caused by exposure to environmental stresses including toxic metals, ozone exposure, and biotic stresses such as insect and pathogen infestation. Plants detoxify toxic metals and metalloids through a GSH-dependent pathway. GSH homeostasis in plants is maintained by the γ-glutamyl cycle, which involves GSH synthesis and degradation and the recycling of component amino acids (Paulose et al. 2013). GSH synthesis is catalyzed by two enzymes – γ-glutamylcysteine synthetase (γ-ECS) and glutathione synthetase (GS). Overexpression of these enzymes enhanced the tolerance to toxic metals and metalloids such as Hg, Cd, Pb, Cr, and As (Dhankher et al. 2002; Li et al. 2005). Once toxic metal is bound to thiol peptides, it is relatively less toxic to plants, and the metal-thiol complexes are then sequestered into vacuoles by glutathione-conjugating pumps, GCPs (Wang et al. 2002; Dhankher et al. 2002; Indriolo et al. 2010; Song et al. 2010). Therefore, enhancing the levels of GSH and its derivatives favors the trapping of toxic metals in thiol complexes resulting in enhanced tolerance and accumulation. This strategy is employed in most of the metal hyperaccumulators (Freeman et al. 2004, 2005).

The role of GSH and PCs in Cd tolerance has been studied in great details. These peptides chelate Cd cations with varying affinities, leading to vacuolar sequestration for metal-peptide complexes (Kneer and Zenk 1992; Ortiz et al. 1992; Howden 1995; Li et al. 1996). Their involvement in detoxifying heavy metals was discovered almost 30 years ago (Grill et al. 1985, 1989). Cadmium stress affects sulfur uptake and assimilation pathway in plants (Zhu et al. 1999), which directly affects the synthesis of amino acids such as cysteine that requires sulfur for its synthesis. Cysteine is a precursor in synthesis of GSH. Zhu et al. (1999) demonstrated the role of γ-ECS in increasing cadmium tolerance and accumulation by overexpressing γ-ECS encoding *E. coli* gene *gsh1* in *Brassica juncea*. Along with transgenics growing better in presence of toxic concentrations of Cd, increased concentrations of PCs, γ-GluCys, and GSH were also seen in the transgenic seedlings, leading to believe that increased production of these compounds lead to increased Cd tolerance and accumulation (Zhu et al. 1999). Metalloid toxicity in non-hypertolerant plants has been observed in *cad1* mutant plants that are either biochemically (Schat 2002) or genetically mutant for PC synthase leading to Cd hypersensitivity (Howden 1995). Arabidopsis engineered to express bacterial GSH1 showed strong tolerance to As and weak tolerance to Hg, whereas it failed to provide any tolerance to Cd stress (Li et al. 2005). Transgenic tobacco (*N. tabacum*) plants expressing serine acetyltransferase (SAT, involved in the production of a cysteine precursor O-acetylserine), GSH1, and PCS, either separately or in combination, have shown increased Cd concentration in roots (Wawrzynski et al. 2006).

Arsenic is present in soil and water in organic forms such as MMA and DMA as well as inorganic forms such as AsV and AsIII, but it is the latter that is a more toxic form and needs remediation (Bentley and Chasteen 2002; Chen et al. 2005). In 2002, Dhankher and coworkers engineered *Arabidopsis* to detoxify As and created a genetics-based strategy for accumulation of As in the aboveground tissue for phytoremediation by co-expressing arsenate reductase, ArsC, and γ-glutamylcysteine synthetase, γ-ECS. The bacterial ArsC reduces AsV to AsIII (Rosen 1999), and γ-ECS is the first step for the synthesis of GSH and PCs and enhances the thiol peptide levels in plants. Transgenic lines expressing light-regulated ArsC were hypersensitive to AsV, whereas the γ-ECS transgenic lines were moderately resistant as compared to wild-type plants. Double transgenic plants made by genetic crossing of As-hypersensitive ArsC plants and moderately resistant γ-ECS overexpressing plants were super-resistant to AsV as compared to the plants expressing γ-ECS alone. The double transgenic ArsC + γ-ECS plants accumulated threefold more As in the aboveground tissues (Dhankher et al. 2002).

Thlaspi species hyperaccumulate Ni up to 3 % of their shoot dry weight (Yadav 2010). The concentrations of GSH, Cys, and O-acetylserine (OAS) in shoot tissue are strongly correlated with the ability to hyperaccumulate Ni in various *Thlaspi* hyperaccumulators collected from serpentine soils. Examples of such hyperaccumulators reported are *T. goesingense*, *T. oxyceras*, and *T. rosulare*, and non-accumulator relatives are *T. perfoliatum*, *T. arvense*, and *A. thaliana* (Kramer et al. 1997; Wenzel and Jockwer 1999; Guerinot and Salt 2001; Peer et al. 2003; Freeman et al. 2004). High concentrations of OAS, Cys, and GSH in Austrian Ni hyperaccumulator *T. goesingense* coincide with constitutively high activity of both SAT and glutathione reductase (GR) enzymes. *Arabidopsis* overproducing SAT from *T. goesingense* has been found to cause accumulation of OAS, Cys, and GSH, mimicking the biochemical changes observed in the Ni hyperaccumulators. In these transgenic *Arabidopsis*, GSH concentration was strongly correlated with increased resistance to Ni-induced growth inhibition and oxidative stress. This observation concluded that high levels of GSH conferred tolerance to Ni-induced oxidative stress in *Thlaspi* Ni hyperaccumulators (Freeman et al. 2004). *Sedum alfredii*, a perennial herb which is also a known Zn/Cd hyperaccumulator, was used as an indicator species by Gupta and coworkers to demonstrate that enzymatic and nonenzymatic antioxidants such as cysteine, nonprotein thiols (NPSH), glutathione (GSH), and PCs have important roles in detoxification of toxicity induced by Pb (Gupta et al. 2010). The abovementioned antioxidants were found in higher concentration in the Pb accumulating *S. alfredii* species when compared with its non-accumulating counterpart suggesting that the capacity of the ecotypes to accumulate different levels of Pb depends upon their ability to detoxify Pb through production of these antioxidants.

In case of mercury, hyperaccumulation can be achieved by overexpression of each of the three enzymes – *γ-ECS*, GS, and PS – involved in the GSH and PC biosynthesis pathway that will increase thiol sinks for Hg (II). Plants transformed with *merB* and $\gamma - ECS$ will trap Hg (II) in the form of thiol-Hg complexes and result in enhanced organic and ionic Hg resistance. Constitutive expression of PCS expression has previously been achieved, which further increases PC sink for Hg (II). However, the second approach of phytoextraction has not been fully achieved yet and more research needs to be done in understanding Hg phytoextraction as such Hg accumulation in plant tissues can be toxic to wildlife. Also, plant tolerance to Hg is generally low, and therefore phytoremediation can be limited by plant tolerance.

7.2.1.2 Chelation with Phytochelatins

Phytochelatins form a family of structures with increasing repetitions of the γ-GluCys dipeptide followed by a terminal Gly, $(\gamma\text{-GluCys})_n$-Gly, where n is generally in the range of 2–8. PCs are structurally related to glutathione (GSH, γ-GluCysGly), and numerous physiological, biochemical, and genetics studies have confirmed that GSH is the substrate for PC biosynthesis (Cobbett 2000; Cobbett and Goldsbrough 2002). PCs are synthesized from GSH by the enzyme phytochelatin synthase, PCS (Grill et al. 1989). *Arabidopsis cad1* mutant was the first PC synthase mutated or inactivated gene. The *cad1* mutants were PC deficient but had wild-type levels of GSH, suggesting a defect in the PCS gene (Howden 1995). The *Arabidopsis cad1/AtPCS1* gene (Howden 1995; Ha 1999; Vatamaniuk et al. 1999) and a similar gene in wheat, *TaPCS1*, have been shown to confer resistance to Cd and Pb when expressed in the yeasts *S. cerevisiae* and *N. glauca*, respectively (Clemens et al. 1999; Cobbett and Meagher 2002; Gisbert et al. 2003). *Arabidopsis cad1-3* mutant was found to be more sensitive to As(V) and Cd and slightly sensitive to Cu, Ag, and Hg, while there was no difference observed when exposed to Zn, selenite, and Ni ions (Ha 1999). Using *B. juncea*, it has been shown that Cd accumulation is accompanied by a rapid induction of PC biosynthesis, and the levels of PC present were theoretically sufficient to chelate all the Cd that is taken up by plants (Haag-Kerwer et al. 1999). A possible role for PCs in Cu tolerance had also been suggested from studies on copper-tolerant *Mimulus guttatus* (Salt et al. 1989). The constitutive overexpression of TaPCS1 in shrub tobacco (*N. glaucum*) substantially increased its tolerance to Pb^{2+} and Cd^{2+} and greatly improved accumulation of Cu^{2+}, Zn^{2+}, Pb^{2+}, and Cd^{2+} in shoots (Martínez et al. 2006).

In the published literature, it has also been observed that excessive levels of PC in transgenic plants lead to increase in heavy metal accumulation without enhancing tolerance and causing hypersensitivity to heavy metals (Pomponi et al. 2006). A similar effect was observed when AtPCS was overexpressed in *Arabidopsis* leading to Cd hypersensitivity (Lee et al. 2003). As far as the role of PCs in copper tolerance/sensitivity, cadmium-tolerant transgenic plants that overexpressed AtPCS1 were not tolerant of copper stress, thereby supporting the hypothesis that PC is not primarily involved in Cu tolerance mechanism. Investigation into Cu tolerance in *cad2-1*, a Cd-sensitive and glutathione (GSH)-deficient *Arabidopsis* mutant, leads to *cad2-1* mutant being more resistant to Cu stress than wild-type plants. This was likely due to the high level of cysteine present in *cad2-1* mutants. However, when the growth medium was supplemented with cysteine, the wild-type plants also exhibited Cu tolerance. Moreover, *S. cerevisiae* that expressed *AtPCS1* showed tolerance to Cd but hypersensitivity to Cu. All these results indicate that PCs are not a major factor in determining Cu tolerance in plants (Lee and Kang 2005). Historically, there has been some disagreement about the role of PCs in metal tolerance and not all studies have supported this role (Steffens 1990; Hall 2002). Although evidence for the role for PCs in Cd detoxification is strong, these peptides may play other important roles in the cell, including metal homeostasis and sulfur metabolism or as antioxidants (Rauser 1995; Dietz et al. 1999; Cobbett 2000; Hall 2002). The participation of PCs in heavy metal detoxification may be a consequence of these other functions (Steffens 1990).

7.2.1.3 Chelation with Metallothioneins

Metallothioneins (MTs) are low-molecular-weight and highly cysteine-rich metal-binding peptides, which play important roles in toxic metal detoxification and metal ion homeostasis. MTs in plants differ considerably from those found in mammals and fungi as they contain mercaptide groups they are able to bind metal ions. Based on arrangement of cysteine residues, metallothioneins from plants can be classified into four subfamilies or classes – MT1 to MT4. Class 1 MTs are characterized by the presence of Cys–X–Cys motifs, whereas in Class 2 MTs both Cys–Cys and Cys–XX–Cys pairs are located toward the N-terminal domain (Robinson et al. 1993). They are able to detoxify metals, achieve homeostasis, and allow metal transport due to their ability to reversibly bind both toxic and essential metal ions.

Human MT2 and mouse MT2 were among the first MT genes expressed in transgenic plants (Lefebvre and Laliberte 1987; Misra and Gedamu 1989). In case of wheat and rice, MTs are not only induced by metal ions, such as Cu and Cd, but also by abiotic stresses such as extreme temperature and nutrient deficiency (Cobbett and Goldsbrough 2002). Plant MTs sequester excess of metals by coordinating metal ions with the multiple cysteine thiol groups (Robinson et al. 1993) and have particular affinity for Zn^{2+}, Cu^{+}, and Cu^{2+} as shown by the expression of the pea gene PsMTa in *E. coli* (Tommey et al. 1991). Overexpression of yeast metallothionein gene (*CUP1*) created Cd-tolerant transgenic cauliflower, which grew well in the presence of 400 μM Cd and accumulated more Cd, especially in the upper leaves (Hasegawa et al. 1997). Also CUP1 overexpression in *N. tabacum* resulted in increased copper accumulation (Thomas et al. 2003). Gene silencing demonstrated that the MT1 class isovariants are required to protect *Arabidopsis* plants from toxic effects of the heavy metal Cd(II) and possibly As. The study (Zimeri et al. 2005) used RNA interference to knock down expression of the *Arabidopsis* class I MT genes: MT1a, MT1b, and MT1c. The MT1 knockdown lines showed greatest sensitivity to Cd(II) and accumulated less Cd, Zn, and As than wild-type plants (Zimeri et al. 2005). When expressed in *N. tabacum*, AtMT2b enhanced root-to-shoot transport of arsenic resulting into decreased As tolerance but increased accumulation in shoots (Grispen et al. 2009). The role of *AtMT2a* and *AtMT3* in Cd detoxification and resistance was studied recently in which the two genes fused to GFP and RFP protected the guard cell chloroplasts of *Vicia faba* from degradation when exposed to high concentrations of cadmium (Lee et al. 2004). It was also identified that the two genes might not be playing a role in cadmium detoxification through vacuole sequestration but are most likely involved in reducing the levels of ROS generated upon cadmium exposure (Lee et al. 2004). Overexpression of CcMT2 from legume *Cajanus cajan* in *Arabidopsis* induced both Cd and Cu tolerance and allowed both metals to accumulate without affecting the expression of endogenous transporters (Sekhar et al. 2011). A heat shock transcription factor A4a (*HsfA4a*) from *Triticum aestivum* when overexpressed in rice enhanced Cd tolerance in rice plants and decreased Cd accumulation in rice plants with knocked-down expression of OsHsfA4a. Under cadmium stress, upregulation of *HsfA4a* along with increased expression of MT genes in wheat and rice such as Os*MT-I-1a* was seen in roots of these plants suggesting that MT has a role to play in cadmium tolerance through *HsfA4a* (Shim et al. 2009). One of the MT genes from sugarcane, *ScMT2-1-3*, not only showed increased tolerance and detoxification to Cd^{2+} and Cu^{2+} when expressed in transgenic *E. coli* but was also upregulated under Cu^{2+} stress and downregulated under Cd^{2+} stress (Guo et al. 2013). A decrease in the levels of peroxidase (POD) activity and malondialdehyde (MDA) accumulation was observed in tobacco plants overexpressing *TaMT3* gene under 35S promoter, leading to believe that *TaMT3* has a role to play in providing increased tolerance to cadmium stress (Zhou et al. 2014). Upon Pb exposure, increased expression of two genes *HiHMA4* and *HiMT2a* coding for a P1B-type ATPase and an MT was observed in roots and leaves of Pb hyperaccumulator plant species *Hirschfeldia incana*, a member of *Brassicaceae* (Auguy et al. 2013). When characterized further, these were seen to play a role in greater lead tolerance and were involved in greater lead accumulation. *OsMT1a* expression was induced specifically by Zn^{2+} treatment. Both transgenic plants and yeasts harboring *OsMT1a* accumulated more Zn^{2+} than wild-type controls, suggesting *OsMT1a* is most likely to be involved in zinc homeostasis. Transgenic rice plants overexpressing *OsMT1a* demonstrated enhanced tolerance to drought. The examination of antioxidant enzyme activities demonstrated that catalase (CAT), peroxidase (POD), and ascorbate peroxidase (APX) were significantly elevated in transgenic plants. Furthermore, the transcripts of several Zn^{2+}-induced CCCH zinc finger transcription factors accumulated in *OsMT1a* transgenic plants, suggesting that OsMT1a not only participates directly in ROS scavenging pathway but also regulates expression of the zinc finger transcription factors via the alteration of Zn^{2+} homeostasis, which leads to improved plant stress tolerance (Yang et al. 2009). Therefore, according to Grennan AK (2011), "although many recent studies have started to reveal the roles of MTs in plants, there is still much more information needed. The large diversity in the metal-binding regions of plant MTs suggests that they have the ability to bind a greater range of metals than their animal counterparts and, consequently, a greater range of function."

7.2.1.4 Other Metal-Binding Peptides

Carboxylic acids such as citrate, malate, oxalate, and amino acids such as histidine and nicotianamine are potential non-thiol ligands for heavy metal detoxification in plants (Rauser 1999). Histidine was first shown to bind a major proportion of Ni in the Ni hyperaccumulator, *Alyssum lesbiacum* (Kramer et al. 1996), and later it was shown to bind Zn in hyperaccumulators as well (Salt et al. 1999; Küpper et al. 2004). Free histidine is an important metal-binding ligand, and in its monodentate form it may not bind to metals very strongly, but multi-histidine residue can make a strong interaction with metals. Nicotianamine (NA), as a polydentate ligand with three carboxyl groups and three nitrogen, makes a strong ligand for Fe^{2+}, Zn^{2+}, Ni^{2+}, and Cu^{2+} (Beneš et al. 1983). They are structurally very similar to the iron-phytosiderophore mugineic acid. NA provides six alternating carboxylate and amine functions; their relative positions favor the formation of six-coordinate metal complexes (Callahan et al. 2006; Leitenmaier and Kupper 2013). They bind to both iron (II) and iron (III) (Pich et al. 1994; Von Wiren 1999). NA has been shown to be important for Cu metabolism (Liao et al. 2000); at the same time they were also found to bind a substantial proportion of total Cu at toxic concentrations in *N. caerulescens* (Mijovilovich et al. 2009). NA has also been shown to bind Zn (Trampczynska et al. 2010), in Cd/Zn hyperaccumulators *Arabidopsis halleri* and *N. caerulescens* (Becher et al. 2004; Weber et al. 2004). An RNAi-mediated knockdown of nicotianamine synthase in *A. halleri* revealed a severe diminishing of Zn and Cd accumulation, which proves importance of NA for hyperaccumulation of these metals (Deinlein et al. 2012). In summary, when present in low concentrations, NA likely is important as a ligand mostly in compartments with low abundance of stronger ligands (e.g., the vacuole, xylem, and phloem). At the very high NA levels in hyperaccumulators, however, it might also take part in binding of these metals in the cytoplasm (Leitenmaier and Kupper 2013).

Anthocyanins have also emerged as important non-thiol ligands in metal binding, which have been shown to be associated with molybdenum (MO) accumulation (Hale et al. 2001). Oxalates were shown to bind Cu in Cu-tolerant lichens and fungi (Fomina et al. 2005), and they have been shown to bind Cu in Cu-sensitive Cd/Zn-hyperaccumulator *N. caerulescens* (Mijovilovich et al. 2009). Organic acids have been shown to facilitate enhanced uptake of metals into the roots of hyperaccumulators (Li et al. 2012). The strategy of binding to oxalate in hyperaccumulators makes sense as Cu and manganese (Mn) oxalates are hardly soluble and much more stable which diminishes their bioavailability as compared to binding to smaller organic acids like malate or citrate (Leitenmaier and Kupper 2013).

7.3 Compartmentation and Sequestration of Heavy Metals

Chelation of metal ions is an important aspect of metal ion detoxification, but a potential mechanism for increased metal tolerance is the sequestration of metal ions and metal-chelate complexes into subcellular compartments. There is considerable evidence that sequestration to the vacuole plays a significant role in detoxifying metal ions in a number of organisms. Manipulation of these sequestration mechanisms may be a necessary to increase plant's tolerance to toxic concentrations of heavy metals (Cobbett and Meagher 2002).

7.3.1 Transport and Storage of GSH- and PC-Bound Metals to Vacuoles

The HMT1 gene encodes a member of the family of ATP-binding cassette (ABC) membrane transport proteins that is located in the vacuolar membrane and is required for the transport of PCs or PC-Cd complexes into vacuolar membrane vesicles (Huang et al. 2012). In fission yeast (*Schizosaccharomyces pombe*) the Cd-sensitive, *hmt1* mutant is unable to accumulate vacuolar PC-Cd (Ortiz et al. 1992). YCF1, a member of ABC family of transporters, is a classical example of transport of both GSH conjugates and $(GSH)_2Cd$ complexes into vacuolar compartment (Li et al. 1996). YCF1 has also been shown to sequester GSH conjugates of Hg (Gueldry et al. 2003) and As(III) (Ghosh et al. 1999) into vacuoles. In mesophyll protoplasts derived from tobacco plants exposed to Cd, almost all of both the Cd and PCs accumulated were confined to the vacuole (Vögeli-Lange and Wagner 1990). AtMRP3 can also transport GS conjugates of Cd (Tommasini et al. 1998). Recently, Guo et al. (2012) confirmed increased tolerance to and accumulation of Cd and As by transgenic *Arabidopsis* overexpressing PCS1 in combination with either YCF1 (Guo et al. 2012) or GSH1 (Guo et al. 2008). It was recently discovered that when *SpHMT1*, which is a PC-Cd transporter, from fission yeast (*S. pombe*) was expressed in Arabidopsis, greater tolerance and accumulation was achieved for Cd, Cu, As, and Zn ions as well as enhanced vacuolar sequestration could also be achieved. Chardonnens et al. (1999) have shown that Zn-tolerant lines of *Silene vulgaris* have increased tonoplast transport of Zn compared with non-tolerant lines. *Arabidopsis* Zn-induced facilitator1 (AtZIF1) is a kind of MFS (major facilitator superfamily) transporter, still part of ABC transporter group, and mRNA of ZIF1 was induced by Zn, and its mutant showed to accumulate more Zn in the shoot and also be sensitive to Cd. Its localization studies indicated it being present in the tonoplast suggesting that it might be playing a role in transport of Zn and its complexes to the vacuoles (Haydon and Cobbett 2007a, b). The CPX or type 1B subclass of P-type ATPase transporters is important in heavy metal detoxification and homeostasis in many organisms, including prokaryotes, fungi, plants, and animals (Cobbett and Meagher 2002). HMA3, a P_{1B}-type heavy metal ATPase, was found to mediate leaf vacuolar storage of Cd, Co, Pb, and Zn in *Arabidopsis* (Morel et al. 2009).

The homologue of yeast efflux transporter *ACR3* was identified in an As hyperaccumulator, *Pteris vittata* (*PvACR3*), and was shown to rescue arsenic-sensitive phenotypes of yeast deficient for *ACR3*. Upregulation of *ACR3* was seen in plant tissues in contact with As, and its knockdown leads to sensitive phenotypes in the presence of As confirming the role of *PvACR3* in As tolerance through vacuolar sequestration of As-bound compounds (Indriolo et al. 2010). The role of ABC transporter family members such as that of ABCC1 in transporting and detoxifying As has also been shown in various species such as yeast, protozoa, and *C. elegans* (Papadopoulou et al. 1994; Broeks et al. 1996; Song et al. 2003; Schwartz et al. 2010; Guo et al. 2012). Two vacuolar PC transporters AtABCC1 and AtABCC2 in *Arabidopsis* were shown to play a role in the tolerance and transport of PC-bound metalloids such as As(III)-PC into the vacuole (Song et al. 2010). They also showed that in the absence of these transporters, *Arabidopsis* plants were extremely sensitive to As. When heterologously expressed, in the presence of PC background, greater tolerance and accumulation of As could be seen. Greater As tolerance in *Arabidopsis* was also achieved when AtABCC1 was overexpressed along with AtPCS1 (Song et al. 2010). Overexpression of AtABCC1 in *Arabidopsis* also resulted in enhanced Cd(II) tolerance and accumulation (Park et al. 2012).

7.4 Heavy Metal Transporters

Metal transporters play a significant role in the uptake and transport of essential and nonessential metals and metalloids across plasma membrane of plant species which are needed for processes such as plant growth and development, signal transduction, and toxic metal detoxification (Krämer et al. 2007; Paulose et al. 2008). With the advances in "omics" in the last few years, a lot of metal transporters have been identified in various plant families. These metal transporters have been broadly categorized into two groups – metal uptake transporters and efflux transporters as described below.

7.4.1 Metal Uptake Transporters

Out of several different types and families of metal uptake transporters, certain metal ions such as Cd, Co, Mn, and Zn get transported through some ZIP (zinc-regulated transporter) transporters because of their nonmetal specific nature. ZAT overexpression in *Arabidopsis* provided strong tolerance to and accumulation of zinc upon exposure to 400 μM $ZnSo_4$ (van der Zaal et al. 1999). Under Zn deficiency, transcripts of ZIP1, ZIP2, and ZIP3 from *A. thaliana* were shown to be induced only in the roots and ZIP4 was found in both roots and shoots (Grotz et al. 1998). When expressed in yeast, another ZIP member from *Arabidopsis*-AtIRT1, which normally uptakes iron, was seen to uptake Zn (Korshunova et al. 1999) and transport Cd leading to Cd-sensitive yeast cells (Rogers et al. 2000). An important role of two ZIP genes, ZNT1 and ZNT2, from Ni hyperaccumulator *Thlaspi japonicum* was also identified in providing tolerance to yeast cells when subjected to greater concentrations of Ni (Mizuno et al. 2005). ZIP proteins from *Medicago truncula̲ta* (MtZIP) were shown to rescue *zrt1/zrt2* mutant in yeast in the presence of Zn and other metals. Under Zn deficiency, ZIP4 from rice was highly expressed in both root and shoots and also complemented a Zn-uptake-deficient mutant strain (Ishimura et al. 2005).

Another class of transporter involved in uptake of toxic metals is the family of natural resistance-associated macrophage protein (NRAMP). Six genes in *Arabidopsis* genome code for NRAMP proteins and have been divided into two groups – group one consisting of *AtNRAMP1* and *6* and group two containing *AtNRAMP2* through *5* (Maser 2001). Cadmium sensitivity and accumulation were increased in yeast cells in which *AtNRAMP 1, 3*, and *4* had been expressed (Thomine et al. 2003). Overexpression and characterization of *AtNRAMP3* and *AtNRAMP4* T-DNA lines further confirmed their role in Cd transport in plants (Thomine et al. 2003). *TjNRAMP4* from *T. japonicum*, which is a homologue of *AtNRAMP4*, also showed increased Ni accumulation and sensitivity when expressed in wild-type yeast cells (Mizuno et al. 2005).

Yellow stripe-like protein family transporters (YSL) allow the uptake of metals that are complexed with secondary amino acids such as phytosiderophores or nicotianamines and got their name from maize mutant, yellow stripe 1, which is unable to uptake iron-phytosiderophore complex (Curie et al. 2001). First YS1 was identified in maize and was termed ZmYS1, which complemented Zn-uptake mutant *zap1* in yeast (Schaaf et al. 2004). Eight genes (AtYSL1 through 8) from *Arabidopsis* were found homologous to ZmYS (Paulose et al. 2008). The role of the transporters from this family in uptake of metals such as Zn, Cd, and other divalent cations is still unclear. OPT3, a member of *Arabidopsis* oligopeptide transporter (OPT) family, was shown to be involved in movement of iron into developing seeds (Stacey et al. 2008). *opt3-2* mutants, which have reduced levels of OPT3, exhibited constitutive expression of iron deficiency responses in roots regardless of iron supplementation, resulting in overaccumulation of iron in leaves, but decreased levels of iron in seeds (Chu 2010, dissertation thesis). Recently, OPT3 has been shown to be a phloem-specific transporter that mediates Fe loading into the phloem. By loading of Fe into the phloem in leaves, OPT3 has been shown to regulate both signaling of Fe demand from shoots to roots and Fe transport to developing tissues (Zhai et al. 2014) and plays a key role regulating Fe, Zn, and Cd distribution within the plant (Mendoza-Cózatl et al. 2014).

7.4.2 Metal Efflux Transporters

Another group of transporters is broadly termed as metal efflux transporters under which the CDF family members (cation diffusion facilitator) have been shown to transport metal cations such as Zn and Cd (Nies 2003). Plant CDFs have been given the name metal tolerance proteins (MTPs) (Delhaize 2003; Kim et al. 2004; Paulose et al. 2008). Twelve CDF members have been identified in *A. thaliana* and the first CDF characterized was termed as AtMTP1. This gene, when overexpressed and knocked down in plants, showed tolerance to Zn with less tissue accumulation of Zn. MTP1 has also been identified in *A. halleri*, a metal accumulator species, and was shown to complement the mutant phenotype of a yeast strain that lacks vacuolar zinc resistance gene (ZRC1) and cobalt transporter gene, COT1 (Dräger et al. 2004). All three MTP1s in *A. halleri* have shown Zn tolerance. Several allelic variants of MTP1 from Ni/Zn-hyperaccumulator plant *T. goesingense*, also a homologue of AtMTP1, have conferred resistance to yeast strains mutant for ZRC1 and COT1. Overexpression of *TgMTP1* also decreases the concentration of Zn in the cells by effluxing it out (Kim et al. 2004).

Arabidopsis HMA2, HMA3, and HMA4 are the members of IB-2 subgroup of P_{IB}-type ATPase transporter family, and it was shown that *AtHMA2* was strongly induced by Zn and Cd exposure and seen to transport Zn outside the cytoplasm. Knocked-down *hma2* mutants, compared to wild-type *Arabidopsis* plants, accumulated higher levels of Zn and Cd (Eren and Argüello 2004). In the presence of higher Cd concentrations, HMA4 from *T. caerulescens* enabled the yeast strain grew better and also effluxed Cd out of the cells through the plasma membrane. High and low concentrations of Zn were also able to induce the expression of *TcHMA4* (Bernard et al. 2004; Papoyan and Kochian 2004). HMA4 gene from *Arabidopsis* is located and expressed at the plasma membrane of the root vascular tissues but its role in metal loading in the xylem was confirmed when its overexpression leads to an increase in the amount of Zn and Cd in the shoots. However, a mutant of this gene showed less translocation of the Zn and Cd from roots to shoots (Verret et al. 2004). The role of HMA4 in root-to-shoot Zn translocation has also been shown earlier by Mills et al. (2005). They also explained the role of *AtHMA4* in metal detoxification by showing that plants' sensitivity to higher concentrations of Cd and Zn increases when *AtHMA4* is disrupted. The role of HMA2 and HMA4 transporters was also studied in *hma2 hma4* double mutants. Decreased amount of Zn in the shoots and nutrient-deficiency phenotype of this mutant, when subjected to zinc-free growth medium, pointed toward the role of these two genes in Zn homeostasis. Promoter analysis of HMA2 and HMA4 revealed that they are located in the vascular tissues of roots, leaves, and stems, and HMA2 was also found to be present at the plasma membrane confirming their role in Zn translocation. These were also thought to be involved in Cd detoxification as in a phytochelatin-deficient mutant background, their sensitivity to Cd increased (Hussain et al. 2004; Wong et al. 2009). *AtHMA3* transporter, a member of the P_{1B-2} subgroup of the P-type ATPase family located in the vacuolar membrane of *Arabidopsis*, has been shown to play a role in transport of heavy metal in plants. When overexpressed in *Arabidopsis*, transgenic plants were more tolerant to Cd, Co, Pb, and Zn. Greater accumulation of Cd was also observed. Sensitive phenotypes of T-DNA knockout lines were also seen in the presence of Cd and Zn. Confocal microscopy reveled participation of *AtHMA3* in vacuolar sequestration of Cd (Morel et al. 2009).

AtATM3, an ATP-binding cassette transporter from *Arabidopsis*, is a mitochondrial protein involved in the biogenesis of iron-sulfur clusters and iron homeostasis in plants. AtATM3-overexpressing plants were shown to exhibit enhanced resistance to Cd, whereas *atatm3* mutant plants were more sensitive to Cd than wild-type controls. The role of AtATM3 in regulating cellular levels of nonprotein thiols (NPSH) was also observed. Increased expression of AtATM3 was seen when GSH biosynthesis was inhibited with increased expression of GSH1 under Cd stress and in the *atatm3* mutant suggesting that it may be involved in the transport of GSH-Cd conjugates across mitochondrial membrane (Kim et al. 2006). Recently, it has been shown that protein with strong similarity to AtATM3 transport Cd conjugates (Hanikenne et al. 2005) and CeHMT1, a close homologue of AtATM3, has been shown to be required for Cd tolerance (Vatamaniuk et al. 2005).

The multidrug and toxin extrusion (MATE) family is the most recently categorized one among five multidrug efflux transporter families (Kuroda and Tsuchiya 2009). Magalhaes et al. (2007) used positional cloning to identify the gene encoding a member of the MATE family, an aluminum-activated citrate transporter, responsible for the major sorghum (*Sorghum bicolor*) aluminum (Al) tolerance. Similarly another study in 2009 showed that expression of an expressed sequence tag, belonging to *MATE* gene family, correlates with the citrate efflux phenotype. This study provided genetic and physiological evidence that citrate efflux is a second mechanism for Al resistance in wheat (Ryan et al. 2009). Iron homeostasis-related FRD3 (ferric reductase defective 3) gene, which encodes a multidrug and toxin efflux (MATE) transporter, is responsible for reduced Zn tolerance in *A. thaliana*. FRD3 works as a multimer and is involved in loading Zn into xylem. Cross-homeostasis between Fe and Zn, therefore, appears to be important for Zn tolerance in *A. thaliana* with FRD3 acting as an essential regulator (Pineau et al. 2012).

MRPs (multidrug resistance-associated proteins) constitute another efflux transporter family belonging to ATP-binding cassette transporter superfamily (ABC) and transport metals that have formed conjugates with GSH. Plant orthologs of yeast cadmium factor 1 (*ScYCF1*) which provides Cd resistance by pumping GSH-Cd conjugates into vacuoles were identified in *Arabidopsis*. The first MRP gene termed as *AtMRP1* showed a similar function as *ScYCF1* (Szczypka et al. 1994; Li et al. 1997). *AtMRP3, 6, 7,* and *14* were subsequently identified and shown to be upregulated by Cd (Bovet et al. 2003), with *AtMRP3* also seen to be involved in transport of Cd. *AtMRP3* promoter was also induced upon metals such as Cd, Ni, As, Co, and Pb exposure in *A. thaliana* and *N. tabacum* (Zientara et al. 2009).

AtPDR12, a member of the pleiotropic drug resistance protein (PDR) family of ABC transporters was found in Arabidopsis. The mRNA level of only AtPDR12 increased in both shoots and roots of Pb(II)-treated Arabidopsis, suggesting that it may be involved in the detoxification of Pb(II). The GFP: AtPDR12 fusion protein at the plasma membrane suggested that AtPDR12 functions as a pump to exclude Pb(II) and/or Pb(II)-containing toxic compounds from the cytoplasm (Lee et al. 2005). Cd tolerance in plants has also been achieved through ABC transporter PDR8 (pleiotropic drug resistance 8) located at the plasma membrane that extrudes Cd out of the cell (Kim et al. 2007).

Apart from these transporter families described above, there are some other transporters that do not fall under these families such as the IREG and PCR1 family from *Arabidopsis*, and have been shown to be involved in Ni tolerance and transport and cadmium tolerance, respectively (Schaaf et al. 2004; Song et al. 2004). Tobacco *NtCBP4* (a calmodulin-binding protein) was isolated from tobacco cDNA library, located at the plasma membrane, and was shown to provide tolerance against heavy metal toxicity. The *NtCBP4* transgenic lines showed improved tolerance to Ni by limiting Ni accumulation and hypersensitivity to Pb due to increased accumulation of Pb. These results lead to believe that *NtCBP4* is involved in the uptake of metals across the plasma membrane (Arazi et al. 1999).

Metalloid such as arsenic is naturally present in soil in the form of arsenate (AsO_4^{-3}). Due to chemical similarity between AsO_4^{-3} and inorganic phosphate, Pi(PO_4^{-3}), As(V) is usually taken up into the plants via phosphate transporters. Out of 9 high-affinity phosphate transporters (PHT1-9) in *Arabidopsis*, PHT1 and PHT4 have been involved in As(V) transport. Overexpression of PHT1 resulted into plants being sensitive to As(V) (Catarecha et al. 2007), while their single and double mutants have been shown to be As(V) tolerant indicating that As(V) uptake is compromised by mutations in these Pi transporters (Shin et al. 2004). Recently, LeBlanc et al. created transgenic *Arabidopsis* overexpressing PHT1 or PHT7 from *Arabidopsis* in combination with or without YCF1, leading to increased As accumulation and tolerance in *Arabidopsis* (LeBlanc et al. 2013).

7.5 Other Genes for Metal Tolerance

Apart from strategies discussed above, plants can be engineered with genes that have other means of dealing with heavy metal toxicity either by volatilization of metals and metalloids or via interacting with proteins in other metabolic pathways as described below.

7.5.1 Phytovolatization

Phytovolatization of a metal or metalloid ion involves the accumulation of metal (loid) species in plant cells and their subsequent conversion to an evaporable, usually less toxic, form such that it can be liberated to atmosphere. The main advantage of phytovolatization is the removal of metal (loid) from a site without the need for plant harvesting and disposal (Kotrba et al. 2009).

Bacterial resistance to Hg^{2+} and organomercurial compounds has been attributed to organomercurial lyase (*MerB*), which converts methylmercury and other organomercurials to Hg^{2+} and mercuric reductase (*MerA)* that reduces elemental Hg^{2+} to nontoxic volatile Hg^0 (Summers 1986). These bacterial mercury resistance genes were cloned and used to genetically engineer plants for methylmercury phytoremediation. Overexpression of *merA, merB*, or a combination of both, in *A. thaliana* (Rugh et al. 1996; Bizily et al. 1999, 2003), *N. tabacum* (Ruiz et al. 2003), and rice (*Oryza sativa*, Heaton et al. 2003), resulted in Hg^{2+} and organomercurial tolerant phenotypes. More than tenfold higher volatilization rate was achieved by the targeting of MerB in the endoplasmic reticulum of *merA/merB*, where MerB exhibited more than 20 times higher specific activity than in plants with cytoplasmically distributed MerB (Bizily et al. 2003).

Selenium (Se) occurs naturally in two forms – selenate and selenite. Due to a similar nature of selenate and sulfate, they are normally taken up and assimilated by the same transporters and pathways (Ng and Anderson 1979; Zayed and Terry 1992; Anderson 1993). One such transporter involved in sulfate transport into plant cells is sulfate permease, and in certain plant species such as *Brassica napus* and *Stylosanthes hamata,* selenate was seen inhibiting the sulfate uptake by this transporter (Hawkesford 2003; Smith et al. 1995). ATP sulfurylase is an enzyme involved in sulfate reduction in plants (Setya et al. 1996). It converts selenate to selenite leading to a decrease in selenate concentration, and recent studies have shown that ATP sulfurylase is also involved in selenate reduction in plants (Pilon-Smits et al. 1999). Overexpression of APS1 in Indian mustard (*B. juncea*) showed that under supplied selenate, activity of ATP sulfurylase in these overexpressed plants was much higher in shoots but not roots compared to wild-type plants. These plants were able to reduce selenate better. Greater selenium accumulation and tolerance was also seen in APS transgenic plants (Pilon-Smits et al. 1999). The biosynthesis of MetSeCys, catalyzed by selenocysteine methyltransferase (SMT) in hyperaccumulating plant species inactivates SeCys for synthesis of SeMet and proteosynthesis. MetSeCys could then be converted to volatile dimethyldiselenide (DMDSe; Terry et al. 2000). Most plants that do not produce DMDSe to convert SeMet from SeCys to volatile dimethylselenide (DMSe), which has been reported to be 500–700 times less toxic than selenate and selenite in soil (Wilber 1980). In addition to an improved Se accumulation in selenocysteine methyltransferase, *B. juncea* overexpressing SMT and ATP sulfurylase showed a higher DMSe production than wild-type control plants and an acquired ability to produce DMDSe (LeDuc et al. 2004). Volatile forms were then efficiently evaporated from the leaves of transgenics grown in culture media and polluted soil (LeDuc et al. 2004; Bañuelos et al. 2007).

7.5.2 Genes Involved in Oxidative Stress Response and Misfolded Protein Repair

As the concentration of heavy metals inside the cell reaches to a point of saturation, the plant as a system begins to experience oxidative stress caused by the production of ROS and the inhibition of metal-dependent antioxidant enzymes (Schützendübel and Polle 2002). In general, heavy metal-induced ROS production causes oxidative damage to photosynthetic pigments; biomolecules such as lipids, proteins, and nucleic acids; and leakage of electrolytes via lipid peroxidation causing dramatic reductions in plant growth and productivity. Plants respond to oxidative stress by production of antioxidative enzymes such as SOD, APX, and GR and nonenzymatic free radical scavengers (Aust et al. 1985). Heavy metal toxicity is reported to increase activity of glucose-6-phosphate dehydrogenase and peroxidase in the leaves of plants grown in polluted soils (Van Assche and Clijsters 1987). In *Nicotiana plumbaginifolia*, leaves exposed to excess Fe have been shown to induce expression of ascorbate peroxidase (APX) and catalases (CAT) (Kampfenkel et al. 1995). Similarly the expression of CAT3 from *B. juncea* was induced upon Cd treatment (Minglin et al. 2005). A brief treatment with low concentration of Al to chickpea seedlings (*Cicer arietinum*) resulted in higher

SOD, APX, and guaiacol peroxidase (GPX) activity (Singh et al. 2012). Peroxiredoxins (PRXes) are peroxide-degrading enzymes with mitochondrial and chloroplastic variants. A T-DNA insertion mutant of *A. thaliana* lacking the expression of mitochondrial PrxIIF (AtPrxIIF knockout) was more sensitive to Cd in terms of root growth than the controls, signifying the involvement of PrxIIF in cellular detoxification of Cd such that root growth is maintained under Cd stress up to a certain threshold (Finkemeier et al. 2005).

Superoxide dismutases (SODs) are enzymes that play a pivotal role in metabolizing singlet O_2, preventing formation of deleterious reactive oxygen species (ROS) including hydrogen peroxide (H_2O_2), hypochlorite (OCl^-), peroxynitrate (ONO_2), and hydroxyl radical (HO^-) (Miller 2012). SOD activity is also induced in tomato seedlings after prolonged Cd treatment (Dong et al. 2006). SOD activity increases significantly in wheat leaves, following exposure to high levels of Cd, probably reflecting the accumulation of superoxide (Lin et al. 2007). On the other hand, Rodríguez-Serrano and coworkers showed a reduced SOD activity in pea plants exposed to Cd. Cd in particular has been shown to reduce the GSH/GSSG ratio and activate antioxidant enzymes such as SOD and GR (Rodríguez-Serrano et al. 2009). Heavy metals also induce the synthesis of stress-related proteins and signaling molecules, such as HSPs, SAPs, salicylic and abscisic acids, and ethylene (Manara 2012). Heat shock proteins (HSPs) are expressed not only in response to elevated temperatures but also in response to other abiotic stresses such as drought and heavy metal stress. Under heavy metal stress, protein-folding mechanism can go haywire resulting in misfolded proteins. Heat shock proteins can act as molecular chaperon ensuring correct folding and repair of misfolded protein (Vierling 1991). Heavy metals induce the expression of low-molecular-weight HSPs in rice (Tseng et al. 1993). *Zea mays* plants exposed to varying soil concentrations of Cu, Ni, Pb, and Zn showed increased chloroplast small HSP contents, with increased time of exposure. These HSPs in turn were shown to protect photosynthesis from heavy metal toxicity (Heckathorn et al. 2004).

Members of stress-associated protein (SAP) family (Vij and Tyagi 2006) were recently shown to provide tolerance to multiple abiotic stress including toxic metals (Mukhopadhyay et al. 2004). OsiSAP8 was shown to be induced in response to heavy metals such as Zn, Cu, Hg, and Cd (Kanneganti and Gupta 2008). Similarly overexpression of AtSAP10 in *Arabidopsis* provided strong tolerance to Ni and Mn. Due to their unique ability to provide tolerance to multiple abiotic stresses, members of SAP family are also the ideal candidates to engineer plants for heavy metal tolerance to tackle the challenges of global climate changes and the effects associated with it (Dixit and Dhankher 2011).

7.6 Heavy Metal Tolerance and Climate Change Adaptations

As global climate change is happening and not in distant future, the burden of environmental stresses including heavy metals and metalloids on crop plants is likely to continue to increase. Flood events can transport heavy metals, cyanide, and hydrocarbons from a contaminated area to a non-contaminated one (Harmon and Wyatt 2008; Hilscherova et al. 2007; Boxall et al. 2009). Climate change is likely to increase frequency of heavy precipitation events worldwide, which would result in transport of historical contaminants from previously undisturbed sediments. This could have implications for residue of toxic metal levels in food crops (Casteel et al. 2006). As irrigation demands may increase because of warmer and drier summers, water of poorer quality, including partially treated wastewater from industries and municipalities, will likely be applied to crops which may result in additional contaminant loadings to crops (Rose et al. 2001). Changes in temperature and precipitation could also increase aerial inputs of volatile and dust-associated contaminants. Finally, changes in bioavailability may occur with less bioavailable forms of contaminant being converted to more bioavailable forms. For example, Booth and Zeller (2005) suggested that increases in temperature could enhance the methylation rate of Hg. The use of composting for treatment of municipal waste is increasing, with a portion of the resulting compost being used in agriculture. This is likely to increase loadings of microbes, heavy metals, and persistent organic pollutants in agricultural land (Déportes et al. 1995; Boxall et al. 2009). Furthermore, with the expansion of crop cultivation on soils that are not optimal for the growth of crop plants such as contaminated lands, development of abiotic stresses including heavy metals stress tolerant plants is becoming increasingly important (Kathuria et al. 2007). Improving the tolerance of major crop plants to heavy metals and other abiotic stresses has been a main goal in agricultural research for a long time. Transgenic approaches offer attractive alternatives to conventional techniques for the genetic improvement of abiotic stress tolerance. The development of new cultivars with enhanced heavy metal tolerance will undoubtedly have an important effect on global food production and food safety.

7.7 Conclusions

Crop production is declining around the world due to several biotic and abiotic stresses, which include heat, cold, drought, heavy metals, etc. (Abedin et al. 2002; Van Nguyen and Ferrero 2006). In the past years, a lot of focus has been on improving plant species and their tolerance toward these stresses but not much has been achieved because of the limited knowledge of the gene/network of genes that might be involved in providing such tolerance to multiple stresses. The research in this direction could have a significant impact on global food security, human health enhancement, and the environment, if more knowledge and information is gained on the hyperaccumulating species and phytoremediation strategy through the use of transgenics. Once this system is fully understood in the model plant species, the knowledge and information gained can be applied on other agricultural crops to engineer crops that will be better able to withstand such abiotic stresses and still produce sustainable yield. This will also help to grow crops for food and biomass production on not so cultivable lands, thus making them more cultivable over time. Developing crops more resilient to heavy metals and other abiotic stresses will enable them to grow on marginal to moderately contaminated soils without losing crop yields.

Acknowledgments A grant (MAS00401) to OPD from the US Department of Agriculture via the University of Massachusetts Amherst supported the writing of this manuscript.

References

Abedin MJ, Feldmann J, Meharg AA (2002) Uptake kinetics of arsenic species in rice plants. Plant Physiol 128:1120–1128

Anderson JW (1993) Selenium interactions in sulfur metabolism. In: De Kok LJ (ed) Sulfur nutrition and assimilation in higher plants-regulatory, agricultural and environmental aspects. SPB Academic Publishing, The Hague, pp 49–60

Arazi T, Sunkar R, Kaplan B, Fromm H (1999) A tobacco plasma membrane calmodulin-binding transporter confers Ni^{2+} tolerance and Pb^{2+} hypersensitivity in transgenic plants. Plant J 20:171–182

Assche F, Clijsters H (1990) Effects of metals on enzyme activity in plants. Plant Cell Environ 13:195–206

Auguy F, Fahr M, Moulin P, Brugel A, Laplaze L, El Mzibri M, Filali-Maltouf A, Doumas P, Smouni A (2013) Lead tolerance and accumulation in Hirschfeldia incana, a Mediterranean *Brassicaceae* from metalliferous mine spoils. PLoS One 8:e61932

Aust SD, Morehouse LA, Thomas CE (1985) Role of metals in oxygen radical reactions. J Free Radic Biol Med 1:3–25

Balogh SJ, Swain EB, Nollet YH (2006) Elevated methylmercury concentrations and loadings during flooding in Minnesota rivers. Sci Total Environ 368:138–148

Bañuelos G, LeDuc DL, Pilon-Smits EAH, Terry N (2007) Transgenic Indian mustard overexpressing selenocysteine lyase or selenocysteine methyltransferase exhibit enhanced potential for selenium phytoremediation under field conditions. Environ Sci Technol 41:599–605

Barceló J, Poschenrieder C (1990) Plant water relations as affected by heavy metal stress: a review. J Plant Nutr 13:1–37

Becher M, Talke IN, Krall L, Krämer U (2004) Cross-species microarray transcript profiling reveals high constitutive expression of metal homeostasis genes in shoots of the zinc hyperaccumulator *Arabidopsis halleri*. Plant J 37:251–268

Beneš I, Schreiber K, Ripperger H, Kircheiss A (1983) Metal complex formation by nicotianamine, a possible phytosiderophore. Experientia 39:261–262

Bentley R, Chasteen TG (2002) Microbial methylation of metalloids: arsenic, antimony, and bismuth. Microbiol Mol Biol Rev 66:250–271

Bernard C, Roosens N, Czernic P, Lebrun M, Verbruggen N (2004) A novel CPx-ATPase from the cadmium hyperaccumulator *Thlaspi caerulescens*. FEBS Lett 569:140–148

Bizily SP, Rugh CL, Summers AO, Meagher RB (1999) Phytoremediation of methylmercury pollution: *merB* expression in *Arabidopsis thaliana* confers resistance to organomercurials. Proc Natl Acad Sci U S A 96:6808–6813

Bizily SP, Kim T, Kandasamy MK, Meagher RB (2003) Subcellular targeting of methylmercury lyase enhances its specific activity for organic mercury detoxification in plants. Plant Physiol 131:463–471

Boening DW (2000) Ecological effects, transport, and fate of mercury: a general review. Chemosphere 40:1335–1351

Bohnert HJ, Nelson DE, Jensen RG (1995) Adaptations to environmental stresses. Plant Cell 7:1099–1111

Booth S, Zeller D (2005) Mercury, food webs, and marine mammals: implications of diet and climate change for human health. Environ Health Perspect 113:521–526

Bovet L, Eggmann T, Meylan-Bettex M, Polier J, Kammer P, Marin E, Feller U, Martinoia E (2003) Transcript levels of AtMRPs after cadmium treatment: induction of AtMRP3. Plant Cell Environ 26:371–381

Boxall ABA, Hardy A, Beulke S, Boucard T, Burgin L, Falloon PD, Haygarth PM, Hutchinson T, Kovats RS, Leonardi G et al (2009) Impacts of climate change on indirect human exposure to pathogens and chemicals from agriculture. Environ Health Perspect 117:508–514

Boyer PD (1954) Spectrophotometric study of the reaction of protein sulfhydryl groups with organic mercurials. J Am Chem Soc 76:4331–4337

Bray EA, Bailey-Serres J, Weretilnyk E (2000) Responses to abiotic stresses. In: Buchanan B, Gruissem W, Jones R (eds) Biochemistry and molecular biology of plants. American Society of Plant Physiologists, Rockville, pp 1158–1203

Broeks A, Gerrard B, Allikmets R, Dean M, Plasterk RH (1996) Homologues of the human multidrug resistance genes MRP and MDR contribute to heavy metal resistance in the soil nematode *Caenorhabditis elegans*. EMBO J 15:6132–6143

Callahan DL, Baker AJM, Kolev SD, Wedd AG (2006) Metal ion ligands in hyperaccumulating plants. J Biol Inorg Chem 11:2–12

Casteel MJ, Sobsey MD, Mueller JP (2006) Fecal contamination of agricultural soils before and after hurricane-associated flooding in North Carolina. J Environ Sci Health A Tox Hazard Subst Environ Eng 41:173–184

Catarecha P, Segura MD, Franco-Zorrilla JM, García-Ponce B, Lanza M, Solano R, Paz-Ares J, Leyva A (2007) A mutant of the Arabidopsis phosphate transporter PHT1;1 displays enhanced arsenic accumulation. Plant Cell 19:1123–1133

Chardonnens A, Koevoets P, van Zanten A, Schat H, Verkleij J (1999) Properties of enhanced tonoplast zinc transport in naturally selected zinc-tolerant *Silene vulgaris*. Plant Physiol 120:779–786

Chatterjee J (2000) Phytotoxicity of cobalt, chromium and copper in cauliflower. Environ Pollut 109:69–74

Chen Z, Zhu YG, Liu WJ, Meharg AA (2005) Direct evidence showing the effect of root surface iron plaque on arsenite and arsenate uptake into rice (*Oryza sativa*) roots. New Phytol 165:91–97

Choi J, Pak C, Lee CW (1996) Micro nutrient toxicity in French marigold. J Plant Nutr 19:901–916

Chu HH (2010) Analyses of Arabidopsis Yellow Stripe-Like (YSL) family of metal transporters. PhD dissertation (Amherst, MA: University of Massachusetts). Paper 159

Clark GT, Dunlop J, Phung HT (2000) Phosphate absorption by *Arabidopsis thaliana*: interactions between phosphorus status and inhibition by arsenate. Funct Plant Biol 27:959–965

Clemens S, Kim EJ, Neumann D, Schroeder JI (1999) Tolerance to toxic metals by a gene family of phytochelatin synthases from plants and yeast. EMBO J 18:3325–3333

Cobbett CS (2000) Phytochelatins and their roles in heavy metal detoxification. Plant Physiol 123:825–832

Cobbett C, Goldsbrough P (2002) Phytochelatins and metallothioneins: roles in heavy metal detoxification and homeostasis. Annu Rev Plant Biol 53:159–182

Cobbett CS, Meagher RB (2002) Arabidopsis and the genetic potential for the phytoremediation of toxic elemental and organic pollutants. Arabidopsis Book 1:e0032

Curie C, Panaviene Z, Loulergue C, Dellaporta SL, Briat JF, Walker EL (2001) Maize yellow stripe1 encodes a membrane protein directly involved in Fe(III) uptake. Nature 409:346–349

Das HK, Mitra AK, Sengupta PK, Hossain A, Islam F, Rabbani GH (2004) Arsenic concentrations in rice, vegetables, and fish in Bangladesh: a preliminary study. Environ Int 30:383–387

De Vries W, Lofts S, Tipping E, Meili M, Groenenberg J, Schütze G (2007) Impact of soil properties on critical concentrations of cadmium, lead, copper, zinc, and mercury in soil and soil solution in view of ecotoxicological effects. Rev Environ Contam Toxicol SE-3. Springer, New York, pp 47–89

Deinlein U, Weber M, Schmidt H, Rensch S, Trampczynska A, Hansen TH, Husted S, Schjoerring JK, Talke IN, Krämer U et al (2012) Elevated nicotianamine levels in *Arabidopsis halleri* roots play a key role in zinc hyperaccumulation. Plant Cell 24:708–723

Delhaize E (2003) Genes encoding proteins of the cation diffusion facilitator family that confer manganese tolerance. Plant Cell Online 15:1131–1142

Déportes I, Benoit-Guyod JL, Zmirou D (1995) Hazard to man and the environment posed by the use of urban waste compost: a review. Sci Total Environ 172:197–222

Dhankher OP, Li Y, Rosen BP, Shi J, Salt D, Senecoff JF, Sashti NA, Meagher RB (2002) Engineering tolerance and hyperaccumulation of arsenic in plants by combining arsenate reductase and gamma-glutamylcysteine synthetase expression. Nat Biotechnol 20:1140–1145

Dhankher OP, Rosen BP, McKinney EC, Meagher RB (2006) Hyperaccumulation of arsenic in the shoots of Arabidopsis silenced for arsenate reductase (ACR2). Proc Natl Acad Sci U S A 103:5413–5418

Dietz KJ, Baier M, Krämer U (1999) Free radicals and reactive oxygen species as mediators of heavy metal toxicity. In: Prasad MNV, Hagemeyer J (eds) Heavy metal stress in plants: from molecules to ecosystems. Springer, Heidelberg, pp 73–97

Dixit AR, Dhankher OP (2011) A novel stress-associated protein "AtSAP10" from Arabidopsis thaliana confers tolerance to nickel, manganese, zinc, and high temperature stress. PLoS One 6:e20921

Dixit V, Pandey V, Shyam R (2002) Chromium ions inactivate electron transport and enhance superoxide generation *in vivo* in pea (*Pisum sativum* L. cv. Azad) root mitochondria. Plant Cell Environ 25:687–693

Dong J, Wu F, Zhang G (2006) Influence of cadmium on antioxidant capacity and four microelement concentrations in tomato seedlings (*Lycopersicon esculentum*). Chemosphere 64:1659–1666

Dräger DB, Desbrosses-Fonrouge AG, Krach C, Chardonnens AN, Meyer RC, Saumitou-Laprade P, Krämer U (2004) Two genes encoding *Arabidopsis halleri* MTP1 metal transport proteins co-segregate with zinc tolerance and account for high MTP1 transcript levels. Plant J 39:425–439

Ebbs SD, Kochian LV (1997) Toxicity of zinc and copper to *Brassica* species: implications for phytoremediation. J Environ Qual 26:776

Ellis DR, Salt DE (2003) Plants, selenium and human health. Curr Opin Plant Biol 6:273–279

Eren E, Argüello JM (2004) Arabidopsis HMA2, a divalent heavy metal-transporting P(IB)-type ATPase, is involved in cytoplasmic Zn^{2+} homeostasis. Plant Physiol 136:3712–3723

Evanko CR, Dzombak DA (1997) Remediation of metals-contaminated soils and groundwater. Technology Evaluation Report(TE-97-01). Ground-Water Remediation Technologies Analysis Center (GWRTAC), Pittsburgh

Falchuk KH, Goldwater LJ, Vallee BL (1977) The biochemistry and toxicology of mercury. In: McAuliffe CA (ed) The biochemical and toxicology mercury. Macmillan, New York, pp 261–284

Finkemeier I, Goodman M, Lamkemeyer P, Kandlbinder A, Sweetlove LJ, Dietz KJ (2005) The mitochondrial type II peroxiredoxin F is essential for redox homeostasis and root growth of *Arabidopsis thaliana* under stress. J Biol Chem 280:12168–12180

Fomina M, Hillier S, Charnock JM, Melville K, Alexander IJ, Gadd GM (2005) Role of oxalic acid overexcretion in transformations of toxic metal minerals by *Beauveria caledonica*. Appl Environ Microbiol 71:371–381

Fontes RLF, Cox FR (1998) Zinc toxicity in soybean grown at high iron concentration in nutrient solution. J Plant Nutr 21:1723–1730

Fordyce FM (2005) Selenium deficiency and toxicity in the environment. In: Selinus O (ed) Essentials of medical geology. Elsevier, Amsterdam, pp 373–415

Freeman JL, Persans MW, Nieman K, Albrecht C, Peer W, Pickering IJ, Salt DE (2004) Increased glutathione biosynthesis plays a role in nickel tolerance in *Thlaspi* nickel hyperaccumulators. Plant Cell 16:2176–2191

Freeman JL, Garcia D, Kim D, Hopf A, Salt DE (2005) Constitutively elevated salicylic acid signals glutathione-mediated nickel tolerance in *Thlaspi* nickel hyperaccumulators. Plant Physiol 137:1082–1091

Ghosh M, Shen J, Rosen BP (1999) Pathways of As(III) detoxification in *Saccharomyces cerevisiae*. Proc Natl Acad Sci 96:5001–5006

Gill SS, Tuteja N (2011) Cadmium stress tolerance in crop plants probing the role of sulfur. Plant Signal Behav 6:215–222

Gisbert C, Ros R, De Haro A, Walker DJ, Pilar Bernal M, Serrano R, Navarro-Aviñó J (2003) A plant genetically modified that accumulates Pb is especially promising for phytoremediation. Biochem Biophys Res Commun 303:440–445

Greger M, Ogren E (1991) Direct and indirect effects of Cd^{2+} on photosynthesis in sugar beet (*Beta vulgaris*). Physiol Plant 83:129–135

Grennan AK (2011) Metallothioneins, a diverse protein family. Plant Physiol 155:1750–1751

Grill E, Winnacker EL, Zenk MH (1985) Phytochelatins: the principal heavy-metal complexing peptides of higher plants. Science 230:674–676

Grill E, Loffler S, Winnacker EL, Zenk MH (1989) Phytochelatins, the heavy-metal-binding peptides of plants, are synthesized from glutathione by a specific-glutamylcysteine dipeptidyl transpeptidase (phytochelatin synthase). Proc Natl Acad Sci 86:6838–6842

Grispen VMJ, Irtelli B, Hakvoort HWJ, Vooijs R, Bliek T, ten Bookum WM, Verkleij JAC, Schat H (2009) Expression of the Arabidopsis metallothionein 2b enhances arsenite sensitivity and root to shoot translocation in tobacco. Environ Exp Bot 66:69–73

Grotz N, Fox T, Connolly E, Park W, Guerinot ML, Eide D (1998) Identification of a family of zinc transporter genes from Arabidopsis that respond to zinc deficiency. Proc Natl Acad Sci 95:7220–7224

Gu JG, Zhao X, Wang X (2003) Reused path of heavy metal pollution in soils and its research advance. J Basic Sci Eng 11:143–151

Gueldry O, Lazard M, Delort F, Dauplais M, Grigoras I, Blanquet S, Plateau P (2003) Ycf1p-dependent Hg(II) detoxification in *Saccharomyces cerevisiae*. Eur J Biochem 270:2486–2496

Guerinot ML, Salt DE (2001) Fortified foods and phytoremediation. Two sides of the same coin. Plant Physiol 125:164–167

Guo J, Dai X, Xu W, Ma M (2008) Overexpressing GSH1 and AsPCS1 simultaneously increases the tolerance and accumulation of cadmium and arsenic in *Arabidopsis thaliana*. Chemosphere 72:1020–1026

Guo J, Xu W, Ma M (2012) The assembly of metals chelation by thiols and vacuolar compartmentalization conferred increased tolerance to and accumulation of cadmium and arsenic in transgenic *Arabidopsis thaliana*. J Hazard Mater 199–200:309–313

Guo J, Xu L, Su Y, Wang H, Gao S, Xu J, Que Y (2013) ScMT2-1-3, a metallothionein gene of sugarcane, plays an important role in the regulation of heavy metal tolerance/accumulation. Biomed Res Int. doi:10.1155/2013/904769

Gupta DK, Huang HG, Yang XE, Razafindrabe BHN, Inouhe M (2010) The detoxification of lead in *Sedum alfredii* H. is not related to phytochelatins but the glutathione. J Hazard Mater 177:437–444

Ha SB (1999) Phytochelatin synthase genes from Arabidopsis and the yeast *Schizosaccharomyces pombe*. Plant Cell Online 11:1153–1164

Haag-Kerwer A, Schafer HJ, Heiss S, Walter C, Rausch T (1999) Cadmium exposure in *Brassica juncea* causes a decline in transpiration rate and leaf expansion without effect on photosynthesis. J Exp Bot 50:1827–1835

Hale KL, McGrath SP, Lombi E, Stack SM, Terry N, Pickering IJ, George GN, Pilon-Smits EA (2001) Molybdenum sequestration in *Brassica* species. A role for anthocyanins? Plant Physiol 126:1391–1402

Hall JLL (2002) Cellular mechanisms for heavy metal detoxification and tolerance. J Exp Bot 53:1–11

Hammerschmidt CR, Fitzgerald WF, Lamborg CH, Balcom PH, Tseng CM (2006) Biogeochemical cycling of methylmercury in lakes and tundra watersheds of Arctic Alaska. Environ Sci Technol 40:1204–1211

Hanikenne M, Motte P, Wu MCS, Wang T, Loppes R, Matagne RF (2005) A mitochondrial half-size ABC transporter is involved in cadmium tolerance in *Chlamydomonas reinhardtii*. Plant Cell Environ 28:863–873

Hänsch R, Mendel RR (2009) Physiological functions of mineral micronutrients (Cu, Zn, Mn, Fe, Ni, Mo, B, Cl). Curr Opin Plant Biol 12:259–266

Harmon SM, Wyatt DE (2008) Evaluation of post-Katrina flooded soils for contaminants and toxicity to the soil invertebrates *Eisenia fetida* and *Caenorhabditis elegans*. Chemosphere 70:1857–1864

Hart BA, Lee CH, Shukla GS, Shukla A, Osier M, Eneman JD, Chiu JF (1999) Characterization of cadmium-induced apoptosis in rat lung epithelial cells: evidence for the participation of oxidant stress. Toxicology 133:43–58

Hasegawa I, Terada E, Sunairi M, Wakita H, Shinmachi F, Noguchi A, Nakajima M, Yazaki J (1997) Genetic improvement of heavy metal tolerance in plants by transfer of the yeast metallothionein gene (CUP1). In: Ando T, Fujita K, Mae T, Matsumoto H, Mori S, Sekiya J (eds) Plant nutrition for sustainable food production and environment SE – 117. Springer, Dordrecht, pp 391–395

Hawkesford MJ (2003) Transporter gene families in plants: the sulphate transporter gene family – redundancy or specialization? Physiol Plant 117:155–163

Haydon MJ, Cobbett CS (2007a) Transporters of ligands for essential metal ions in plants. New Phytol 174:499–506

Haydon MJ, Cobbett CS (2007b) A novel major facilitator superfamily protein at the tonoplast influences zinc tolerance and accumulation in Arabidopsis. Plant Physiol 143:1705–1719

Heaton ACP, Rugh CL, Kim T, Wang NJ, Meagher RB (2003) Toward detoxifying mercury-polluted aquatic sediments with rice genetically engineered for mercury resistance. Environ Toxicol Chem 22:2940–2947

Heckathorn SA, Mueller JK, Laguidice S, Zhu B, Barrett T, Blair B, Dong Y (2004) Chloroplast small heat-shock proteins protect photosynthesis during heavy metal stress. Am J Bot 91:1312–1318

Hegedüs A, Erdei S, Horváth G (2001) Comparative studies of H2O2 detoxifying enzymes in green and greening barley seedlings under cadmium stress. Plant Sci 160:1085–1093

Hilscherova K, Dusek L, Kubik V, Cupr P, Hofman J, Klanova J, Holoubek I (2007) Redistribution of organic pollutants in river sediments and alluvial soils related to major floods. J Soils Sediments 7:167–177

Howden R (1995) Cadmium-sensitive, cad1 mutants of *Arabidopsis thaliana* are phytochelatin deficient. Plant Physiol 107:1059–1066

Huang TL, Huang HJ (2008) ROS and CDPK-like kinase-mediated activation of MAP kinase in rice roots exposed to lead. Chemosphere 71:1377–1385

Huang B, Hatch E, Goldsbrough PB (1987) Selection and characterization of cadmium tolerant cells in tomato. Plant Sci 52:211–221

Huang J, Zhang Y, Peng J-S, Zhong C, Yi H-Y, Ow DW, Gong J-M (2012) Fission yeast HMT1 lowers seed cadmium through phytochelatin-dependent vacuolar sequestration in Arabidopsis. Plant Physiol 158:1779–1788

Hussain D, Haydon MJ, Wang Y, Wong E, Sherson SM, Young J, Camakaris J, Harper JF, Cobbett CS (2004) P-type ATPase heavy metal transporters with roles in essential zinc homeostasis in Arabidopsis. Plant Cell 16:1327–1339

Indriolo E, Na G, Ellis D, Salt DE, Banks JA (2010) A vacuolar arsenite transporter necessary for arsenic tolerance in the arsenic hyperaccumulating fern *Pteris vittata* is missing in flowering plants. Plant Cell 22:2045–2057

Ishimaru Y, Suzuki M, Kobayashi T, Takahashi M, Nakanishi H, Mori S, Nishizawa NK (2005) OsZIP4, a novel zinc-regulated zinc transporter in rice. J Exp Bot 56:3207–3214

Kaiser J (2001) Environmental health. Second look at arsenic finds higher risk. Science 293:2189

Kampfenkel K, Van Montagu M, Inze D (1995) Effects of iron excess on *Nicotiana plumbaginifolia* plants (implications to oxidative stress). Plant Physiol 107:725–735

Kanneganti V, Gupta AK (2008) Overexpression of OsiSAP8, a member of stress associated protein (SAP) gene family of rice confers tolerance to salt, drought and cold stress in transgenic tobacco and rice. Plant Mol Biol 66:445–462

Kathuria H, Giri J, Tyagi H, Tyagi AK (2007) Advances in transgenic rice biotechnology. CRC Crit Rev Plant Sci 26:65–103

Keating M, Mahaffey K, Schoeny R, Rice GE, Bullock OR (1997) Mercury study report to congress, vol 1. Executive summary, EPA-452/R-97-003, United States Environmental Protection Agency, Research Triangle Park, NC

Kim D, Gustin JL, Lahner B, Persans MW, Baek D, Yun DJ, Salt DE (2004) The plant CDF family member TgMTP1 from the Ni/Zn hyperaccumulator Thlaspi goesingense acts to enhance efflux of Zn at the plasma membrane when expressed in *Saccharomyces cerevisiae*. Plant J 39:237–251

Kim DY, Bovet L, Kushnir S, Noh EW, Martinoia E, Lee Y (2006) AtATM3 is involved in heavy metal resistance in Arabidopsis. Plant Physiol 140:922–932

Kim DY, Bovet L, Maeshima M, Martinoia E, Lee Y (2007) The ABC transporter AtPDR8 is a cadmium extrusion pump conferring heavy metal resistance. Plant J 50:207–218

Kneer R, Zenk MH (1992) Phytochelatins protect plant enzymes from heavy metal poisoning. Phytochemistry 31:2663–2667

Korshunova YO, Eide D, Clark WG, Guerinot ML, Pakrasi HB (1999) The IRT1 protein from Arabidopsis thaliana is a metal transporter with a broad substrate range. Plant Mol Biol 40:37–44

Kotrba P, Najmanova J, Macek T, Ruml T, Mackova M (2009) Genetically modified plants in phytoremediation of heavy metal and metalloid soil and sediment pollution. Biotechnol Adv 27:799–810

Krämer U, Clemens S (2006) Functions and homeostasis of zinc, copper, and nickel in plants. In: Tamas M, Martinoia E (eds) Molecular biology of metal homeostasis and detoxification SE-96. Springer, Berlin/Heidelberg, pp 216–271

Kramer U, Cotter-Howells JD, Charnock JM, Baker AJM, Smith JAC (1996) Free histidine as a metal chelator in plants that accumulate nickel. Nature 379:635–638

Kramer U, Smith RD, Wenzel WW, Raskin I, Salt DE (1997) The role of metal transport and tolerance in nickel hyperaccumulation by *Thlaspi goesingense* Halacsy. Plant Physiol 115:1641–1650

Krämer U, Talke IN, Hanikenne M (2007) Transition metal transport. FEBS Lett 581:2263–2272

Krupa Z (1988) Cadmium-induced changes in the composition and structure of the light-harvesting chlorophyll a/b protein complex II in radish cotyledons. Physiol Plant 73:518–524

Krupa Z, Oquist G, Huner NPA (1993) The effects of cadmium on photosynthesis of *Phaseolus vulgaris* – a fluorescence analysis. Physiol Plant 88:626–630

Küpper H, Mijovilovich A, Meyer-Klaucke W, Kroneck PMH (2004) Tissue- and age-dependent differences in the complexation of cadmium and zinc in the cadmium/zinc hyperaccumulator *Thlaspi caerulescens* (Ganges ecotype) revealed by x-ray absorption spectroscopy. Plant Physiol 134:748–757

Kuroda T, Tsuchiya T (2009) Multidrug efflux transporters in the MATE family. Biochim Biophys Acta 1794:763–768

Larsen E, Moseholm L, Nielsen M (1992) Atmospheric deposition of trace elements around point sources and human health risk assessment. II: uptake of arsenic and chromium by vegetables grown near a wood preservation factory. Sci Total Environ 126:263–275

LeBlanc MS, McKinney EC, Meagher RB, Smith AP (2013) Hijacking membrane transporters for arsenic phytoextraction. J Biotechnol 163:1–9

LeDuc DL, Tarun AS, Montes-Bayon M, Meija J, Malit MF, Wu CP, AbdelSamie M, Chiang CY, Tagmount A, deSouza M et al (2004) Overexpression of selenocysteine methyltransferase in Arabidopsis and Indian mustard increases selenium tolerance and accumulation. Plant Physiol 135:377–383

LeDuc DL, AbdelSamie M, Montes-Bayon M, Wu CP, Reisinger SJ, Terry N (2006) Overexpressing both ATP sulfurylase and selenocysteine methyltransferase enhances selenium phytoremediation traits in Indian mustard. Environ Pollut 144:70–76

Lee S, Kang BS (2005) Phytochelatin is not a primary factor in determining copper tolerance. J Plant Biol 48:32–38

Lee CW, Choi JM, Pak CH (1996) Micronutrient toxicity in seed geranium (Pelargonium x hortorum Bailey). J Am Soc Hort Sci 121:77–82

Lee S, Moon JS, Ko TS, Petros D, Goldsbrough PB, Korban SS (2003) Overexpression of Arabidopsis phytochelatin synthase paradoxically leads to hypersensitivity to cadmium stress. Plant Physiol 131:656–663

Lee J, Shim D, Song WY, Hwang I, Lee Y (2004) Arabidopsis metallothioneins 2a and 3 enhance resistance to cadmium when expressed in *Vicia faba* guard cells. Plant Mol Biol 54:805–815

Lee M, Lee K, Lee J, Noh EW, Lee Y (2005) AtPDR12 contributes to lead resistance in Arabidopsis. Plant Physiol 138:827–836

Lefebvre DD, Laliberte JF (1987) Mammalian metallothionein functions in plants. In: Verma D, Brisson N (eds) Molecular genetic of plant-microbe interactions SE-7. Springer, Dordrecht, pp 32–34

Leitenmaier B, Küpper H (2013) Compartmentation and complexation of metals in hyperaccumulator plants. Front Plant Sci 4:1–13

Li ZS, Szczypka M, Lu YP, Thiele DJ, Rea PA (1996) The yeast cadmium factor protein (YCF1) is a vacuolar glutathione S-conjugate pump. J Biol Chem 271:6509–6517

Li ZS, Lu YP, Zhen RG, Szczypka M, Thiele DJ, Rea PA (1997) A new pathway for vacuolar cadmium sequestration in *Saccharomyces cerevisiae*: YCF1-catalyzed transport of bis (glutathionato) cadmium. Proc Natl Acad Sci U S A 94:42–47

Li Y, Dhankher OP, Carreira L, Balish RS, Meagher RB (2005) Arsenic and mercury tolerance and cadmium sensitivity in Arabidopsis plants expressing bacterial gamma-glutamylcysteine synthetase. Environ Toxicol Chem 24:1376

Li T, Xu Z, Han X, Yang X, Sparks DL (2012) Characterization of dissolved organic matter in the rhizosphere of hyperaccumulator *Sedum alfredii* and its effect on the mobility of zinc. Chemosphere 88:570–576

Liao MT, Hedley MJ, Woolley DJ, Brooks RR, Nichols MA (2000) Copper uptake and translocation in chicory (*Cichorium intybus* L. cv Grasslands Puna) and tomato (*Lycopersicon esculentum* Mill. cv Rondy) plants grown in NFT system. II. The role of nicotianamine and histidine in xylem sap copper transport. Plant Soil 223:245–254

Lin R, Wang X, Luo Y, Du W, Guo H, Yin D (2007) Effects of soil cadmium on growth, oxidative stress and antioxidant system in wheat seedlings (*Triticum aestivum* L.). Chemosphere 69:89–98

Liu Y, Cotgreave I, Atzori L, Grafström RC (1992) The mechanism of Hg^{2+} toxicity in cultured human oral fibroblasts: the involvement of cellular thiols. Chem Biol Interact 85:69–78

Magalhaes JV, Liu J, Guimaraes CT, Lana UG, Alves VM, Wang YH, Kochian LV (2007) A gene in the multidrug and toxic compound extrusion (MATE) family confers aluminum tolerance in sorghum. Nat Genet 39:1156–1161

Manara A (2012) Plant responses to heavy metal toxicity. In: Furini A (ed) Plants and heavy metals, Briefs in molecular science. Springer, Dordrecht, pp 27–53

Marin AR, Pezeshki SR, Masschelen PH, Choi HS (1993) Effect of dimethylarsenic acid (DMAA) on growth, tissue arsenic, and photosynthesis of rice plants. J Plant Nutr 16:865–880

Martínez M, Bernal P, Almela C, Vélez D, García-Agustín P, Serrano R, Navarro-Aviñó J (2006) An engineered plant that accumulates higher levels of heavy metals than *Thlaspi caerulescens,* with yields of 100 times more biomass in mine soils. Chemosphere 64:478–485

Maser P (2001) Phylogenetic relationships within cation transporter families of Arabidopsis. Plant Physiol 126:1646–1667

Meharg AA, Macnair MR (1992) Suppression of the high affinity phosphate uptake system: a mechanism of arsenate tolerance in *Holcus lanatus* L. J Exp Bot 43:519–524

Mendoza-Cózatl DG, Xie Q, Akmakjian GZ, Jobe TO, Patel A, Stacey MG, Song L, Demoin DW, Jurisson SS, Stacey G et al (2014) OPT3 is a component of the iron-signaling network between leaves and roots and misregulation of OPT3 leads to an over-accumulation of cadmium in seeds. Mol Plant 7:1455–1469

Mijovilovich A, Leitenmaier B, Meyer-Klaucke W, Kroneck PMH, Götz B, Küpper H (2009) Complexation and toxicity of copper in higher plants. II. Different mechanisms for copper versus cadmium detoxification in the copper-sensitive cadmium/zinc hyperaccumulator *Thlaspi caerulescens* (Ganges Ecotype). Plant Physiol 151:715–731

Miller AF (2012) Superoxide dismutases: ancient enzymes and new insights. FEBS Lett 586:585–595

Mills RF, Francini A, Ferreira da Rocha PSC, Baccarini PJ, Aylett M, Krijger GC, Williams LE (2005) The plant P1B-type ATPase AtHMA4 transports Zn and Cd and plays a role in detoxification of transition metals supplied at elevated levels. FEBS Lett 579:783–791

Minamata Disease Research Group (1968) Minamata Disease Research Group: Minamata Disease. Medical School of Kumamoto University, Kumamoto

Minglin L, Yuxiu Z, Tuanyao C (2005) Identification of genes up-regulated in response to Cd exposure in Brassica juncea L. Gene 363:151–158

Misra S, Gedamu L (1989) Heavy metal tolerant transgenic *Brassica napus* L. and *Nicotiana tabacum* L. plants. Theor Appl Genet 78:161–168

Mizuno T, Usui K, Horie K, Nosaka S, Mizuno N, Obata H (2005) Cloning of three ZIP/Nramp transporter genes from a Ni hyperaccumulator plant *Thlaspi japonicum* and their Ni^{2+}-transport abilities. Plant Physiol Biochem 43:793–801

Morel M, Crouzet J, Gravot A, Auroy P, Leonhardt N, Vavasseur A, Richaud P (2009) AtHMA3, a P1B-ATPase allowing Cd/Zn/Co/Pb vacuolar storage in Arabidopsis. Plant Physiol 149:894–904

Mukhopadhyay A, Vij S, Tyagi AK (2004) Overexpression of a zinc-finger protein gene from rice confers tolerance to cold, dehydration, and salt stress in transgenic tobacco. Proc Natl Acad Sci U S A 101:6309–6314

Ng BH, Anderson JW (1979) Light-dependent incorporation of selenite and sulphite into selenocysteine and cysteine by isolated pea chloroplasts. Phytochemistry 18:573–580

Nies DH (2003) Efflux-mediated heavy metal resistance in prokaryotes. FEMS Microbiol Rev 27:313–339

Ortiz DF, Kreppel L, Speiser DM, Scheel G, McDonald G, Ow DW (1992) Heavy metal tolerance in the fission yeast requires an ATP-binding cassette-type vacuolar membrane transporter. Embo J 11:3491–3499

Papadopoulou B, Roy G, Dey S, Rosen B, Ouellette M (1994) Contribution of the *Leishmania* P-glycoprotein-related gene ltpgpA to oxyanion resistance. J Biol Chem 269:11980–11986

Papoyan A, Kochian LV (2004) Identification of *Thlaspi caerulescens* genes that may be involved in heavy metal hyperaccumulation and tolerance. Characterization of a novel heavy metal transporting ATPase. Plant Physiol 136:3814–3823

Park J, Song W-Y, Ko D, Eom Y, Hansen TH, Schiller M, Lee TG, Martinoia E, Lee Y (2012) The phytochelatin transporters AtABCC1 and AtABCC2 mediate tolerance to cadmium and mercury. Plant J 69:278–288

Patra M, Sharma A (2000) Mercury toxicity in plants. Bot Rev 66:379–422

Patra M, Bhowmik N, Bandopadhyay B, Sharma A (2004) Comparison of mercury, lead and arsenic with respect to genotoxic effects on plant systems and the development of genetic tolerance. Environ Exp Bot 52:199–223

Paulose B, Jaiwal PK, Dhankher OP (2008) Heavy metal transporters in plants. In: Jaiwal S, Dhankher OP (eds) Plant membrane and vacuolar transporters. CABI, Wallingford/Cambridge, pp 213–238

Paulose B, Chhikara S, Coomey J, Jung HI, Vatamaniuk O, Dhankher OP (2013) A γ-glutamyl cyclotransferase protects Arabidopsis plants from heavy metal toxicity by recycling glutamate to maintain glutathione homeostasis. Plant Cell 25:4580–4595

Peer WA, Mamoudian M, Lahner B, Reeves RD, Murphy AS, Salt DE (2003) Identifying model metal hyperaccumulating plants: germplasm analysis of 20 *Brassicaceae* accessions from a wide geographical area. New Phytol 159:421–430

Pich A, Scholz G, Stephan UW (1994) Iron-dependent changes of heavy metals, nicotianamine, and citrate in different plant organs and in the xylem exudate of two tomato genotypes. Nicotianamine as possible copper translocator. Plant Soil 165:189–196

Pilon-Smits EAHE, Hwang S, Mel Lytle C, Zhu Y, Tai J, Bravo R, Chen Y, Leustek T, Terry N (1999) Overexpression of ATP sulfurylase in Indian mustard leads to increased selenate uptake, reduction, and tolerance. Plant Physiol 119:123–132

Pineau C, Loubet S, Lefoulon C, Chalies C, Fizames C, Lacombe B, Ferrand M, Loudet O, Berthomieu P, Richard O (2012) Natural variation at the FRD3 MATE transporter locus reveals cross-talk between Fe homeostasis and Zn tolerance in *Arabidopsis thaliana*. PLoS Genet 8:e1003120

Pomponi M, Censi V, Di Girolamo V, De Paolis A, di Toppi LS, Aromolo R, Costantino P, Cardarelli M (2006) Overexpression of Arabidopsis phytochelatin synthase in tobacco plants enhances Cd^{2+} tolerance and accumulation but not translocation to the shoot. Planta 223:180–190

Prasad MNV (1995) Cadmium toxicity and tolerance in vascular plants. Environ Exp Bot 35:525–545

Rauser WE (1995) Phytochelatins and related peptides: structure, biosynthesis, and function. Plant Physiol 109:1141–1149

Rauser WE (1999) Structure and function of metal chelators produced by plants: the case for organic acids, amino acids, phytin, and metallothioneins. Cell Biochem Biophys 31:19–48

Robinson NJ, Tommey AM, Kuske C, Jackson PJ (1993) Plant metallothioneins. Biochem J 295(Pt 1):1–10

Rodríguez-Serrano M, Romero-Puertas MC, Pazmiño DM, Testillano PS, Risueño MC, Del Río LA, Sandalio LM (2009) Cellular response of pea plants to cadmium toxicity: cross talk between reactive oxygen species, nitric oxide, and calcium. Plant Physiol 150:229–243

Rogers EE, Eide DJ, Guerinot ML (2000) Altered selectivity in an Arabidopsis metal transporter. Proc Natl Acad Sci U S A 97:12356–12360

Rose JB, Epstein PR, Lipp EK, Sherman BH, Bernard SM, Patz JA (2001) Climate variability and change in the United States: potential impacts on water- and foodborne diseases caused by microbiologic agents. Environ Health Perspect 109(Suppl):211–221

Rosen BP (1999) Families of arsenic transporters. Trends Microbiol 7:207–212

Rugh CL (2001) Mercury detoxification with transgenic plants and other biotechnological breakthroughs for phytoremediation. Vitro Cell Dev Biol Plant 37:321–325

Rugh CL, Wilde HD, Stack NM, Thompson DM, Summers AO, Meagher RB (1996) Mercuric ion reduction and resistance in transgenic Arabidopsis thaliana plants expressing a modified bacterial merA gene. Proc Natl Acad Sci U S A 93:3182–3187

Ruiz ON, Hussein HS, Terry N, Daniell H (2003) Phytoremediation of organomercurial compounds via chloroplast genetic engineering. Plant Physiol 132:1344–1352

Ryan PR, Raman H, Gupta S, Horst WJ, Delhaize E (2009) A second mechanism for aluminum resistance in wheat relies on the constitutive efflux of citrate from roots. Plant Physiol 149:340–351

Salt DE, Thurman DA, Tomsett AB, Sewell AK (1989) Copper phytochelatins of *Mimulus guttatus*. Proc R Soc B Biol Sci 236:79–89

Salt DE, Prince RC, Pickering IJ, Raskin I (1995) Mechanisms of cadmium mobility and accumulation in Indian mustard. Plant Physiol 109:1427–1433

Salt DE, Prince RC, Baker AJM, Raskin I, Pickering IJ (1999) Zinc ligands in the metal hyperaccumulator Thlaspi caerulescens as determined using x-ray absorption spectroscopy. Environ Sci Technol 33:713–717

Schaaf G, Ludewig U, Erenoglu BE, Mori S, Kitahara T, von Wirén N (2004) ZmYS1 functions as a proton-coupled symporter for phytosiderophore- and nicotianamine-chelated metals. J Biol Chem 279:9091–9096

Schat H (2002) The role of phytochelatins in constitutive and adaptive heavy metal tolerances in hyperaccumulator and non-hyperaccumulator metallophytes. J Exp Bot 53:2381–2392

Schmoger MEV (2000) Detoxification of arsenic by phytochelatins in plants. Plant Physiol 122:793–802

Schützendübel A, Polle A (2002) Plant responses to abiotic stresses: heavy metal-induced oxidative stress and protection by mycorrhization. J Exp Bot 53:1351–1365

Schwartz MS, Benci JL, Selote DS, Sharma AK, Chen AGY, Dang H, Fares H, Vatamaniuk OK (2010) Detoxification of multiple heavy metals by a half-molecule ABC transporter, HMT-1, and coelomocytes of *Caenorhabditis elegans*. PLoS One 5:e9564

Scoccianti V, Crinelli R, Tirillini B, Mancinelli V, Speranza A (2006) Uptake and toxicity of Cr(III) in celery seedlings. Chemosphere 64:1695–1703

Sekhar K, Priyanka B, Reddy VD, Rao KV (2011) Metallothionein 1 (CcMT1) of pigeonpea (*Cajanus cajan* L.) confers enhanced tolerance to copper and cadmium in Escherichia coli and Arabidopsis thaliana. Environ Exp Bot 72:131–139

Setya A, Murillo M, Leustek T (1996) Sulfate reduction in higher plants: molecular evidence for a novel 5′-adenylylsulfate reductase. Proc Natl Acad Sci 93:13383–13388

Shanker AK, Cervantes C, Loza-Tavera H, Avudainayagam S (2005) Chromium toxicity in plants. Environ Int 31:739–753

Sharma D, Sharma C, Tripathi R (2003) Phytotoxic lesions of chromium in maize. Chemosphere 51:63–68

Shim D, Hwang JU, Lee J, Lee S, Choi Y, An G, Martinoia E, Lee Y (2009) Orthologs of the class A4 heat shock transcription factor HsfA4a confer cadmium tolerance in wheat and rice. Plant Cell 21:4031–4043

Shin H, Shin HS, Dewbre GR, Harrison MJ (2004) Phosphate transport in Arabidopsis: Pht1;1 and Pht1;4 play a major role in phosphate acquisition from both low- and high-phosphate environments. Plant J 39:629–642

Shu X, Yin L, Zhang Q, Wang W (2012) Effect of Pb toxicity on leaf growth, antioxidant enzyme activities, and photosynthesis in cuttings and seedlings of *Jatropha curcas* L. Environ Sci Pollut Res Int 19:893–902

Singh S, Verma A, Dubey VK (2012) Effectivity of anti-oxidative enzymatic system on diminishing the oxidative stress induced by aluminium in chickpea (*Cicer arietinum* L.) seedlings. Braz J Plant Physiol 24:47–54

Smith FW, Ealing PM, Hawkesford MJ, Clarkson DT (1995) Plant members of a family of sulfate transporters reveals functional subtypes. Proc Natl Acad Sci U S A 92:9373–9377

Song WY, Sohn EJ, Martinoia E, Lee YJ, Yang YY, Jasinski M, Forestier C, Hwang I, Lee Y (2003) Engineering tolerance and accumulation of lead and cadmium in transgenic plants. Nat Biotechnol 21:914–919

Song WY, Martinoia E, Lee J, Kim D, Kim DY, Vogt E, Shim D, Choi KS, Hwang I, Lee Y (2004) A novel family of cys-rich membrane proteins mediates cadmium resistance in Arabidopsis. Plant Physiol 135:1027–1039

Song WY, Park J, Mendoza-Cózatl DG, Suter-Grotemeyer M, Shim D, Hörtensteiner S, Geisler M, Weder B, Rea PA, Rentsch D et al (2010) Arsenic tolerance in Arabidopsis is mediated by two ABCC-type phytochelatin transporters. Proc Natl Acad Sci U S A 107:21187–21192

Stacey MG, Patel A, McClain WE, Mathieu M, Remley M, Rogers EE, Gassmann W, Blevins DG, Stacey G (2008) The Arabidopsis AtOPT3 protein functions in metal homeostasis and movement of iron to developing seeds. Plant Physiol 146:589–601

Stadtman ER, Oliver CN (1991) Metal-catalyzed oxidation of proteins. Physiological consequences. J Biol Chem 266:2005–2008

Steffens JC (1990) The heavy metal-binding peptides of plants. Annu Rev Plant Physiol Plant Mol Biol 41:553–575

Summers AO (1986) Organization, expression, and evolution of genes for mercury resistance. Annu Rev Microbiol 40:607–634

Szczypka M, Wemmie J, Moye-Rowley W, Thiele D (1994) A yeast metal resistance protein similar to human cystic fibrosis transmembrane conductance regulator (CFTR) and multidrug resistance-associated protein. J Biol Chem 269:22853–22857

Terry N, Zayed AM, De Souza MP, Tarun AS (2000) Selenium in higher plants. Annu Rev Plant Physiol Plant Mol Biol 51:401–432

Thomas JC, Davies EC, Malick FK, Endreszl C, Williams CR, Abbas M, Petrella S, Swisher K, Perron M, Edwards R et al (2003) Yeast metallothionein in transgenic tobacco promotes copper uptake from contaminated soils. Biotechnol Prog 19:273–280

Thomine S, Lelièvre F, Debarbieux E, Schroeder JI, Barbier-Brygoo H (2003) AtNRAMP3, a multispecific vacuolar metal transporter involved in plant responses to iron deficiency. Plant J 34:685–695

Tommasini R, Vogt E, Fromenteau M, Hörtensteiner S, Matile P, Amrhein N, Martinoia E (1998) An ABC-transporter of *Arabidopsis thaliana* has both glutathione-conjugate and chlorophyll catabolite transport activity. Plant J 13:773–780

Tommey AM, Shi J, Lindsay WP, Urwin PE, Robinson NJ (1991) Expression of the pea gene PSMTA in *E. coli*. Metal-binding properties of the expressed protein. FEBS Lett 292:48–52

Trampczynska A, Küpper H, Meyer-Klaucke W, Schmidt H, Clemens S (2010) Nicotianamine forms complexes with Zn(II) *in vivo*. Metallomics 2:57–66

Tseng TS, Tzeng SS, Yeh KW, Yeh CH, Chang FC, Chen YM, Lin CY (1993) The heat-shock response in rice seedlings: isolation and expression of cDNAs that encode Class I low-molecular-weight heat-shock proteins. Plant Cell Physiol 34:165–168

Van Assche F, Clijsters H (1987) Enzyme analysis in plants as a tool for assessing phytotoxicity of heavy metal polluted soils. Med Fac Landbouww Rijksuniv Gent 52:1819–1824

Van der Zaal BJ, Neuteboom LW, Pinas JE, Chardonnens AN, Schat H, Verkleij JA, Hooykaas PJ (1999) Overexpression of a novel Arabidopsis gene related to putative zinc-transporter genes from animals can lead to enhanced zinc resistance and accumulation. Plant Physiol 119:1047–1055

Van Huysen T, Abdel-Ghany S, Hale KL, LeDuc D, Terry N, Pilon-Smits EAH (2003) Overexpression of cystathionine-gamma-synthase enhances selenium volatilization in *Brassica juncea*. Planta 218:71–78

Van Nguyen N, Ferrero A (2006) Meeting the challenges of global rice production. Paddy Water Environ 4:1–9

Vatamaniuk OK, Mari S, Lu YP, Rea PA (1999) AtPCS1, a phytochelatin synthase from Arabidopsis: isolation and *in vitro* reconstitution. Proc Natl Acad Sci U S A 96:7110–7115

Vatamaniuk OK, Bucher EA, Sundaram MV, Rea PA (2005) CeHMT-1, a putative phytochelatin transporter, is required for cadmium tolerance in *Caenorhabditis elegans*. J Biol Chem 280:23684–23690

Verret F, Gravot A, Auroy P, Leonhardt N, David P, Nussaume L, Vavasseur A, Richaud P (2004) Overexpression of AtHMA4 enhances root-to-shoot translocation of zinc and cadmium and plant metal tolerance. FEBS Lett 576:306–312

Vierling E (1991) The roles of heat shock proteins in plants. Annu Rev Plant Physiol Plant Mol Biol 42:579–620

Vij S, Tyagi AK (2006) Genome-wide analysis of the stress associated protein (SAP) gene family containing A20/AN1 zinc-finger(s) in rice and their phylogenetic relationship with Arabidopsis. Mol Genet Genomics 276:565–575

Vögeli-Lange R, Wagner GJ (1990) Subcellular localization of cadmium and cadmium-binding peptides in tobacco leaves: implication of a transport functions for cadmium-binding peptides. Plant Physiol 92:1086–1093

Von Wiren N (1999) Nicotianamine chelates both FeIII and FeII. Implications for metal transport in plants. Plant Physiol 119:1107–1114

Wang J, Zhao FJ, Meharg AA, Raab A, Feldmann J, McGrath SP (2002) Mechanisms of arsenic hyperaccumulation in *Pteris vittata*. Uptake kinetics, interactions with phosphate, and arsenic speciation. Plant Physiol 130:1552–1561

Wawrzynski A, Kopera E, Wawrzynska A, Kaminska J, Bal W, Sirko A (2006) Effects of simultaneous expression of heterologous genes involved in phytochelatin biosynthesis on thiol content and cadmium accumulation in tobacco plants. J Exp Bot 57:2173–2182

Weber M, Harada E, Vess C, Roepenack-Lahaye EV, Clemens S (2004) Comparative microarray analysis of Arabidopsis thaliana and Arabidopsis halleri roots identifies nicotianamine synthase, a ZIP transporter and other genes as potential metal hyperaccumulation factors. Plant J 37:269–281

Wenzel W, Jockwer F (1999) Accumulation of heavy metals in plants grown on mineralised soils of the Austrian Alps. Environ Pollut 104:145–155

Wilber CG (1980) Toxicology of selenium: a review. Clin Toxicol 17:171–230

Wilber C (1983) Selenium: a potential environmental poison and a necessary food constituent. Charles C Thomas Pub Ltd, Springfield

Williams PN, Price AH, Raab A, Hossain SA, Feldmann J, Meharg AA (2005) Variation in arsenic speciation and concentration in paddy rice related to dietary exposure. Environ Sci Technol 39:5531–5540

Wong CKE, Jarvis RS, Sherson SM, Cobbett CS (2009) Functional analysis of the heavy metal binding domains of the Zn/Cd-transporting ATPase, HMA2, in Arabidopsis thaliana. New Phytol 181:79–88

Xiong XZ, Li PJ, Wang YS, Ten H, Wang LP, Song SH (1987) Environmental capacity of arsenic in soil and mathematical model. Huanjing Kexue 8:8–14

Yadav SK (2010) Heavy metals toxicity in plants: an overview on the role of glutathione and phytochelatins in heavy metal stress tolerance of plants. South Afr J Bot 76:167–179

Yang Z, Wu Y, Li Y, Ling HQ, Chu C (2009) OsMT1a, a type 1 metallothionein, plays the pivotal role in zinc homeostasis and drought tolerance in rice. Plant Mol Biol 70:219–229

Zayed AM, Terry N (1992) Selenium volatilization in broccoli as influenced by sulfate supply. J Plant Physiol 140:646–652

Zayed A, Lytle CM, Terry N (1998) Accumulation and volatilization of different chemical species of selenium by plants. Planta 206:284–292

Zhai Z, Gayomba SR, Jung HI, Vimalakumari NK, Piñeros M, Craft E, Rutzke MA, Danku J, Lahner B, Punshon T et al (2014) OPT3 is a phloem-specific iron transporter that is essential for systemic iron signaling and redistribution of iron and cadmium in Arabidopsis. Plant Cell 26:2249–2264

Zhou B, Yao W, Wang S, Wang X, Jiang T (2014) The metallothionein gene, TaMT3, from *Tamarix androssowii* confers Cd^{2+} tolerance in tobacco. Int J Mol Sci 15:10398–10409

Zhu YL, Pilon-Smits EA, Tarun AS, Weber SU, Jouanin L, Terry N (1999) Cadmium tolerance and accumulation in Indian mustard is enhanced by overexpressing gamma-glutamylcysteine synthetase. Plant Physiol 121:1169–1178

Zhu YG, Williams PN, Meharg AA (2008) Exposure to inorganic arsenic from rice: a global health issue? Environ Pollut 154:169–171

Zientara K, Wawrzyńska A, Lukomska J, López-Moya JR, Liszewska F, Assunção AGL, Aarts MGM, Sirko A (2009) Activity of the AtMRP3 promoter in transgenic *Arabidopsis thaliana* and *Nicotiana tabacum* plants is increased by cadmium, nickel, arsenic, cobalt and lead but not by zinc and iron. J Biotechnol 139:258–263

Zimeri AM, Dhankher OP, McCaig B, Meagher RB (2005) The plant MT1 metallothioneins are stabilized by binding cadmiums and are required for cadmium tolerance and accumulation. Plant Mol Biol 58:839–855

Prospects of Genetic Manipulation for Enhanced Heavy Metal Tolerance and Bioremediation in Relation to Climate Change

8

Sudhakar Srivastava and P. Suprasanna

Abstract

"Climate change" is indicative of changes in mean annual temperature, precipitation patterns, and atmospheric CO_2 and greenhouse gas concentrations. These parameters have been observed to alter significantly in the past few decades and are expected to affect the plant and animal health in the coming years. The ability of plants to take up heavy metals and metalloids depends, for example, on their bioavailability in the medium, competition for uptake by other elements, and plant habit and growth characteristics. These factors are influenced by the prevailing climatic conditions of an area. Thus, climate change is expected to indirectly alter mineral composition of plants. Metal accumulation ability of plants also holds promise for the cleanup of contaminated sites, the technology described as phytoremediation. Climate change might thus affect proposed remediation abilities of plants. This review focuses on the effect of climatic parameters on the metal accumulation and tolerance of plants and prospective strategies of genetic manipulation for achieving enhanced remediation potential in relation to climate change.

Keywords

Bioremediation • Climate change • Genetic manipulation • Global warming • Metal tolerance

S. Srivastava
Nuclear Agriculture and Biotechnology Division, Bhabha Atomic Research Centre, Mumbai 400085, Maharashtra, India

Institute of Environment & Sustainable Development, Banaras Hindu University, Varanasi 221005, India
e-mail: sudhakar.srivastava@gmail.com

P. Suprasanna (✉)
Nuclear Agriculture and Biotechnology Division, Bhabha Atomic Research Centre, Mumbai 400085, Maharashtra, India
e-mail: prasanna@barc.gov.in

8.1 Climate Change: Introduction

Continued worldwide industrialization has induced dramatic changes in the global biological, chemical, and physical environment and overall climate. In recent terms, the climate change is referred to the ongoing changes in modern climate, including increase in green-

P.K. Jaiwal et al. (eds.), *Genetic Manipulation in Plants for Mitigation of Climate Change*,
DOI 10.1007/978-81-322-2662-8_8

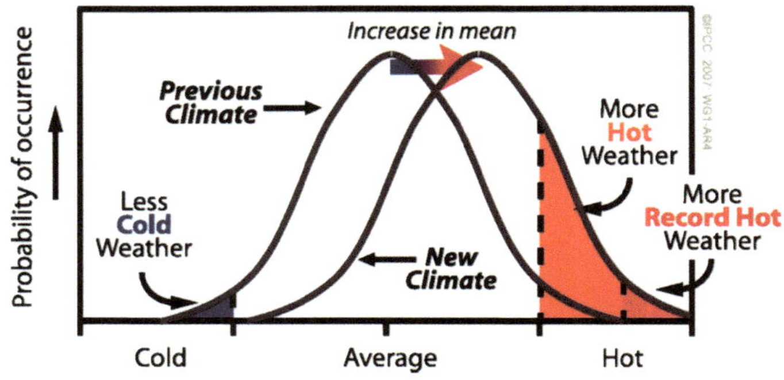

Fig. 8.1 Schematic showing the effect on extreme temperatures when the mean temperature increases, for a normal temperature distribution

Source: www.ipcc.ch/publications_and_data/ar4/wg1/en/box-ts-5-figure-1.html

house gas (atmospheric CO_2 and CH_4) concentrations, increase in mean annual temperature (global warming; Fig. 8.1), changes in precipitation patterns, melting of ice caps and glaciers and reduced snow cover, and increases in ocean temperatures and ocean acidity (due to seawater absorbing heat and carbon dioxide from the atmosphere) (Intergovernmental Panel on Climate Change (IPCC) 2007). The increase of climatic variability due to global climate change exerts climatic stress on plants that is novel in magnitude and frequency (Hegerl et al. 2011).

The rise in the level of greenhouse gases causes a rise in the Earth's mean temperature because the thermal radiation from the Earth's surface is absorbed by greenhouse gases and reradiated in all directions including back to the surface. This effect is pronounced as greenhouse effect. Over the last century, atmospheric concentrations of CO_2 increased from a preindustrial value of 278–379 ppm in 2005, and the average global temperature rose by 0.74 °C. The atmospheric CO_2 concentration is predicted to increase to as high as 500–1000 ppm by the year 2100 (IPCC 2007) while the global average temperature may increase by 1.8–4.0 °C (Noyes et al. 2009). Several predictions of effects of climate change range in the extent of effect but do suggest serious impacts on ecosystems and plants (Noyes et al. 2009; Thomson et al. 2010). Climate change may have an impact on growth and productivity of crops (Lobell and Field 2007; Thomson et al. 2010), mineral and biochemical composition, and response to environmental constraints.

Plants are used in remediation of various pollutants like organics and heavy metals and in eco-sensitive zones for the prevention of erosion, water flow and land degradation management. Climate-mediated effects on the plant growth would also affect the efficiency of these applications. Heavy metal contamination is caused by anthropogenic activities (mining, metal processing, fossil fuel combustion) and agricultural practices (application of fertilizers, fungicides, and sewage sludge dispersal). In the case of As, natural geochemical weathering of rocks and biogeochemical and microbial reactions have resulted in widespread contamination (Srivastava et al. 2012). Climate change may also have an impact on the level of heavy metal contamination due to effects on the bioavailability of metals and alter the environmental fate and behavior/toxicity of pollutants through altering physicochemical–biological properties of atmosphere, water, and soils (Dowdall et al. 2008; Noyes et al. 2009).

This review focuses on the issue of climate change with the perspective of phytoremediation potential of plants and highlights the directions for future to maintain/increase plants' potential for metal accumulation.

8.2 Effect of Climate Change on Growth Responses of Plants

The major contributor of climate change is the rapid increase in CO_2 concentration, which impacts directly the agricultural productivity and crop quality as well as indirectly through changes in global temperature and precipitation patterns. Carbon source–sink relationships are believed to play a major role in determining the ability of a plant to utilize e[CO_2] and avoid downward acclimation of photosynthesis upon prolonged e[CO_2] exposure (Tausz et al. 2013). CO_2 enrichment to 550–600 μL L^{-1} is well established to increase grain yield by 10–16 % (Kimball 2006; Long et al. 2006; Högy et al. 2009). However, the extent of the CO_2 response on grain yield may vary among crop cultivars and according to environmental conditions and nutrient availability (Ainsworth and Long 2005). In the presence of e[CO_2], increased photosynthetic rate and biomass production and enhanced water and nutrient use efficiency due to changes in root morphology are observed (Korner et al. 1997; Phillips et al. 2009; Prior et al. 2011). Wang et al. (2003) found significant increases in ascorbate, glutathione (GSH), anthocyanin, and phenolic contents as well as oxygen radical absorbance activity in strawberry fruit under e[CO_2]. Effect of elevated CO_2 on amino acid composition has also been reported (Högy et al. 2009) with significant effect reported for glycine and glutamic acid, the component amino acids of glutathione. In this study, minerals such as K, Mo, and Pb increased, while Mg, Fe, Cd, and Si decreased, suggesting that elevated CO_2 may affect chemical composition of grains. Högy et al. (2013) grew wheat in a FACE and found an increase in thousand grain weight with size distribution shifted toward larger grains but no significant impact on biomass and grain yield. They also observed a change in protein composition and mineral concentrations. Tausz-Posch et al. (2013) suggest that growth under eCO_2 is influenced by climatic factors (e.g., high temperature and water supply) and that there are intraspecific variation in response toward e[CO_2] and climatic factors with respect to photosynthetic performance and antioxidant levels. Ainsworth et al. (2006) have identified CO_2-responsive transcripts through microarray analysis of soybean plants grown in SoyFACE facility. [CO_2]-responsive genes largely suggest that elevated [CO_2] stimulates the respiratory breakdown of carbohydrates, which provides increased energy and biochemical precursors for leaf expansion and growth at elevated [CO_2]. Tausz et al. (2013) identified important traits for improved crop performance under elevated CO_2, which may be utilized for non-crop plants too for the purpose of developing heavy metal accumulating varieties.

Tropospheric ozone (O_3) is an important secondary pollutant resulting from emission of volatile organic compounds and nitrogen oxides and is recognized as a greenhouse gas (Serengil et al. 2011). Ground-level O_3 concentrations have doubled since the nineteenth century, with concentrations in major crop growing regions of the United States, India, and China increasing by up to 10 % yr^{-1} (Vingarzan 2004; Ashmore et al. 2006, Galant et al. 2012). Climate models predict that mean surface O_3 concentrations may rise 20–25 % globally by 2050, with concentrations in India and South Asia reaching comparable values by 2020 (IPCC 2001; Dentener et al. 2005; Van Dingenen et al. 2009). Acute exposure to ozone can result in chlorosis and necrosis. Long-term exposure leads to reduced photosynthetic efficiency and mobilization of energy reserves (Ahsan et al. 2010) and accelerates senescence of plant cell. In FACE trials with soybean, a 13 ppb increase in O_3 from 56 to 69 ppb resulted in a 20 % decrease in crop yield (Morgan et al. 2004, 2006). Proteomic studies on rice, wheat, soybean, and poplar under continuous (2–5 d) high O_3 (120–200 ppb) stress revealed reductions in the photosynthetic and carbon metabolism proteins and induction of defense/stress-related proteins (Cho et al. 2008; Feng et al. 2008; Renaut et al. 2009; Ahsan et al. 2010; Sarkar et al. 2010; Galant et al. 2012). Glutathione S-transferases, ascorbate peroxidase, methionine sulfoxide reductase, carotenoids, and amino acid-associated proteins were found to be altered either in oxidation state or in abundance in response to O_3 treatment in experiments conducted by Galant et al. (2012) on soybean FACE system. In another experiment with soybean exposed to O_3 at ambient and elevated CO_2 (Gillespie et al. 2012), antioxidant enzymes superoxide dismutase (SOD) and monodehydroascorbate reductase (MDHAR) and genes of various enzymes were significantly altered. Hoshika et al. (2013) reported that ambient ozone increase affects the response of stomata and delays their closure and thus increases water loss under both well-watered and drought conditions.

High temperatures increase the transpiration and stomatal conductance, but decrease the photosynthesis resulting in significant reduction in the plant biomass yield (Djanaguiraman et al. 2010; Qaderi et al. 2012). Shahbaz et al. (2013) found that in Chinese cabbage, UV-A+B had no significant impact on biomass production, dry matter, and pigment content. However, combined stress of Cu and UV caused decline in biomass and pigment content. The reason behind the toxicity was UV-mediated increase in Cu accumulation in both roots and shoots. Drought and salinity stresses are also outcome of climate change due to altered precipitation patterns and warming. These stresses decrease growth rate by reducing photochemical efficiency and photosynthetic rates (Chaves et al. 2009). The series of events leading to the perturbation of cellular metabolism under drought and salt stress are suggested to be as follows: less water availability, stomatal closure, altered gaseous exchange, inhibition of photosynthesis, effect on electron flow in ETC in chloroplast and mitochondria, and increase in the production of reactive oxygen species (Munns and Tester 2008). Hence, the imposition of oxidative stress and altered energy reactions caused by ETC over-reduction affect general metabolism and growth. Through the above discussion, it becomes evident that individual components of climate produce a wide array of morphological, biochemical, and molecular effects on plants. Hence, climate change, as a whole, is bound to cause serious impacts on plant growth and, therefore, on their applications as crop or phytoremediator in the coming future.

8.3 Effect of Climate/Geographical Conditions on Metal Composition/Bioavailability

Metal composition means total metal concentrations in soil or water, which do not necessarily correspond with metal bioavailability, which is the proportion of total metals that are available for incorporation into biota (bioaccumulation). Metal bioavailability is controlled by many factors including concentration and speciation of metals, mineralogy, pH, redox potential, temperature, total organic content, and suspended particulate content, as well as volume of water, water velocity, and duration of water availability. Climate (viz., humid, arid, tropical, temperate, etc.) strongly influences soil types, and both these factors control element mobility and availability. Schiedek et al. (2007) have discussed the impacts of climate change on heavy metal contamination for marine ecosystems. The predictions of increased flooding and hydrological changes pose risk of mobilization of metals due to altered chemistry (Visser et al. 2012). Arsenic contamination in groundwater of Bangladesh and West Bengal, India, has been associated to the depth and age of groundwater (Fendorf et al. 2010). As released from flooded contaminated soils has been found to be temperature dependent (Weber et al. 2010). Warming can enhance P release by the increase of soil phosphatase activity and may thus facilitate As mobilization and higher plant capture (Cao and Ma 2004). Soil warming can alter release of soluble metal ions into soil solution via decomposition of organic matter (leading to loss of cation-exchange capacity; Sardans et al. 2008a; van Gestel 2008), change of microbial communities (Rajkumar et al. 2013), and the destruction of soil aggregates and through change in soil enzymatic activity (Cao and Ma 2004; Sardans and Penuelas 2006).

Meharg et al. (2009) compiled an extensive data set of total As analysis for 901 polished (white) grain samples, originating from 10 countries from 4 continents. They found that median total As contents of rice varied sevenfold, with Egypt (0.04 mg kg^{-1}) and India (0.07 mg kg^{-1}) having the lowest As content while the United States (0.25 mg kg^{-1}) and France (0.28 mg kg^{-1}) having the highest content. Although this study highlighted the differential ability of varieties for taking up As, it may also indicate how climate/geographical conditions prevailing in different countries/locations affect the As level in rice grains.

Thawing of permafrost and the subsequent export of carbon are a widespread phenomenon, and the projection is that it will increase even more in the near future. Rydberg et al. (2010) suggest that this may lead to release of substantial amounts of Hg into sensitive arctic freshwater and marine systems due to thawing of northern peatlands. Klaminder et al. (2010) reconstructed the pollution history of three high-latitude lakes situated in a region where a recent climatic shift has occurred on the basis of 206 Pb/207 Pb ratio between atmospheric contaminants (206 Pb/207 Pb ratio 1.16) and geogenic Pb in the catchment soil (206 Pb/207 Pb ratio 1.22) and concluded that soil processes stimulated by the ongoing climate change at high latitudes might work counteractive to efforts to reduce contaminant levels in subarctic lakes. Stern et al. (2012) reviewed the impact of climate change on Hg levels, chemistry, and dynamics in arctic region. They concluded that the effects of recent climate warming on the Arctic's physical environment have been profound. From Hg perspective, the most important impacts have occurred in precipitation rates and type (rain vs. snow), riverine discharge and seasonality, and lake ice and sea-ice seasonality, etc.

Since climatic conditions and geographical characteristics affect metal composition and availability in soils and waters, climate change would affect plants' performance for phytoremediation in two contrasting ways, viz., plant growth and metal bioavailability. Plant growth is dependent on mineral and heavy metal bioavailability and plants themselves affect metal concentrations in rhizosphere. Further, individual climatic components (CO_2, temperature, ozone, water availability) have different effects on plant growth and metal bioavailability. Hence, these interactions need careful and in-depth analyses to predict future scenarios.

8.4 Effect of Climate Change on Metal Accumulation

8.4.1 Effect of Temperature

Effect of high temperature on metal accumulation is attributable to several influences, viz., increased water loss and hence transpiration, changes in lipid composition of plasma membrane and its fluidity (Lynch and Steponkus 1987), effect on optimal biochemical processes of plants and associated microbes including the metal uptake, and altered metal bioavailability. Vaughan et al. (2009) reported that Fe-reducing conditions in three flooded soils developed within 2 days of water saturation when temperature was >9 °C, but only after 20 days when temperature was <3.9 °C. The equilibrium between the cell wall exchange sites and the metal in solution changes with temperature. Intracellular ion accumulation may thus increase with the increasing cation-exchange capacity (CEC) of cell walls due to the ion gradient established around the plasma membrane. As higher temperatures lead to increased extracellular concentrations of heavy metals, the cell wall exchange properties impact the ion availability for uptake, ion diffusion rates in the apoplast, and membrane transporters. Several authors have documented effect of temperature on metal accumulation profile of plants.

Ekvall and Greger (2003) found that two ecotypes of *Pinus sylvestris* reacted differently to temperature in their Cd uptake translocation. Fritioff et al. (2005) found that heavy metal accumulation increased in two submersed plant species (*Elodea Canadensis* and *Potamogeton natans*) as the temperature increased from 5 to 20 °C. Milner et al. (2007) studied the effects of plant growth at 10 °C and 25 °C on GST-specific activity and glutathione (GSH) pool size in herbicide-resistant and susceptible *Alopecurus myosuroides* biotypes and found that levels of GST and GSH do change as per the growth temperatures. Sardans et al. (2008b) found that warming treatment increased the accumulation of Al in *Erica multiflora* and *Globularia alypum*; As, Cr, and Pb in *E. multiflora*; and Sb and Zn in *G. alypum*. Lu et al. (2009) showed that low-temperature treatment (4 °C) significantly inhibited Cd uptake and reduced upward translocation of Cd to shoots by up to 90 % in one ecotype of *Sedum alfredii*, whereas no such effect was observed in the other ecotypes investigated.

Chen et al. (2010) carried out pot experiments to investigate the effects of increased soil temperature on shoot uptake of heavy metals by *Zea mays* and *Vigna radiata* from heavy metal-contaminated soils. They found that timing of heating treatment and chelant application significantly affected heavy metal accumulation. Irrigation with 100 °C water after the chelant addition or with 100 °C chelant solutions directly resulted in significantly higher phytoextraction of metals in the two crops compared with 25 °C chelant solutions. In addition, a novel application method to increase soil temperature using underground polyvinyl chloride tubes increased the chelant-assisted extraction efficiency of Cu approximately 10- to 14-fold in corn and fivefold in mung bean. Hawrylak-Nowak et al. (2010) investigated the effect of Se supply in *Cucumis sativus* grown under short-term low-temperature stress (a day/night temperature of 10 °C/5 °C for the first 24 h and then at 20 °C/15 °C for the next 24 h followed by 25/20 °C for 7 days). Compared with the control, the Se-treated plants showed significant increase of proline content in leaves immediately after chilling than after rewarming. However, the resistance of plants to low temperature was not clearly enhanced, as concluded on the basis of FW and photosynthetic pigment accumulation. Li D et al. (2011) found that higher temperature increased Cd accumulation in roots of wheat plants in Cd-polluted soils, but reduced root elongation. Rofkar and Dwyer (2011) assessed seasonal effects on As uptake in *Carex stricta* and *Spartina pectinata* in conditions representative of spring (15/5 °C) and summer (28/17 °C) temperatures and light regimes. Both species had comparable rates of As uptake into roots in summer conditions, but *C. stricta* had a higher maximum net influx rate in spring conditions than *S. pectinata*.

Li et al. (2012) also found that the high temperature increased Cu, Zn, and Fe leaf concentrations in *Solanum tuberosum*; however, the treatment decreased Cd, Pb, Fe, Zn, and Cu concentrations in tubers. Pourghasemian et al. (2013) investigated the effect of temperature (18 and 23 °C) on Cd uptake, translocation, as well as tolerance in wild and cultivated species of safflower. Net accumulation of Cd via root increased with an increase in temperature for some varieties but not for all; however, Cd translocation to shoots significantly increased in all genotypes.

8.4.2 Effect of Carbon Dioxide

In the past decade, the responses of plants to e[CO_2] under heavy metals have been investigated by several researchers. The e[CO_2] has been found to increase the uptake of essential micronutrients, such as Cu, Fe, Mn, and Zn (Lieffering et al. 2004; Jia et al. 2007; Yang et al. 2007; Zheng et al. 2008; Högy and Fangmeier 2008). In a greenhouse pot experiment, a significant increase in frond biomass and accumulation of Cu was reported in *Pteridium revolutum* and *P. aquilinum* (Zheng et al. 2008) in e[CO_2] conditions. The study pointed out that phytoextraction efficiency of potential candidates can be increased by increasing the ambient CO_2 concentrations. In *sorghum* and *trifolium*, both increased biomass and higher accumulation of Cs were observed in the presence of e[CO_2] (Wu et al. 2009), which was attributed to CO_2-mediated decrease in the rhizosphere soil pH and subsequent greater root exudation of carbonic acid. Cs uptake was also found to increase in response to e[CO_2] in *Phytolacca americana* (C_3 species) and *Amaranthus cruentus* (C4 species) (Tang et al. 2011) Similarly, Kim and Kang (2011) observed an increase in biomass production and Pb uptake in pine seedlings in Pb-contaminated soils as a result of e[CO_2]. In fly ash-amended soils also, e[CO2] significantly increased the total biomass production and concentrations of Br, Co, Cu, Fe, Mn, Ni, Pb, and Zn in roots, stems, and seeds of soybean plants (Rodriguez et al. 2011). Song et al. (2012) further evaluated the effects of e[CO_2] on Cs uptake in *Phytolacca americana* and *Amaranthus cruentus* and found that shoot and root biomass and Cs accumulation were higher under e[CO_2] than under ambient CO_2. In addition, positive influences of e[CO_2] were noticed on the populations of bacteria, actinomycetes, and fungi and the microbial C and N in the rhizosphere soils of both species.

In contrast, some results showed that e[CO_2] had either no effect or reduced metal accumulation in plants. Jia et al. (2007) found that e[CO_2] decreased leaf Cu concentrations of a Japonica rice growing in Cu-contaminated soil to close to that growing in control soil. While at low CO_2 concentrations, Cu concentrations in leaves were significantly higher in plants growing at Cu-contaminated site than that at control site. Jia et al. (2010) observed that the e[CO_2] reduced Cd accumulation in *L. mutiforum* and *L. perenne* grown in Cd-amended soils. Jia et al. (2011) further demonstrated that at e[CO_2], *L. perenne* may be better protected against Cd stress due to higher synthesis of high-molecular-weight PCs (PC_4, PC_5, PC_6) in shoots and roots. Jia et al. explained that the decreased Cd accumulation in plant tissues can be attributed to dilution effects as a result of increased plant biomass due to e[CO_2] levels due to increased photosynthesis and enhanced antioxidant enzyme activities. In addition in various experiments of Jia et al., root length, surface area, volume, tip number, and fine roots all decreased under Cd exposure, while by contrast, elevated levels of CO_2 significantly increased all those parameters in the presence of Cd, compared to the CO_2 control. High-CO_2-grown green microalga *Scenedesmus armatus* has also been found to produce significantly more PCs and better tolerate Cd toxicity than low-CO_2-grown one (Tukaj et al. 2007); however, growth of alga was decreased under both low and high CO_2 conditions.

Natali et al. (2009) observed significant increase in Mn concentrations in leaves of forest trees (*Quercus chapmanii*, *Q. geminata*, and *Q. myrtifolia*) under e[CO_2], which was attributed to increased photosynthetic demand for Mn. However, other heavy metals with similar chemical properties to Mn and with higher bioavailabilities did not increase. Li et al. (2010) demonstrated that e[CO_2] significantly increased the total biomass of six Chinese rice in the presence of metals (Cu and Cd); however, it differently affected the accumulation patterns of Cd. The variable Cd accumulation pattern was ascribed to genetic differences among the rice varieties and to the changes in exudation rates and spectrum of organic acids. Guo et al. (2011) also reported variable effects of e[CO_2] on Cd and Cu uptake by rice and wheat growing in metal-contaminated soils with an increase in Cd concentration but decline in Cu concentration in both plants. Guo et al. (2011) suggested that differential binding affinity of Cd and Cu to organic matter and hence variable mobilization in soil might be responsible for observed accumulation pattern. Van den Berge et al. (2011) questioned that will species that are sensitive/tolerant to Zn pollution still have the same sensitivity/tolerance in a future climate? They analyzed the response of constructed grassland communities to Zn supply, under a current climate and a future climate (elevated CO_2 and warming). Zn concentrations increased in roots and shoots with Zn addition, but this increase did not differ between climates. Light-saturated net CO_2 assimilation rate (Asat) of the species, on the other hand, responded differently to Zn addition depending on climate. They suggested that if Asat in a future climate does not compromise productivity and survival, sensitivity will not be altered by climate change.

8.4.3 Effect of Ozone and Other Environmental Factors

Shukla et al. (2008) studied interactive effects of Cd and UV-B exposure on essential nutrient (Ca, Mg, K) uptake and biomass and chlorophyll content in *Brassica campestris* seedlings. They found that Ca, Mg, and K uptake was reduced in roots under high doses of Cd and simultaneous exposure to UV-B for 10 days. Cd uptake was significantly enhanced by 33 % in the presence of UV-B. Li Y et al. (2011) studied the combined effects of O_3 (<10 and 80 ppb, 7 h day^{-1}) and Cd on growth and physiology of *Triticum aestivum.* O_3 considerably depressed light-saturated net photosynthetic rate, stomatal conductance, chlorophyll content, and total biomass without or with Cd. Ozone significantly increased the activity of superoxide dismutase, catalase, and peroxidase but also the malondialdehyde content. Ozone can reduce the stomatal conductance and thus limit transpiration-driven metal transport from root to shoot and metal concentration in grains. Pleijel (2011) stated that as O_3 negatively affects grain yield, it would effectively increase metal concentration. Pleijel (2012) studied the effect of ozone on Cd and Zn in wheat grain. Grain Zn concentration was significantly enhanced by ozone, while Zn yield was not affected. Cd concentration was unaffected by ozone, but Cd yield was significantly negatively affected.

Guo et al. (2012) studied the effects of elevated O_3 (50 % higher than the ambient O_3) on oxidative stress and bioaccumulation of Cd in wheat using FACE system. The study revealed higher concentrations of Cd in wheat tissues (shoots, husk, and grains) under elevated O3 supply. In addition, the combined exposure to Cd and elevated O_3 levels strongly affected the antioxidant enzymes and accelerated oxidative stress in wheat leaves. In experiments of Castagna et al. (2013), two poplar clones (I-214 and Eridano), differently sensitive to O_3, were grown for 5 weeks in pots supplied with Cd and then exposed to 15-d O_3 fumigation. Cadmium induced a reduction in stomatal conductance and a significant accumulation of H_2O_2 and NO in both clones and negatively affected the carotenoid content in I-214. Ozone, on the other hand, counteracted Cd accumulation in the aboveground organs and significantly increased the xanthophyll deep-oxidation state, indicating photoinhibition in O_3-treated plants. Surprisingly, O_3 alone or in combination with Cd decreased H_2O_2 accumulation in I-214. The NO production was generally stimulated by Cd, while it decreased following O_3 exposure in I-214. The overall data indicated that Cd and O_3 induced clone-specific responses. Moreover, when they applied in combination, antagonistic rather than synergistic effects were observed. Thus, various studies indicate either increase or decrease in metal accumulation in the presence of O_3. These variations may be attributed to experimental conditions, plant types, and metal bioavailability.

Nutrient availability is also a climate- and geography-regulated phenomenon as discussed above. It will be interesting and significant to study and analyze how plants manage during nutrient-limiting conditions to maintain abiotic stress tolerance. In a recent study, Bashir et al. (2013) investigated the importance of the sulfur and sulfur-containing compounds for the plants' ability to perform under Cd stress. In this study, the authors studied the effects of sulfur deprivation in Cd-stressed *Arabidopsis thaliana* by measuring changes in magnitude of oxidative stress, key components of ascorbate–glutathione antioxidant pathway and sulfur-assimilation pathway, photosynthetic efficiency, chloroplast ultrastructure, and biomass accumulation. They concluded that under S-deprived conditions, plants shift toward non-sulfur tolerance mechanisms and thus could maintain equilibrium between growth and defense.

8.4.4 Effect of Drought and Salinity

Several studies have evaluated the interaction of metal accumulation with soil moisture/water levels because the water levels directly influence metal bioavailability as well as plant growth. Variable results of positive or negative influence of water status on metal accumulation have been found, and further metal accumulation has been either linked to or not to water stress tolerance. An increase in Se accumulation was observed under low soil moisture in tall fescue by Tennant and Wu (2000), which was explained by the decreased plant biomass. Bhatia et al. (2005) also found that the Ni concentrations increased in shoots of *Stackhousia tryonii* as the soil moisture levels decreased from 100 to 20 % field capacity. In contrast, Angle et al. (2003) established that uptake of Zn or Ni in *Alyssum murale*, *Berkheya coddii*, and *Thlaspi caerulescens* increased at higher soil moisture levels rather than at low soil moisture levels. Further, as plants produced more biomass at higher soil moisture levels, this further enhanced the amount of metals extracted from soil. Similarly, Pascual et al. (2004) reported that bioavailable metal concentration in soils and thereby metal accumulation in *Lolium multiflorum* decreased under water-deficit condition. Whiting et al. (2003) compared the responses of non-hyperaccumulator and hyperaccumulator species to Ni and water stress conditions and found that *Alyssum montanum* and *Lepidium heterophyllum* had reduced survival rate and shoot growth as compared to that of hyperaccumulator species (*Alyssum murale* and *Thlaspi caerulescens*). However, metal hyperaccumulation itself did not confer resistance to drought stress. Instead, higher expressions of more general stress tolerance genes were probably the reason behind increased tolerance. However, when Disante et al. (2011) evaluated whether high Zn concentrations in soil solution impart water stress resistance in *Quercus suber*, they found that morpho-physiological responses induced by Zn stress improved water conservation strategies (via stomatal closure, decrease in transpiratory losses, and slow decrease of C fixation) and imparted short-term drought tolerance. The increased metal accumulation may also alter drought stress tolerance of plants through effects on the uptake of K+, turgor maintenance, and osmotic adjustment (Rajkumar et al. 2013). Hofacker et al. (2013) demonstrated that temperature controls trace metal dynamics during soil flooding via its influence on microbial reduction of terminal electron acceptors. Even at low temperatures, soil flooding may trigger the release of chalcophile metals from contaminated floodplain soils by sorbent reduction, competitive sorption, and formation of nanoparticulate metal-bearing colloids.

Raziuddin et al. (2011) conducted hydroponic experiment to investigate the sole and combined effects of Cd and salinity on growth and photosynthesis of *Brassica napus* and *B. juncea*. Cd and NaCl treatments imposed significant negative effects on shoot and root length, weight, water content, number of plant leaves, photosynthesis, and stomatal conductance compared to control, with the effects being more pronounced under combined treatments than in single ones. Bauddh and Singh (2012) found that *Ricinus communis* produced more biomass and was more tolerant to salinity and drought in the presence of Cd and removed more Cd in a given time than Indian mustard. Though the amount of Cd accumulated in the roots and shoots of Indian mustard was higher per unit biomass than that in castor, total removal of the metal from soil was much higher in castor on per plant basis in the same period in presence of the stresses. Salinity alone enhanced Cd uptake, whereas drought stress reduced its uptake in both the plants. Zhang BL et al. (2013) investigated the effect of NaCl on Cd uptake, translocation, and oxidative stress using two tobacco cultivars differing in Cd tolerance. The growth inhibition of the tobacco plants exposed to Cd toxicity and oxidative stress was alleviated by moderate addition of NaCl through recovered or enhanced activities of antioxidative enzymes catalase and glutathione peroxidase. Cd concentration of shoots and roots in the two cultivars was also reduced by addition of NaCl. Hence, similar to other stresses, salt and metal interactions appear to be genus and species specific whereby salt treatment either decreased or increased metal uptake by the plants and that too variably in different plants.

8.5 Prospective Strategies for Genetic Manipulation for Enhanced Phytoremediation in Changing Climate

Phytoremediation is the use of plants to remediate contaminated soils and water and is a promising approach. It is environmentally friendly and maintains the biological properties and physical structure of the soil (Srivastava et al. 2012). However, phytoremediation applicability is dependent on several factors like metal accumulation potential, metal bioavailability in soils/water, and plant growth in metal-contaminated environment. These factors are affected by climate change. A number of transgenics have been generated with the view of developing tolerance against abiotic stresses like drought, heat, and salinity. These transgenic approaches should be viewed with the perspective of their utilization in the climate change with the predicted increases in temperature and changes in water availability. No transgenic plant developed for abiotic stress has yet been tested for performance under elevated CO_2 to our knowledge. However, at elevated CO_2 concentrations, plant allocation to defensive compounds will be affected enough to impact plant–herbivore interactions as indicated in a study with transgenic cotton (Coviella et al. 2002). Further, there are also no studies to test the effectiveness of a transgenic developed, say, for drought tolerance against metal stress to our knowledge. However, the basic mechanisms of abiotic stress tolerance involve some common transcription factors, hormones, and strategies. Hence, it is likely that these transgenics would also tolerate the stress of increasing greenhouse gases and sustain growth and hence may perform better for remediation.

Plant exposure to almost all of the abiotic stresses leads to an accumulation of reactive oxygen species (ROS) with the concomitant increase in antioxidant defense mechanisms. Therefore, one common strategy has been to strengthen antioxidant machinery to augment ROS scavenging or to devise ways to reduce the production of ROS per se. Alternative oxidase (AOX) constitutes a branch of the mitochondrial ETC, whose activity gets induced under abiotic stresses due to its importance to avoid ROS generation by the chain. Wang and Vanlerberghe (2013) found that mild to moderate drought resulted in increase in AOX amount, accompanied by a progressive increase in expression of different ROS-scavenging components in *Nicotiana tabacum*. However, RNAi transgenic plants with suppressed AOX amount were able to manage ROS load. Under severe drought stress, however, plants lacking AOX suffered significantly more cellular damage than WT. Zhang Z et al. (2013) demonstrated that rice ascorbate peroxidase (*OsAPX2*) loss of function mutants had lower APX activity and were sensitive to abiotic stresses (drought, salt, and cold) while overexpression of *OsAPX2* increased APX activity and enhanced stress tolerance. CrPrx and CrPrx1 are class III peroxidases from *Catharanthus roseus*, which were expressed in *Nicotiana tabacum*. Increased oxidative stress tolerance was observed in transgenics when treated with H_2O_2 under strong light conditions. However, against abiotic stress, CrPrx exhibited better cold tolerance while CrPrx1 showed improved germination under salt and dehydration stresses (Kumar et al. 2012).

Tocopherols are a group of powerful antioxidants having additional roles in signaling and gene expression, with α-tocopherol being the most potent form. In a study, α-tocopherol-enriched transgenic *Brassica juncea* plants were assessed against salt, heavy metal, and osmotic stress (Kumar D et al. 2013). Transgenic plants showed lesser oxidative damage and higher activities and transcript levels of antioxidant enzymes than WT plants under stress. Tocopherol cyclase (TC; encoded by gene VTE1) catalyzes the penultimate step of tocopherol synthesis. Guo et al. (2009) used wild-type and transgenic tobacco plants overexpressing VTE1 from *Arabidopsis* to examine the role of tocopherol in ozone sensitivity. Wild-type plants responded to ozone by severe leaf necrosis while the transgenic lines exhibited limited injury. Compared with the wild type, VTE1-overexpressing plants had lower increase in hydrogen peroxide, malondialdehyde contents, and ion leakage and lower decrease of net photosynthetic rate 48 h following the ozone exposure. Transgenic plants also better maintained the structural integrity of the photosynthetic apparatus. Ascorbate is a major antioxidant and free-radical scavenger in plants. Monodehydroascorbate reductase (MDHAR) is crucial for ascorbate regeneration and essential for maintaining a reduced pool of ascorbate. To examine whether an overexpressed level of MDHAR could minimize the deleterious effects of environmental stresses, Eltayeb et al. (2007) developed transgenic tobacco plants overexpressing *Arabidopsis thaliana* MDHAR gene in the cytosol. The transgenic plants exhibited up to 2.1-fold higher MDHAR activity and 2.2-fold higher level of reduced AsA compared to non-transformed control plants. The transgenic plants showed enhanced stress tolerance in term of significantly higher net photosynthesis rates under ozone, salt, and PEG stresses.

Carotenoids are essential components of the photosynthetic apparatus involved in plant photoprotection. To investigate the protective role of zeaxanthin under high light and UV stress, Gotz et al. (2002) transformed tobacco plants with a heterologous carotenoid gene encoding beta-carotene hydroxylase (crtZ; responsible for the conversion of beta-carotene into zeaxanthin) from *Erwinia uredovora*. The transformants synthesized zeaxanthin more rapidly and in larger quantities than controls upon transfer to high-intensity white light. The UV-exposed transformed plants demonstrated a greater tolerance with a higher biomass under UV-stressed conditions. Thus, enhancement of zeaxanthin formation imparted UV stress protection and prevented damage. Nakajima et al. (2002) introduced an antisense DNA for an ozone-inducible 1-aminocyclopropane-1-carboxylate (ACC) synthase from tomato into tobacco. In transgenic plants, ozone-inducible accumulation of ACC and ethylene production was reduced and the visible damage was attenuated. No significant differences in growth and morphology were observed between transgenic and wild-type plants; however, stomatal conductance of transgenic plants was higher than that of wild-type plants. These findings indicated that the suppression of ACC synthase can improve the ozone tolerance of plants without reducing their gas absorption and productivity.

It is thought that manipulation of the expression of critical stress-responsive genes should ultimately provide increased protection against abiotic stress. The regulation of gene expression is a key factor in plant acclimation to stress. Hence, altered expression of specific regulatory proteins, such as transcription factors, protein kinases, and ubiquitin ligases, can be used to regulate the expression of downstream genes to affect metabolic responses to stress. Mukhopadhyay et al. (2004) isolated and characterized a rice gene (OsiSAP1), encoding a zinc-finger protein, which is expressed under a variety of abiotic stresses. This discovery led to the identification of a new family of stress-associated proteins (SAPs) that plays an important role in regulating plant responses to a wide range of abiotic stress conditions. Expression of the *AlSAP* gene from *Aeluropus littoralis* was responsive to salt, osmotic, heat, and cold stress, as well as abscisic acid and salicylic acid, and its overexpression in tobacco enhanced the tolerance to salt, drought, heat, and freezing stress (Ben Saad et al. 2010). Similarly, the overexpression of OsiSAP8 in both tobacco and rice improved the tolerance to salt, drought, and cold stress (Kanneganti and Gupta 2008). Hozain et al. (2012) ectopically expressed *Arabidopsis thaliana* AtSAP5 gene in cotton (*Gossypium hirsutum*). Under non-stressful conditions, the rate of net CO_2 assimilation for transgenic lines was less sensitive to rapidly developing water deficit over 4 d than WT plants. In addition, enhanced protection of seedling growth and leaf viability was associated with the expression of AtSAP5. Babitha et al. (2013) expressed AtbHLH17 (basic helix-loop-helix) and AtWRKY28 TFs, which are known to be upregulated under drought and oxidative stress, respectively, in *Arabidopsis*. The transgenic lines exhibited enhanced tolerance to NaCl, mannitol, and oxidative stress. Growth under stress and recovery growth were substantially superior in transgenics exposed to gradual long-term desiccation stress conditions. Hence, coexpressing two or more TFs may substantially improve the stress tolerance of the plants. Vadez et al. (2013) assessed how root traits are influenced under water stress in groundnut (*Arachis hypogaea*) in WT and transgenics (transformed with DREB1A, dehydration-responsive element-binding protein 1A) and demonstrated that water stress promotes rooting growth more strongly in transgenic events than in the WT, especially in deep soil layers, and this leads to increased water extraction.

Morran et al. (2011) generated transgenic wheat and barley plants with constitutive and drought-inducible expression of the TaDREB2 and TaDREB3 transcription factors. Transgenic populations with constitutive overexpression showed slower growth, delayed flowering, and lower grain yields relative to the WT controls. However, both transgenics showed improved survival under severe drought conditions. The undesired changes in plant development could be alleviated by using a drought-inducible promoter. In addition to drought tolerance, both transgenic plants with constitutive overexpression of the transgene showed a significant improvement in frost tolerance. Nuclear factor-Y (NF-Y) is a heterotrimeric transcription factor composed of NF-YA, NF-YB, and NF-YC proteins. Ni et al.

(2013) identified and characterized a gene GmNFYA3 from *Glycine max*. Overexpression of GmNFYA3 in *Arabidopsis* resulted in reduced leaf water loss and enhanced drought tolerance; however, the transgenic *Arabidopsis* exhibited increased sensitivity to high salinity and exogenous ABA. Moreover, the transcript levels of ABA biosynthesis, signaling, and stress-responsive genes were generally higher in GmNFYA3 plants than in WT controls under normal conditions. MicroRNA319 (miR319) is one of the first characterized and conserved microRNA families in plants and has been demonstrated to target TCP (for Teosinte Branched/Cycloidea/Proliferating Cell Factors [PCF]) genes encoding plant-specific transcription factors. Zhou et al. (2013) investigated the role that miR319 plays in the plant response to abiotic stress using transgenic *Agrostis stolonifera* overexpressing a rice (*Oryza sativa*) miR319 gene. Transgenic plants overexpressing Osa-miR319a displayed morphological changes and enhanced drought and salt tolerance associated with increased leaf wax content and water retention. A rice stress-responsive transcription factor encoded by the rice NAC1 gene (SNAC1) plays an important role in drought stress tolerance. Saad et al. (2013) introduced the SNAC1 gene in wheat and transgenics displayed significantly enhanced tolerance to drought and salinity as compared to wild type. In addition, the expression of genes involved in abiotic stress/ABA signaling was effectively regulated by the alien SNAC1 gene.

Calcium-binding proteins that contain EF-hand motifs have been reported to play important roles in transduction of signals associated with biotic and abiotic stresses. Wang et al. (2013) found that transgenic *Arabidopsis* seedlings expressing an EF-hand family gene (*MtCaMP1*) from *Medicago truncatula* had higher survival rate and greater tolerance than WT under drought and salt stress. This was attributed to effective osmoregulation due to greater accumulation of proline and by minimizing toxic Na+ accumulation under drought and salinity stress, respectively. Protein phosphorylation/dephosphorylation is a major signaling event induced by osmotic stress in higher plants. Sucrose non-fermenting 1-related protein kinase 2 (SnRK2) family members play essential roles in the response to hyperosmotic stresses in plants. Tian et al. (2013) cloned TaSnRK2.3 gene from common wheat, overexpressed in *Arabidopsis*, and subjected to severe abiotic stresses. Overexpression of TaSnRK2.3 resulted in an improved root system and significantly enhanced tolerance to drought, salt, and freezing stresses, simultaneously demonstrated by enhanced expression of abiotic stress-responsive genes and ameliorative physiological indices.

One of the approaches employed for the decontamination of environment includes identification and overexpression of genes involved in the detoxification mechanism of plants. Glutathione S-transferases (GSTs) are a superfamily of enzymes, principally known for their role in detoxification reactions. Different classes of GSTs have been used to develop plants with improved detoxification potential. Kumar S et al. (2013) found differential expression of OsGSTLs genes in As-sensitive and As-tolerant genotypes. They expressed one member of Lambda class OsGSTL2 in *Arabidopsis*. OsGSTL2 provided tolerance for heavy metals and other abiotic stresses like cold, osmotic stress, and salt.

Dehydrins are a type of late embryogenesis abundant (LEA) protein. Accumulation of dehydrins enhances the drought, cold, and salt tolerances of transgenic plants, although the underlying mechanism is unclear. MtCAS31 (*Medicago truncatula* cold-acclimation-specific protein 31) is a dehydrin that was isolated and overexpressed in *Arabidopsis* by Xie et al. (2012). MtCAS31 overexpression dramatically reduced stomatal density through interaction with a gene involved in stomatal development (AtICE1: inducer of CBF expression 1) and markedly enhanced the drought tolerance of transgenics. Plants synthesize a set of evolutionary conserved proteins called heat shock proteins (HSPs) upon heat stress, and many groups have produced thermotolerant plants by overexpressing these HSPs (Khurana et al. 2013). The HSP family has been classified into five groups depending on their molecular weight: HSP100, HSP90, HSP70, HSP60, and small HSPs. Jiang et al. (2009) expressed RcHSP17.8, a cytosolic class I sHSP, from *Rosa chinensis* in *Arabidopsis thaliana* plants, which exhibited increased tolerance to multiple abiotic stresses such as heat, salt, osmotic, and drought stress. Chauhan et al. (2012) characterized wheat chloroplastic sHSP26. Transgenic *Arabidopsis* plants were substantially tolerant under continuous high-temperature regimen than WT plants, as measured by higher biomass and seed yield. Zou et al. (2012) generated transgenic rice plants overexpressing OsHsp17.0 and OsHsp23.7. Both transgenic lines demonstrated higher germination ability compared to WT plants when subjected to mannitol and NaCl. Transgenic rice lines displayed a higher tolerance to drought and salt stress compared to WT plants.

Lu et al. (2013) overexpressed *Arabidopsis* molybdenum cofactor sulfurase gene (LOS5) in maize that markedly enhanced the expression of aldehyde oxidase (ZmAO) and aldehyde oxidase (AO) activity, leading to ABA accumulation and increased drought tolerance. Transgenic maize exhibited reductions in stomatal aperture, which led to decreased water loss. The increased drought tolerance in transgenic plants was associated with ABA accumulation via activated AO and expression of stress-related gene via ABA induction, which sequentially induced a set of favorable stress-related physiological and biochemical responses.

Singh et al. (2012) characterized a rice protein of unknown function, OsCBSX4 (cystathionine β-synthase domain-containing protein), which was upregulated under high salinity, heavy metal, and oxidative stresses at seedling stage. Transgenic tobacco plants overexpressing OsCBSX4 exhibited improved tolerance toward salinity, heavy metal, and oxidative stress and could grow and set seeds under continuous presence of 150 mM NaCl.

8.6 Conclusions and Future Directions

The studies conducted do indicate that metal bioavailability and hence the metal uptake by the plants as well as plant growth per se are intricately controlled by various climatic and geographical factors including water, salt, temperature, CO_2, and other gases. To date a few studies have tried to evaluate the interaction of some components like CO_2 and temperature, temperature and metal, CO_2 and metal, etc. But this seems unlikely to study the effect of all parameters on growth and metal uptake of plants. For a better evaluation of future climate effects on growth and metal accumulation profile of plants, one approach might be to conduct experiments in different natural field locations. More comprehensive statistical modeling can be another suitable approach to deal with the issue of climate change to predict future effects. The need exists to develop transgenics for better remediation of metal-contaminated areas presently as well as in future climatic conditions. Laboratory and field studies have been done to develop transgenics for increasing the metal accumulation potential of plants (Srivastava et al. 2012) and to tolerate climatic adverse conditions like salinity, drought (Aguado-Santacruz 2006), and temperature (Grover et al. 2013). However, no transgenic has yet been tested for metal removal potential in field with a view of changing climate. This needs timely attention so that application of phytoremediation technology continues in future.

References

Aguado-Santacruz GA (2006) Genetic manipulation of plants for increased drought tolerance. In: Guevara-Gonzalez GG, Torres-Pacheco I (eds) Advances in agricultural and food biotechnology. Research Signpost, Trivandrum, pp 71–97

Ahsan N, Nanjo Y, Sawada H, Kohno Y, Komatsu S (2010) Ozone stress-induced proteomic changes in leaf total soluble and chloroplast proteins of soybean reveal that carbon allocation is involved in adaptation in the early developmental stage. Proteomics 10:2605–2619

Ainsworth EA, Long SP (2005) What have we learned from 15 years of free-air CO_2 enrichment (FACE)? A meta-analytic review of the responses of photosynthesis, canopy properties and plant production to rising CO_2. New Phytol 165:351–372

Ainsworth EA, Rogers A, Vodkin LO, Walter A, Schurr U (2006) The effects of elevated CO_2 concentration on soybean gene expression. An analysis of growing and mature leaves. Plant Physiol 142:135–147

Angle JS, Baker AJM, Whiting SN, Chaney RL (2003) Soil moisture effects on uptake of metals by *Thlaspi*, *Alyssum*, and *Berkheya*. Plant Soil 256:325–332

Ashmore M, Toet S, Emberson L (2006) Ozone – a significant threat to future world food production? New Phytol 170:201–204

Babitha KC, Ramu SV, Pruthvi V, Mahesh P, Nataraja KN, Udayakumar M (2013) Co-expression of *AtbHLH17* and *AtWRKY28* confers resistance to abiotic stress in Arabidopsis. Transgenic Res 22:327–341

Bashir H, Ahmad J, Bagheri R, Nauman M, Irfan Qureshi M (2013) Limited sulfur resource forces *Arabidopsis thaliana* to shift towards non-sulfur tolerance under cadmium stress. Environ Exp Bot 94:19–32

Bauddh K, Singh RP (2012) Growth, tolerance efficiency and phytoremediation potential of *Ricinus communis* (L.) and *Brassica juncea* (L.) in salinity and drought affected cadmium contaminated soil. Ecotoxicol Environ Saf 85:13–22

Ben Saad R, Zouari N, Ben Ramdhan W, Azaza J, Meynard D, Guiderdoni E, Hassairi A (2010) Improved drought and salt stress tolerance in transgenic tobacco overexpressing a novel A20/AN1 zinc-finger "AlSAP" gene isolated from the halophyte grass *Aeluropus littoralis*. Plant Mol Biol 72:171–190

Bhatia N, Baker AJM, Walsh KB, Midmore DJ (2005) A role for nickel in osmotic adjustment in drought-stressed plants of the nickel hyperaccumulator *Stackhousia tryonii* Bailey. Planta 223:134–139

Cao XD, Ma LQ (2004) Effects of compost and phosphate on plant arsenic accumulation from soils near pressure-treated wood. Environ Pollut 132:435–442

Castagna A, Di Baccio D, Tognetti R, Ranieri A, Sebastiani L (2013) Differential ozone sensitivity interferes with cadmium stress in poplar clones. Biol Plant 57:313–324

Chauhan H, Khurana N, Nijhavan A, Khurana JP, Khurana P (2012) The wheat chloroplastic small heat shock protein (sHSP26) is involved in seed maturation and germination and imparts tolerance to heat stress. Plant Cell Environ 35:1912–1931

Chaves MM, Flexas J, Pinheiro C (2009) Photosynthesis under drought and salt stress: regulation mechanisms from whole plant to cell. Ann Bot 103:551–560

Chen Y, Wang C, Wang G, Luo C, Mao Y, Shen Z, Li X (2010) Heating treatment schemes for enhancing chelant-assisted phytoextraction of heavy metals from contaminated soils. Environ Toxicol Chem 27:888–896

Cho K, Shibato J, Agrawal GK, Jung YH, Kubo A, Jwa NS, Tamogami S, Satoh K, Kikuchi S, Higashi T, Kimura S, Saji H, Tanaka Y, Iwahashi H, Masuo Y, Rakwal R (2008) Integrated transcriptomics, proteomics, and metabolomics analyses to survey ozone responses in the leaves of rice seedling. J Proteome Res 7:2980–2998

Coviella CE, Stipanovic RD, Trumble JT (2002) Plant allocation to defensive compounds: interactions between elevated CO_2 and nitrogen in transgenic cotton plants. J Exp Bot 53:323–331

Dentener F, Stevenson M, Cofala J, Mechler R, Amann M, Bergamaschi P, Raes F, Derwent R (2005) The impact of air pollutants and methane emission controls on tropospheric ozone and radiative forcing: CTM calculations for the period of 1990–2030. Atmos Chem Phys 5:1731–1755

Disante KB, Fuentes D, Cortina J (2011) Response to drought of Zn-stressed *Quercus suber* L. seedlings. Environ Exp Bot 70:96–103

Djanaguiraman M, Prasad PVV, Seppanen M (2010) Selenium protects sorghum leaves from oxidative damage under high temperature high stress by enhancing antioxidant defense system. Plant Physiol Biochem 48:999–1007

Dowdall M, Standring M, Shaw G, Strand P (2008) Will global warming affect soil-to-plant transfer of radionuclides? J Environ Radioact 99:1736–1745

Ekvall L, Greger M (2003) Effects of environmental biomass-producing factors on Cd uptake in two Swedish ecotypes of *Pinus sylvestris*. Environ Pollut 121:401–411

Eltayeb AE, Kawano N, Badawi GH, Kaminaka H, Sanekata T, Shibahara T, Inanaga S, Tanaka K (2007) Overexpression of monodehydroascorbate reductase in transgenic tobacco confers enhanced tolerance to ozone, salt and polyethylene glycol stresses. Planta 225:1255–1264

Fendorf S, Michael HA, van Geen A (2010) Spatial and temporal variations of groundwater arsenic in South and Southeast Asia. Science 328:1123–1127

Feng YW, Komatsu S, Furukawa T, Koshiba T, Kohno Y (2008) Proteome analysis of proteins responsive to ambient and elevated ozone in rice seedlings. Agric Ecosyst Environ 125:255–265

Fritioff A, Kautsky L, Greger M (2005) Influence of temperature and salinity on heavy metal uptake by submersed plants. Environ Pollut 133:265–274

Galant A, Koester RP, Ainsworth EA, Hicks LM, Jez JM (2012) From climate change to molecular response: redox proteomics of ozone-induced responses in soybean. New Phytol 194:220–229

Gillespie KM, Xu F, Richter KT, McGrath JM, Markelz RJC, Ort DR, Leakey ADB, Ainsworth EA (2012) Greater antioxidant and respiratory metabolism in field-grown soybean exposed to elevated O_3 under both ambient and elevated CO_2. Plant Cell Environ 35:169–184

Gotz T, Sandmann G, Romer S (2002) Expression of a bacterial carotene hydroxylase gene (crtZ) enhances UV tolerance in tobacco. Plant Mol Biol 50:129–142

Grover A, Mittal D, Negi M, Lavania D (2013) Generating high temperature tolerant transgenic plants: achievements and challenges. Plant Sci 205–206:38–47

Guo J, Li XF, Qi DM, Chen SY, Li ZQ, Nijs I, Li YG, Liu GS (2009) Effects of ozone on wild type and transgenic tobacco. Biol Plant 53:670–676

Guo H, Zhu J, Zhou H, Sun Y, Yin Y, Pei D, Ji R, Wu J, Wang X (2011) Elevated CO_2 levels affects the concentrations of copper and cadmium in crops grown in soil contaminated with heavy metals under fully open-air field conditions. Environ Sci Technol 45:6997–7003

Guo H, Tian R, Zhu J, Zhou H, Pei D, Wang X (2012) Combined cadmium and elevated ozone affect concentrations of cadmium and antioxidant systems in wheat under fully open-air conditions. J Hazard Mater 209–210:27–33

Hawrylak-Nowak B, Matraszek R, Szymanska M (2010) Selenium modifies the effect of short-term chilling stress on cucumber plants. Biol Trace Elem Res 138:307–315

Hegerl GC, Hanlon H, Beierkuhnlein C (2011) Elusive extremes. Nat Geosci 4:142–143

Hofacker AF, Voegelin A, Kaegi R, Weber F-A, Kretzschmar R (2013) Temperature-dependent formation of metallic copper and metal sulfide nanoparticles during flooding of a contaminated soil. Geochim Cosmochim Acta 103:316–332

Högy P, Fangmeier A (2008) Effects of elevated atmospheric CO_2 on grain quality of wheat. J Cereal Sci 48:580–591

Högy P, Wieser H, Kohler P, Schwadorf K, Breuer J, Franzaring J, Muntifering R, Fangmeier A (2009) Effects of elevated CO_2 on grain yield and quality of wheat: results from a 3-year free-air CO_2 enrichment experiment. Plant Biol 11:60–69

Högy P, Brunnbauer M, Koehler P, Schwadorf K, Breuer J, Franzaring J, Zhunusbayeva D, Fangmeier A (2013) Grain quality characteristics of spring wheat (*Triticum aestivum*) as affected by free-air CO_2 enrichment. Environ Exp Bot 88:11–18

Hoshika Y, Omasa K, Paoletti E (2013) Both ozone exposure and soil water stress are able to induce stomatal sluggishness. Environ Exp Bot 88:19–23

Hozain M, Abdelmageed H, Lee J, Kang M, Fokar M, Allen RD, Holaday AS (2012) Expression of AtSAP5 in cotton up-regulates putative stress-responsive genes and improves the tolerance to rapidly developing water deficit and moderate heat stress. J Plant Physiol 169:1261–1270

IPCC (2001) Atmospheric chemistry and greenhouse gases. In: Houghton JT et al (eds) Climate change 2001: the scientific basis – contribution of Working Group I to the third assessment report of the Intergovernmental Panel on Climate Change. Cambridge University Press, Cambridge, pp 239–288

IPCC (2007) Climate change: synthesis report. In: Core Writing Team, Pachauri RK, Reisinger A (eds) Contribution of Working Groups I, II and III to the fourth assessment report of the Intergovernmental Panel on Climate Change. IPCC, Geneva

Jia HX, Guo HY, Yin Y, Wang Q, Sun Q, Wang XR, Zhu JG (2007) Responses of rice growth to copper stress under free-air CO_2 enrichment (FACE). Chin Sci Bull 52:2636–2641

Jia Y, Tang SR, Wang RG, Ju XH, Ding YZ, Tu SX, Smith DL (2010) Effects of elevated CO_2 on growth, photosynthesis, elemental composition, antioxidant level, and phytochelatin concentration in *Lolium mutiforum* and *Lolium perenne* under Cd stress. J Hazard Mater 180:384–394

Jia Y, Ju X, Liao S, Song Z, Li Z (2011) Phytochelatin synthesis in response to elevated CO_2 under Cd stress in *Lolium perenne* L. J Plant Physiol 168:1723–1728

Jiang C, Xu J, Zhang H, Zhang X, Shi J, Li M, Ming F (2009) A cytosolic class I small heat shock protein, RcHSP17.8, of *Rosa chinensis* confers resistance to a variety of stresses to *Escherichia coli*, yeast and *Arabidopsis thaliana*. Plant Cell Environ 32:1046–1059

Kanneganti V, Gupta AK (2008) Overexpression of OsiSAP8, a member of stress associated protein (SAP) gene family of rice confers tolerance to salt, drought and cold stress in transgenic tobacco and rice. Plant Mol Biol 66:445–462

Khurana N, Chauhan H, Khurana P (2013) Wheat chloroplast targeted sHSP26 promoter confers heat and abiotic stress inducible expression in transgenic *Arabidopsis* plants. PLoS One 8:e54418

Kim S, Kang H (2011) Effects of elevated CO_2 and Pb on phytoextraction and enzyme activity. Water Air Soil Pollut 219:365–375

Kimball BA (2006) The effect of free-air [CO_2] enrichment of cotton, wheat, and sorghum. In: Nösberger J, Long SP, Norby RJ, Stitt M, Hendrey GR, Blum H (eds) Managed ecosystems and CO_2. Springer, Berlin, pp 47–70

Klaminder J, Hammarlund D, Kokfelt U, Vonk JE, Bigler C (2010) Lead contamination of subarctic lakes and its response to reduced atmospheric fallout: can the recovery process be counteracted by the ongoing climate change? Environ Sci Technol 44:2335–2340

Korner C, Diemer M, Schappi B, Niklaus P, Arnone J III (1997) The responses of alpine grass and to four seasons of CO_2 enrichment: a synthesis. Acta Oecol 18:165–175

Kumar S, Jaggi M, Sinha AK (2012) Ectopic overexpression of vacuolar and apoplastic *Catharanthus roseus* peroxidases confers differential tolerance to salt and dehydration stress in transgenic tobacco. Protoplasma 249:423–432

Kumar D, Yusuf MA, Singh P, Sardar M, Sarin NB (2013a) Modulation of antioxidant machinery in α-tocopherol-enriched transgenic *Brassica juncea* plants tolerant to abiotic stress conditions. Protoplasma 250(5):1079–1089. doi:10.1007/s00709-013-0484-0

Kumar S, Asif MH, Chakrabarty D, Tripathi RD, Dubey RS, Trivedi PK (2013b) Expression of a rice Lambda class of glutathione S-transferase, *OsGSTL2*, in *Arabidopsis* provides tolerance to heavy metal and other abiotic stresses. J Hazard Mater 248–249:228–237

Li ZY, Tang SR, Deng XF, Wang RG, Song ZG (2010) Contrasting effects of elevated CO_2 on Cu and Cd uptake by different rice varieties grown on contaminated soils with two levels of metals: implication for phytoextraction and food safety. J Hazard Mater 177:352–361

Li D, Zhou D, Wang P, Li L (2011a) Temperature affects cadmium-induced phytotoxicity involved in subcellular cadmium distribution and oxidative stress in wheat roots. Ecotoxicol Environ Saf 74:2029–2035

Li Y, Li C, Zheng Y, Wu G, Wuyun T, Xu H, He X, Jiang G (2011b) Cadmium pollution enhanced ozone damage to winter wheat: biochemical and physiological evidences. J Environ Sci (China) 23:255–265

Li Y, Zhang Q, Wang R, Gou X, Wang H, Wang S (2012) Temperature changes the dynamics of trace element accumulation in *Solanum tuberosum* L. Clim Chang 112:655–672

Lieffering M, Kim HY, Kobayashi K, Okada M (2004) The impact of elevated CO_2 on the elemental concentrations of field-grown rice grains. Field Crop Res 88:279–286

Lobell DB, Field CB (2007) Global scale climate—crop yield relationships and the impacts of recent warming. Environ Res Lett 2:014002

Long SP, Zhu XG, Naidu SL, Ort DR (2006) Can improvement in photosynthesis increase crop yields. Plant Cell Environ 29:315–330

Lu LL, Tian SK, Yang XE, Li TQ, He ZL (2009) Cadmium uptake and xylem loading are active processes in the hyperaccumulator *Sedum alfredii*. J Plant Physiol 166:579–587

Lu Y, Li Y, Zhang J, Xiao Y, Yue Y, Duan L, Zhang M, Li Z (2013) Overexpression of *Arabidopsis* molybdenum cofactor sulfurase gene confers drought tolerance in maize (*Zea mays* L.). PLoS One 8:e52126

Lynch DV, Steponkus PL (1987) Plasma membrane lipid alterations associated with cold acclimation of winter rye seedlings (*Secale cereale* L. cv Puma). Plant Physiol 83:761–767

Meharg AA, Williams PN, Adomako E, Lawgali YY, Deacon C, Villada A, Cambell RCJ, Sun G, Zhu Y-G, Feldmann J, Raab A, Zhao F-J, Islam R, Hossain S, Yanai J (2009) Geographical variation in total and inorganic arsenic content of polished (white) rice. Environ Sci Technol 43:1612–1617

Milner LJ, Reade JPH, Cobb AH (2007) The effect of temperature on glutathione S-transferase activity and glutathione content in *Alopecurus myosuroides* (black grass) biotypes susceptible and resistant to herbicides. Weed Res 47:106–112

Morgan PB, Bernacchi CJ, Ort DR, Long SP (2004) An *in vivo* analysis of the effect of season-long open-air elevation of ozone to anticipated 2050 levels on photosynthesis in soybean. Plant Physiol 135:2348–2357

Morgan PB, Miles TA, Bollero GA, Nelson RA, Long SP (2006) Season-long elevation of ozone concentration to projected 2050 levels under fully open-air conditions substantially decreases the growth and production of soybean. New Phytol 170:333–343

Morran S, Eini O, Pyvovarenko T, Parent B, Singh R, Ismagul A, Eliby S, Shirley N, Langridge P, Lopato S (2011) Improvement of stress tolerance of wheat and barley by modulation of expression of DREB/CBF factors. Plant Biotechnol J 9:230–249

Mukhopadhyay A, Vij S, Tyagi AK (2004) Overexpression of a zinc-finger protein gene from rice confers tolerance to cold, dehydration, and salt stress in transgenic tobacco. Proc Natl Acad Sci U S A 101:6309–6314

Munns R, Tester M (2008) Mechanisms of salinity tolerance. Annu Rev Plant Biol 59:651–681

Nakajima N, Itoh T, Takikawa S, Asai N, Tamaoki M, Aono M, Kubo A, Azumi Y, Kamada H, Saji H (2002) Improvement in ozone tolerance of tobacco plants with an antisense DNA for 1-aminocyclopropane-1-carboxylate synthase. Plant Cell Environ 25:727–735

Natali SM, Sanudo-Wilhelmy SA, Lerdau MT (2009) Plant and soil mediation of elevated CO_2 impacts on trace metals. Ecosystems 12:715–727

Ni Z, Hu Z, Jiang Q, Zhang H (2013) *GmNFYA3*, a target gene of miR169, is a positive regulator of plant tolerance to drought stress. Plant Mol Biol 82:113–129

Noyes PD, McElwee MK, Miller HD, Clark BW, Van Tiem LA, Walcott KC, Erwin KN, Levin ED (2009) The toxicology of climate change: environmental contaminants in a warming world. Environ Int 35:971–986

Pascual I, Antolin MC, Garcia C, Polo A, Sanchez-Diaz M (2004) Plant availability of heavy metals in a soil amended with a high dose of sewage sludge under drought conditions. Biol Fertil Soils 40:291–299

Phillips RP, Bernhardt ES, Schlesinger WH (2009) Elevated CO_2 increases root exudation from loblolly pine (*Pinus taeda*) seedlings as an N-mediated response. Tree Physiol 29:1513–1523

Pleijel H (2011) Reduced ozone by air filtration consistently improved grain yield in wheat. Environ Pollut 159:897–902

Pleijel H (2012) Effects of ozone on zinc and cadmium accumulation in wheat-dose-response functions and relationships with protein, grain yield, and harvest index. Ecol Evol 2:3186–3194

Pourghasemian N, Ehsanzadeh P, Greger M (2013) Genotypic variation in safflower (*Carthamus* spp.) cadmium accumulation and tolerance affected by temperature and cadmium levels. Environ Exp Bot 87:218–226

Prior SA, Runion GB, Marble SC, Rogers HH, Gilliam CH, Torbert HA (2011) A review of elevated atmospheric CO2 effects on plant growth and water relations: implications for horticulture. Hortic Sci 46:158–162

Qaderi MM, Kurepin LV, Reid DM (2012) Effects of temperature and watering regime on growth, gas exchange and abscisic acid content of canola (*Brassica napus*) seedlings. Environ Exp Bot 75:107–113

Rajkumar M, Prasad MNV, Swaminathan S, Freitas H (2013) Climate change driven plant-metal-microbe interactions. Environ Int 53:74–86

Raziuddin F, Hassan G, Akmal M, Shah SS, Mohammad F, Shafi M, Bakht J, Zhou W (2011) Effects of cadmium and salinity on growth and photosynthesis parameters of *Brassica* species. Pak J Bot 43:333–340

Renaut J, Bohler S, Hausman JF, Hoffmann L, Sergeant K, Ahsan N, Jolivet Y, Dizengremel P (2009) The impact of atmospheric composition on plants: a case study of ozone and poplar. Mass Spectrom Rev 28:495–516

Rodriguez JH, Klumpp A, Fangmeier A, Pignata ML (2011) Effects of elevated CO_2 concentrations and fly ash amended soils on trace element accumulation and translocation among roots, stems and seeds of *Glycine max* (L.) Merr. J Hazard Mater 187:58–66

Rofkar JR, Dwyer DF (2011) Effects of light regime, temperature, and plant age on uptake of arsenic by *Spartina pectinata* and *Carex stricta*. Int J Phytoremediation 13:528–537

Rydberg J, Klaminder J, Rosen P, Bindler R (2010) Climate driven release of carbon and mercury from permafrost mires increases mercury loading to subarctic lakes. Sci Total Environ 408:4778–4783

Saad ASI, Li X, Li H-P, Huang T, Gao C-S, Guo M-W, Cheng W, Zhao G-Y, Liao Y-C (2013) A rice stress-responsive NAC gene enhances tolerance of transgenic wheat to drought and salt stresses. Plant Sci 203–204:33–40

Sardans J, Penuelas J (2006) Introduction of the factor of partitioning in the lithogenic enrichment factors of trace element bioaccumulation in plant tissues. Environ Monit Assess 115:473–498

Sardans J, Penuelas J, Prieto P, Estiarte M (2008a) Changes in Ca, Fe, Mg, Mo, Na, and S content in a Mediterranean shrub land under warming and drought. J Geophys Res 13:1–11

Sardans J, Penuelas J, Estiarte M (2008b) Warming and drought change trace element bioaccumulation patterns in a Mediterranean shrub land. Chemosphere 70:874–885

Sarkar A, Rakwal R, Bhushan Agrawal S, Shibato J, Ogawa Y, Yoshida Y, Kumar Agrawal G, Agrawal M (2010) Investigating the impact of elevated levels of ozone on tropical wheat using integrated phenotypical, physiological, biochemical, and proteomics approaches. J Proteome Res 9:4565–4584

Schiedek D, Sundelin B, Readman JW, Macdonald RW (2007) Interactions between climate change and contaminants. Mar Pollut Bull 54:1845–1856

Serengil Y, Augustaitis A, Bytnerowicz A, Grulke N, Kozovitz AR, Matyssek R, Müller-Starck G, Schaub M, Wieser G, Coskun AA, Paoletti E (2011) Adaptation of forest ecosystems to air pollution and climate change: a global assessment on research priorities. iForest – Biogeosci. Forestry 4:44–48

Shahbaz M, Parmar S, Stuiver CEE, Hawkesford MJ, De Kok LJ (2013) Copper toxicity and sulfur metabolism in Chinese cabbage are affected by UV radiation. Environ Exp Bot 88:60–70

Shukla UC, Murthy RC, Kakkar P (2008) Combined effect of ultraviolet-B radiation and cadmium contamination on nutrient uptake and photosynthetic pigments in *Brassica campestris* L. seedlings. Environ Toxicol 23:712–719

Singh AK, Kumar R, Pareek A, Sopory SK, Singla-Pareek S (2012) Overexpression of rice CBS domain containing protein improves salinity, oxidative, and heavy metal tolerance in transgenic tobacco. Mol Biotechnol 52:205–216

Song N, Zhang X, Wang F, Zhang C, Tang S (2012) Elevated CO_2 increases Cs uptake and alters microbial communities and biomass in the rhizosphere of *Phytolacca americana* Linn (pokeweed) and *Amaranthus cruentus* L. (purple amaranth) grown on soils spiked with various levels of Cs. J Environ Radioact 112:29–37

Srivastava S, Suprasanna P, D'Souza SF (2012) Mechanisms of arsenic tolerance and detoxification in plants and their application in transgenic technology: a critical appraisal. Int J Phytoremediation 14:506–517

Stern GA, Macdonald RW, Outridge PM, Wilson S, Chetelat J, Cole A, Hintelmann H, Loseto LL, Steffen A, Wang F, Zdanowicz C (2012) How does climate change influence arctic mercury? Sci Total Environ 414:22–42

Tang S, Liao S, Guo J, Song Z, Wang R, Zhou X (2011) Growth and cesium uptake responses of *Phytolacca americana* Linn. and *Amaranthus cruentus* L. grown on cesium contaminated soil to elevated CO_2 or inoculation with a plant growth promoting rhizobacterium *Burkholderia* sp. D54, or in combination. J Hazard Mater 198:188–197

Tausz M, Tausz-Posch S, Norton RM, Fitzgerald GJ, Nicolas ME, Seneweera S (2013) Understanding crop physiology to select breeding targets and improve crop management under increasing atmospheric CO_2 concentrations. Environ Exp Bot 88:71–80

Tausz-Posch S, Borowiak K, Dempsey RW, Norton RM, Seneweera S, Fitzgerald GJ, Tausz M (2013) The effect of elevated CO_2 on photochemistry and antioxidative defence capacity in wheat depends on environmental growing conditions – a FACE study. Environ Exp Bot 88:81–92

Tennant T, Wu L (2000) Effects of water stress on selenium accumulation in tall fescue (*Festuca arundinacea* Schreb) from a selenium-contaminated soil. Arch Environ Contam Toxicol 38:32–39

Thomson LJ, Macfadyen S, Hoffmann AA (2010) Predicting the effects of climate change on natural enemies of agricultural pests. Biol Control 52:296–306

Tian S, Mao X, Zhang H, Chen S, Zhai C, Yang S, Jing R (2013) Cloning and characterization of TaSnRK2.3, a novel SnRK2 gene in common wheat. J Exp Bot 64:2063–2080

Tukaj Z, Bascik-Remisiewicz A, Skowronski T, Tukaj C (2007) Cadmium effect on the growth, photosynthesis, ultrastructure and phytochelatin content of green microalga *Scenedesmus armatus*: a study at low and elevated CO_2 concentration. Environ Exp Bot 60:291–299

Vadez V, Rao JS, Bhatnagar-Mathur P, Sharma KK (2013) DREB1A promotes root development in deep soil layers and increases water extraction under water stress in groundnut. Plant Biol 15:45–52

Van den Berge J, Naudts K, Janssens IA, Ceulemans R, Nijs I (2011) Does the stress tolerance of mixed grassland communities change in a future climate? A test with heavy metal stress (zinc pollution). Environ Pollut 159:3294–3301

Van Dingenen R, Dentener FJ, Raes F, Krol MC, Emberson L, Cofala J (2009) The global impact of ozone on agricultural crop yields under current and future air quality legislation. Atmos Environ 43:604–618

van Gestel CAM (2008) Physico-chemical and biological parameters determine metal bioavailability in soils. Sci Total Environ 406:385–395

Vaughan KL, Rabenhorst MC, Needelman BA (2009) Saturation and temperature effects on the development of reducing conditions in soils. Soil Sci Soc Am J 73:663–667

Vingarzan R (2004) A review of surface ozone background levels and trends. Atmos Environ 38:3431–3442

Visser A, Kroes J, van Vliet MTH, Blenkinsop S, Fowler HJ, Broers HP (2012) Climate change impacts on the leaching of a heavy metal contamination in a small lowland catchment. J Contam Hydrol 127:47–64

Wang J, Vanlerberghe GC (2013) A lack of mitochondrial alternative oxidase compromises capacity to recover from severe drought stress. Physiol Plant 149(4):461–473. doi:10.1111/ppl.12059

Wang SY, Bunce JA, Maas JL (2003) Elevated carbon dioxide increases contents of antioxidant components in field-grown strawberries. J Agric Food Chem 51:4315–4320

Wang T-Z, Zhang J-L, Tian Q-Y, Zhao M-G, Zhang W (2013) A *Medicago truncatula* EF-Hand family gene, *MtCaMP1*, is involved in drought and salt stress tolerance. PLoS One 8:e58952

Weber FA, Hofacker AF, Voegelin A, Kretzschmar R (2010) Temperature dependence and coupling of iron and arsenic reduction and release during flooding of a contaminated soil. Environ Sci Technol 44:116–122

Whiting SN, Neumann PM, Baker AJM (2003) Nickel and zinc hyperaccumulation by *Alyssum murale* and *Thlaspi caerulescens* (Brassicaceae) do not enhance survival and whole-plant growth under drought stress. Plant Cell Environ 26:351–360

Wu H, Tang S, Zhang X, Guo J, Song Z, Tian S, Smith DL (2009) Using elevated CO2 to increase the biomass of a *Sorghum vulgare* × *Sorghum vulgare* var. sudanense hybrid and *Trifolium pratense* L. and to trigger hyperaccumulation of cesium. J Hazard Mater 170:861–870

Xie C, Zhang R, Qu Y, Miao Z, Zhang Y, Shen X, Wang T, Dong J (2012) Overexpression of MtCAS31 enhances drought tolerance in transgenic Arabidopsis by reducing stomatal density. New Phytol 195:124–135

Yang LX, Wang YL, Dong GC, Gu H, Huang JY, Zhu JG, Yang HJ, Liu G, Han Y (2007) The impact of free-air CO_2 enrichment (FACE) and nitrogen supply on grain quality of rice. Field Crop Res 102:128–140

Zhang BL, Shang SH, Zhang HT, Jabeen Z, Zhang GP (2013a) Sodium chloride enhances cadmium tolerance through reducing cadmium accumulation and increasing anti-oxidative enzyme activity in tobacco. Environ Toxicol Chem 32:1420–1425

Zhang Z, Zhang Q, Wu J, Zheng X, Zheng S, Sun X, Qiu Q, Lu T (2013b) Gene knockout study reveals that cytosolic ascorbate peroxidase 2 (OsAPX2) plays a critical role in growth and reproduction in rice under drought, salt and cold stresses. PLoS One 8:e57472

Zheng J, Wang H, Li Z, Tang S, Chen Z (2008) Using elevated carbon dioxide to enhance copper accumulation in *Pteridium revolutum*, a copper-tolerant plant, under experimental conditions. Int J Phytorem 10:161–172

Zhou M, Li D, Li Z, Hu Q, Yang C, Zhu L, Luo H (2013) Constitutive expression of a *miR319* gene alters plant development and enhances salt and drought tolerance in transgenic creeping bentgrass. Plant Physiol 161:1375–1391

Zou J, Liu C, Liu A, Zou D, Chen X (2012) Overexpression of *OsHsp17.0* and *OsHsp23.7* enhances drought and salt tolerance in rice. J Plant Physiol 169:628–635

9 Biotechnological Approaches to Mitigate Adverse Effects of Extreme Climatic Factors on Plant Productivity

Kuldeep Bauddh, Manish Sainger, Sanjeev Kumar, Poonam Ahlawat Sainger, Pawan Kumar Jaiwal, and Rana Pratap Singh

Abstract

Elevated carbon dioxide (CO_2), high temperature, drought, cold and freezing, and buildup of the greenhouse gases and suspended particulate matters (SPMs) in the air, the major consequences of climate change and climate variability, affect plant productivity in various ways. To mitigate the negative effects of these climatic factors on plants, a clear understanding of complex tolerance mechanisms involved in making plants more tolerant to these multiple stresses is necessary. The various metabolic and molecular pathways involved in inducing tolerance to these odd climatic conditions indicate that multiple genetic regulatory systems are involved in mitigating these stresses. However, the share of various metabolic shifts in providing tolerance to the plants is not yet clear. The conventional breeding is unable to manage such complex genetic traits; therefore, scientists are looking for Gene Technologies to handle this problem. The genetic transformation of plants with regulatory genes, e.g., transcription factors, has paved a way to produce improved plant varieties possessing tolerance to the multiple stresses that originated from the climate change. The pertinent literature indicates that a lot more is to be done to overcome the challenges in developing transgenic plants suitable for the future climate. It needs an urgent attention to resolve the availability of plant-based products in the recent future.

K. Bauddh
Centre for Environmental Sciences, Central University of Jharkhand, Brambe, Ranchi 835205, India

M. Sainger • P.K. Jaiwal
Centre for Biotechnology, Maharshi Dayanand University, Rohtak 12400, Haryana, India

S. Kumar • R.P. Singh (✉)
Department of Environmental Science, Babasaheb Bhimrao Ambedkar University (A Central University), Lucknow 226025, Uttar Pradesh, India
e-mail: cceseditor@gmail.com

P.A. Sainger
Department of Environmental Sciences, Maharishi Dayanand University, Rohtak 124001, India

P.K. Jaiwal et al. (eds.), *Genetic Manipulation in Plants for Mitigation of Climate Change*,
DOI 10.1007/978-81-322-2662-8_9

Keywords
Cold • Drought • Freezing • Heat • *Oryza sativa* L. • Stress responsive gene • Transcription factor • *Triticum aestivum* L.

9.1 Introduction

There is an increasing acceptance in various sectors that climatic variability and global warming observed during the past few decades are a reality and not a myth. The extreme hot and cold seasons and subsequent drought and flood and the elevated carbon dioxide (CO_2) and other greenhouse gases of anthropogenic origin are generally considered to be the major factors resulting from such climatic variability and extreme weathers throughout the world. Agriculture, horticulture, forestry, availability of potable and irrigation water, and energy are considered to be more sensitive areas especially in the tropical and subtropical countries including India. All these areas, directly or indirectly, are related to the plant productivity and biodiversity in different ecosystems. Though the application of inorganic fertilizers has been found to be well correlated with the crop yield (Bakht et al. 2009; Wang et al. 2010; Zaman et al. 2010), excessive use of these fertilizers has been reported not only to contaminate the groundwater and surface water bodies (Chandna et al. 2011; Rawat et al. 2010, 2012) but also responsible for the emissions of greenhouse gases, e.g., nitrogen oxides (NOx) and volatilization of NH_3, which accelerate the process of global warming (Fig. 9.1) (Galloway et al. 2008; Foulkes et al. 2009; Yang et al. 2011; Cui et al. 2012; Gogoi and Baruah 2012). Therefore, there is an increasing concern to find out ways to mitigate the negative effects of climate change on biodiversity and plant productivity in general and crop productivity in particular.

Climate change is generally related to the abiotic stresses due to extreme environmental changes, and most of the abiotic stress tolerance mechanisms in plants are complex due to the involvement of multiple metabolic pathways. Hence,. manipulating these characters through conventional breeding remains a big challenge.

New tools and techniques related to the molecular breeding and genetic engineering in plants have emerged as significant way to identify, select, and transfer the specific key regulatory genes or transcription factors of synthetic nature as well as those isolated from the living organisms across the taxonomic boundaries, e.g., microorganism and distantly related plants and animals, etc. using gene technologies. The works related to the biotechnological strategies to mitigate the problems of climate change are, however, in the initial stage and the efforts made in this direction may be limited to one (Ewert et al. 2002; Zhang et al. 2011; Xu-rong et al. 2013) or a few stress (Waditee et al. 2005), whereas the mitigation measures for these complex climatic factors need handling of multiple stresses in one go. Some success stories have been reported for tolerance to multiple stresses in plants using the regulatory genes, e.g., transcription factor, which opens up new hopes in this area (Xiong and Yang 2003; Su and Wu 2004; Shabala and Pottosin 2014).

This chapter critically assesses the potentials and limitations of biotechnological approaches for the mitigation of the multiple stress-related multigenic control of extreme temperature, and water extremes and other climate change-related stresses occur along with these stresses.

9.2 Climate Change Causes and Consequences

Before the industrial era, the concentration of atmospheric CO_2 was almost 280 ppm. The concentration of CO_2 has been increased dramatically with the industrialization and has reached more than 360 ppm by 2000 (Fig. 9.1a). The rate of this increased CO_2 over the past century is unmatched with the previous period and at least during the past 20,000 years. The enhanced

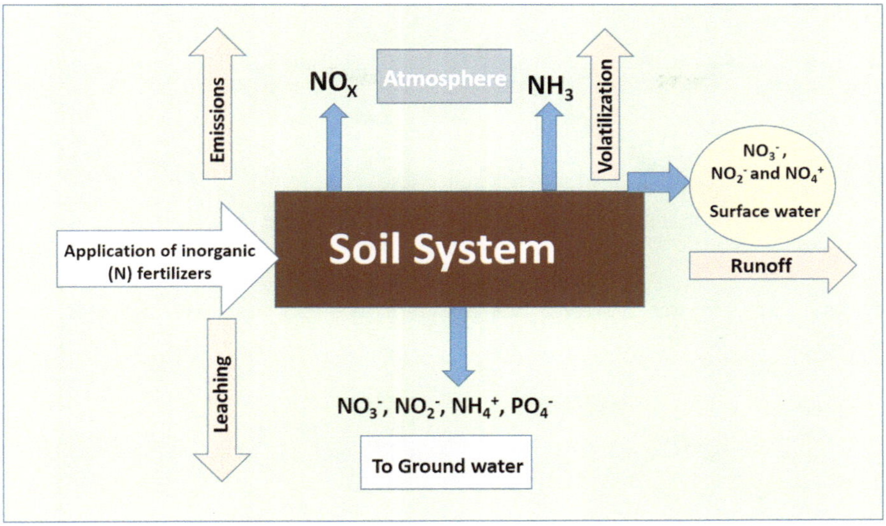

Fig. 9.1 Environmental degradation due to the application of inorganic fertilizers in crop fields. Effects of application of inorganic fertilizers (Based on Galloway et al. 2008; Foulkes et al. 2009; Rawat et al. 2010, 2012; Yang et al. 2011; Cui et al. 2012; Gogoi and Baruah 2012)

buildup of greenhouse gases during the recent past has been observed as a consequence of rapid urbanization, globalization, industrialization, deforestation, and indiscriminate chemical fertilizers during the green revolution phase of the agriculture (Figs. 9.1 and 9.2a–d). Climate change is represented by the global warming, rapid climate variability, and unscheduled expansion or shrinkage of the extreme climates, i.e., heat, cold, and rain, etc. The carrying capacity of the various climatic conditions is diminishing as sudden high rain fall causing flood and water logging may be followed by extreme temperature variations (Webb et al. 2012). The hot days are increasing in tropics and subtropics and cold days are decreasing. Similarly, in temperate regions, freezing, ice storms, earthquakes, etc. are increasing. It appears, therefore, that the greenhouse gases and air pollution causing global warming are not the only causes and consequence of the climate change, but many known and unknown factors are also involved. The environmental system is an open and huge system, where simulations and modeling of data may not give the exact picture of the expected consequences.

9.3 Metabolic Changes in Plants Due to Climate Variability and Climate Change

The major blame for the recent global warming has been primarily designated to the elevated CO_2 levels of the recent past, which has been correlated with increase in population, urbanization, and other anthropogenic activities (Fig. 9.3).

Many studies reported on the effects of elevated CO_2 on the plant productivity claims that the availability of high CO_2 levels enhances photosynthetic fixation; hence, the global warming may be beneficial to plant productivity (Morison and Gifford 1983; Prior et al. 2003; Prior and Runion 2011) (Fig. 9.4).

However, most of these studies have been performed with the isolated crop plants in simulated laboratory conditions with the supply of elevated CO_2. This will not mimic exactly multiple stress causing natural agroclimatic conditions of the global warming, which are expected to occur simultaneously in the natural ecosystems in the climate change regime. It is evident that the elevated CO_2 will delimit the carbon fixation by the green plants, provided all other relevant factors

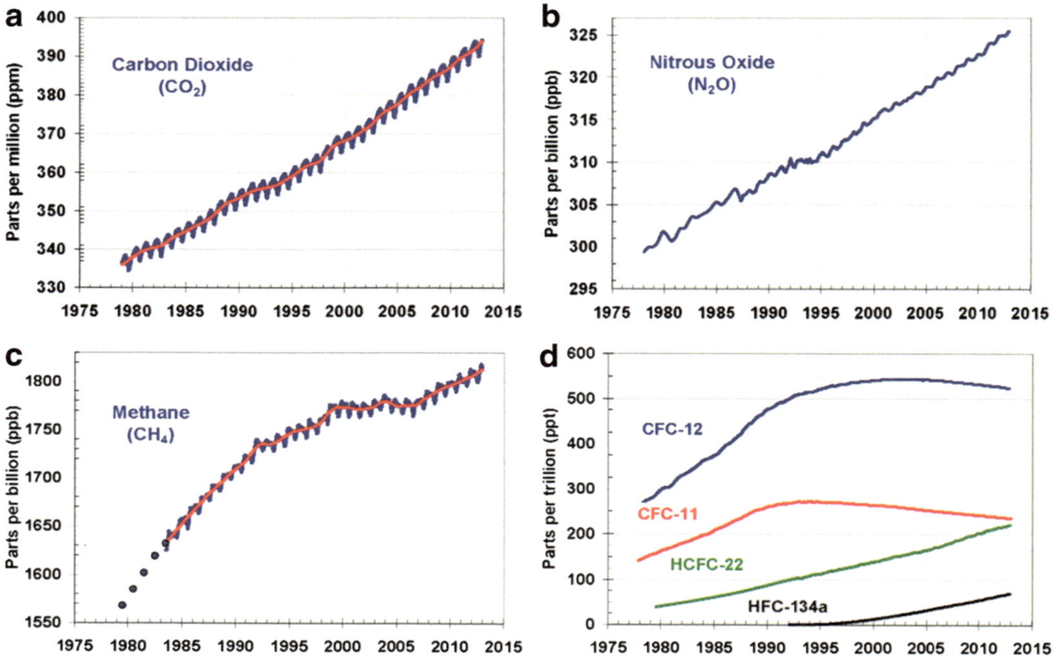

Fig. 9.2 Status of major greenhouses gases CO_2 (**a**) nitrous oxide (**b**) methane (**c**) and chlorofluorocarbons (**d**) in the atmosphere (Source: http://www.esrl.noaa.govgmdaggi)

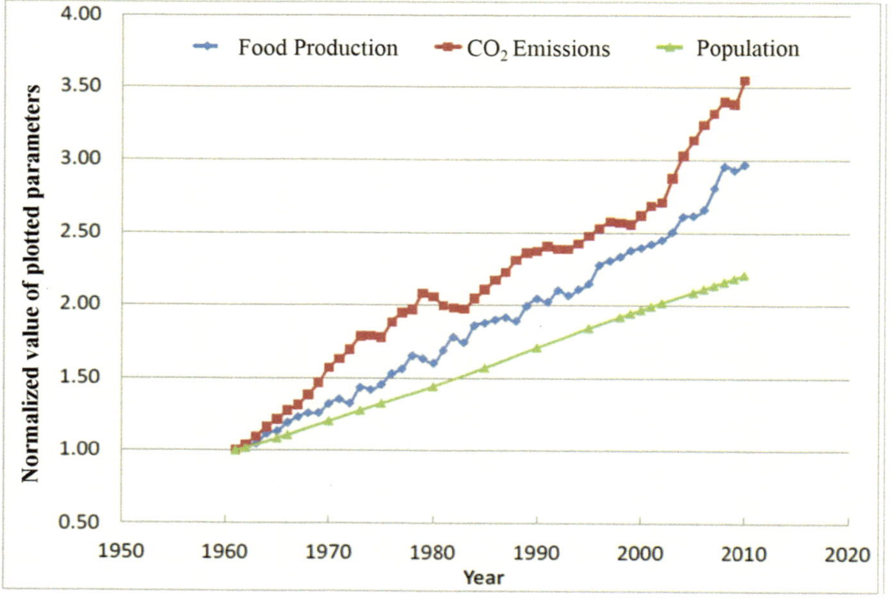

Fig. 9.3 Global population, CO_2 emissions, and food production data over the period 1961–2010 (Source: www.co2science.org)

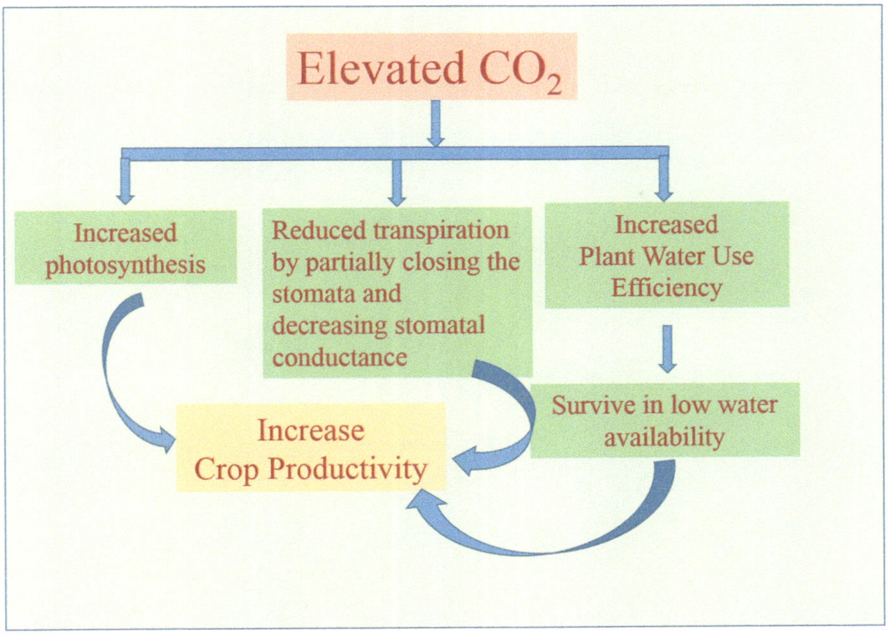

Fig. 9.4 Direct effect of elevated CO_2 on the crop productivity (Morison and Gifford 1983; Prior et al. 2003; Prior and Runion 2011; Miura et al. 2013; Zhou et al. 2013)

do not get limited. Many other factors like the synthesis of transporters, enzymes, precursors, and energy-carrying compounds, etc. will also be required, for enhanced productivity which is dependent on the metabolism of nitrogen and other macro- and micronutrients which may get limited due to CO_2 enhanced high temperature and drought.

The elevated CO_2 will increase temperature regimes and water stress to the plants which will cause metabolic shifts to increase over the production of reactive oxygen species, higher osmolyte synthesis, protein denaturation, reduced water availability, increased cellular injury, and apoptosis and will also reduce the microbial richness of the soil and cause reduction in the growth and productivity of plants (Sung et al. 2003a, b; Mittler et al. 2004; Howarth 2005; Ferrandino and Lovisolo 2014; Minocha et al. 2014; Shi and Chan 2014) (Fig. 9.5).

The extreme temperature is reported to cause a number of metabolic changes in plants which directly or indirectly reduced its growth and productivity (Wahid et al. 2007; Hanumappa and Nguyen 2010; Bernard et al. 2014; Valifard et al. 2014) (Fig. 9.6).

It appears, therefore, that the effects of the elevated CO_2 and other increased greenhouse gases need a revisit in relation to their complex interactions with other primary and secondary stresses produced in the natural ecosystems as climate change, climatic variability, and global warming. Further studies are required to explore the up- and down regulations of productivity-related genes and transcription factor under the influence of multiple stresses. In most of the studies increase in CO_2 fixation and related productivity enhancement have been reported in controlled conditions by manipulating single stress or the related genes, however, in natural fields the conditions of multiple stresses and their complex interactions occur. These studies are hence, unable to explain the regulatory procedures at molecular level which are responsible for the complex plant responses under the complex agroclimatic changes.

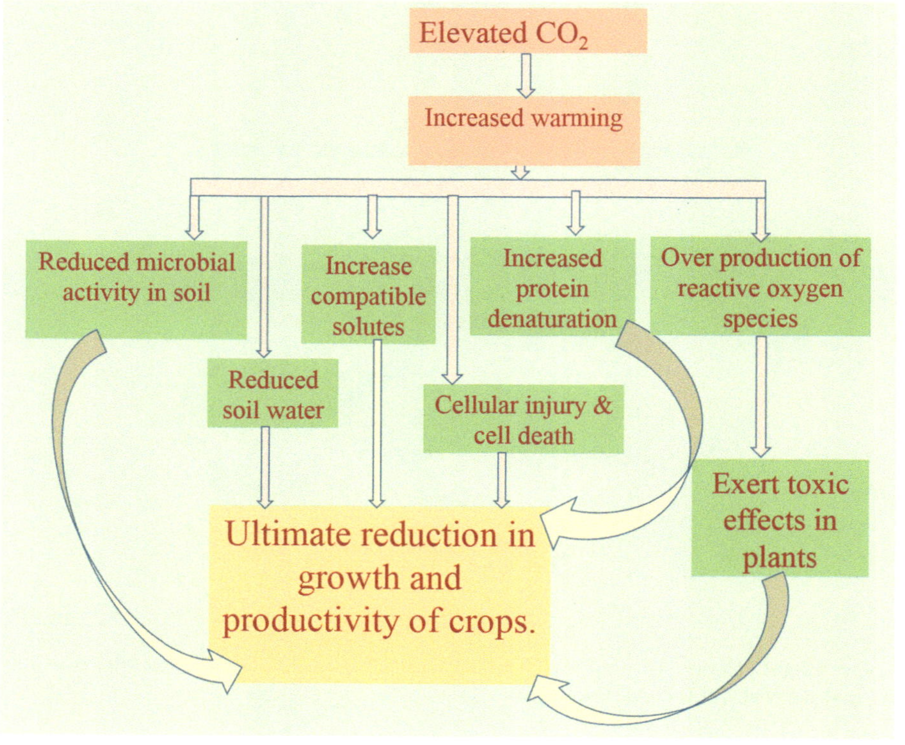

Fig. 9.5 Indirect effect of elevated CO_2 on the crop productivity (Mittler et al. 2004; Howarth 2005; Theocharis et al. 2012; Ferrandino and Lovisolo 2014; Minocha et al. 2014)

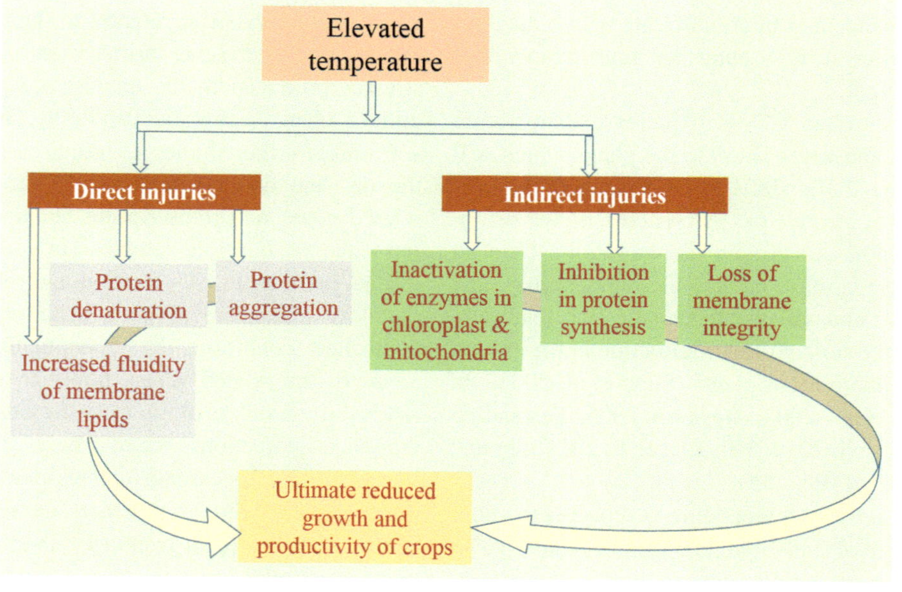

Fig. 9.6 Effect of extreme temperature on the growth and productivity of crops (Knight 2000; Larkindale and Knight 2002; Wahid et al. 2007; Hanumappa and Nguyen 2010; Evrard et al. 2013; Bernard et al. 2014; Valifard et al. 2014)

9.4 Gene and Loci Responsive to Major Climatic Factors and Assessment of Genetic Improvement in Plants to Mitigate Extreme Climatic Factors

Generally plants respond to the abiotic stresses through a wide range of reactions from morphological changes to alterations in the patterns of protein expressions, production of secondary metabolites and antioxidant compounds, etc. (Srivastava and Singh 1987; Liu et al. 2003; Gill et al. 2013; Bernard et al. 2014; Valifard et al. 2014) (Fig. 9.7).

Biotechnological approaches has made some success in improving water-use efficiency (WUE) and drought resistance in the crops like *Zea mays*, *Triticum aestivum*, *Carica papaya*, *Oryza sativa*, *Brassica napus*, *Gossypium hirsutum*, and *Saccharum officinarum* (Table 9.1) (Haake et al. 2002; Almeida et al. 2007; Chen et al. 2007; Ratnakumar et al. 2009; Zhang et al. 2011; de Paiva Rolla et al. 2013; Xu-rong et al. 2013; Tripathi et al. 2014). Haake et al. (2002) have reported that the transcription factor *CBF4* is a

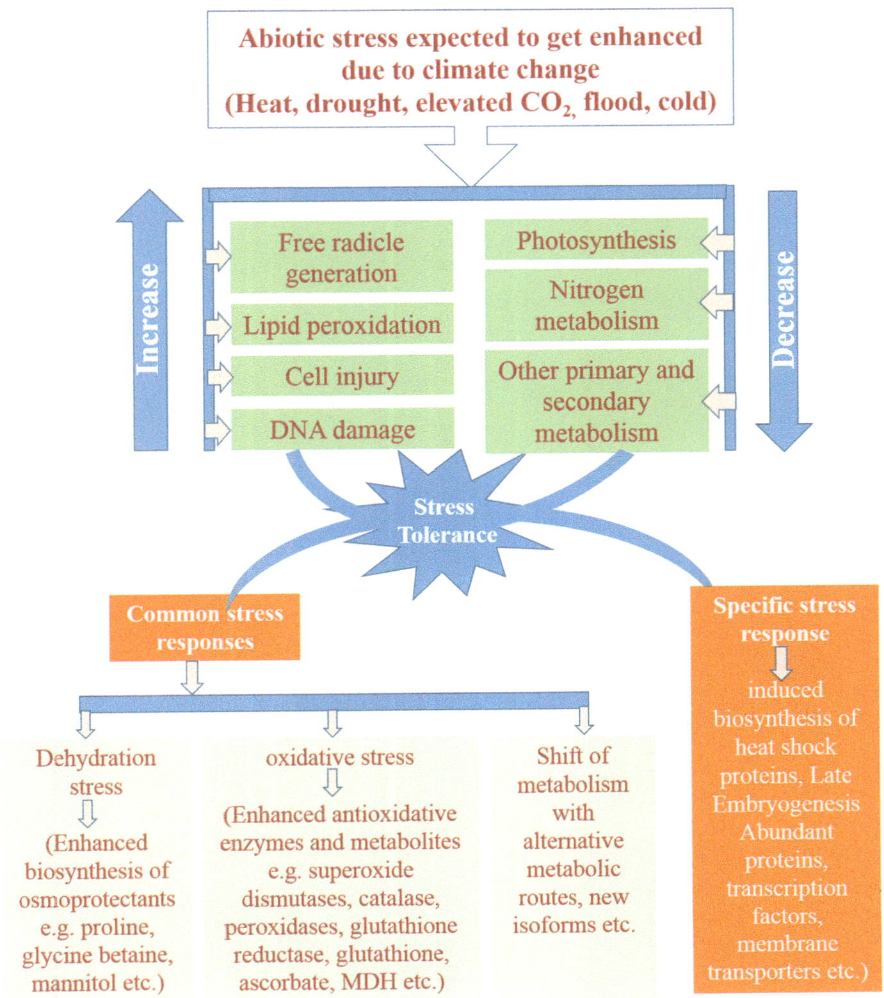

Fig. 9.7 Natural mechanisms in plants to defend the abiotic stresses (Srivastava and Singh 1987; Gill et al. 2013; Bernard et al. 2014; Danquah et al. 2014; Valifard et al. 2014)

Table 9.1 Stress-responsive genes contributed to tolerance of climate change-related extreme climatic conditions in transgenic plants

Plant species	Gene source	Gene	Type and intensity of stress	Outcome	References
Triticum aestivum L.	*Triticum aestivum* L.	*RcaB*	Heat; 35 °C	Increased photosynthesis, improvement of light use efficiency, improved antioxidant capacity	Wang et al. (2014)
Carica papaya L.	*Arabidopsis thaliana* L.	*CBF (CpCBF2)*	Heat; 37 °C	Temperature stress tolerance	Zhu et al. (2013)
Nicotiana tabacum L.	*Zea mays* L.	*AChE*	Heat; 45 °C	Maize AChE activity enhanced heat tolerance	Yamamoto et al. (2011)
Chrysanthemum pacificum L.	*Arabidopsis thaliana* L.	*AtDREB1A*	Heat; 45 °C	High RuBisCO and sucrose phosphate synthase activity; higher photosynthetic activity	Hong et al. (2009)
Zea mays L.	*Arabidopsis thaliana* L.	*GASA4 T-DNA* (SALK_042431)	Heat; 50 °C	Elevated heat tolerance	Ko et al. (2007)
Arabidopsis thaliana L.	*Arabidopsis thaliana* L.	*AtDREB2A*	Heat; 37 °C	High plant survival	Sakuma et al. (2006)
Oryza sativa L.	*Hordeum vulgare* L.	*BADH*	Heat; 45 °C	Develop heat tolerance	Kishitani et al. (2000)
Carica papaya L.	*Arabidopsis thaliana* L.	*CBF (CpCBF2)*	Cold; 7 °C	Temperature stress tolerance	Zhu et al. (2013)
Arabidopsis thaliana L.	*Oryza sativa* L.	*OsDREB1B*	Freezing; –4 °C	High plant survival under low temperature	Qin et al. (2007)
Arabidopsis thaliana L.	*Arabidopsis thaliana* L.	*ApGSMT* and *ApDMT*	Cold; 0 °C	Develop cold tolerance	Waditee et al. (2005)
Brassica napus L.	*Arabidopsis thaliana*	*BNCBF5/BNCBF17*	Cold; –2 °C	Higher photosynthetic activity and enhanced freezing tolerance	Savitch et al. (2005)
Arabidopsis thaliana L.	*Zea mays* L.	*ZmDREB1A*	Cold; 4 °C	Enhanced cold tolerance	Qin et al. (2004)
Zea mays L.	*Nicotiana tabacum* L.	*NPK1*	Freezing; –1 °C	Enhance freezing tolerance in transgenic maize	Shou et al. (2004)
Arabidopsis thaliana L.	*Capsicum annuum* L.	*CaPF1*	Cold; 4 °C	Enhance cold tolerance	Yi et al. (2004)
Oryza sativa L.	*Oryza sativa* L.	*OsMAPK5*	Cold; 4 °C	Develop cold tolerance	Xiong and Yang (2003)
Nicotiana tabacum L.	*Citrus unshiu* L.	*CuCOR19*	Freezing –4 °C	Enhancement of cold tolerance and inhibition of lipid peroxidation	Hara et al. (2003)
Oryza sativa L.	*Hordeum vulgare* L.	*BADH*	Cold; 5 °C	Develop cold tolerance	Kishitani et al. (2000)

Nicotiana tabacum L.	*Arabidopsis thaliana* L.	*fad7*	Cold; 1 °C	Enhancement of cold tolerance	Kodama et al. (1994)
Oryza sativa L.	*Arthrobacter globiformis* L.	*codA*	Drought; water regimes of 23 days	Transgenic plants showed better physiological performance at seedling, vegetative, and reproductive stage	Kathuria et al. (2009)
Gossypium hirsutum L.	*Escherichia coli*	*betA*	Drought; 4-week water-deficit stress	Glycine betaine accumulation was positively correlated with drought stress	Lv et al. (2007)
Nicotiana tabacum L.	*Arabidopsis thaliana* L.	*TPS1*	Drought; water regimes of 21 days	Transgenic plants showed sustained photosynthesis under stress	Almeida et al. (2007)
Arabidopsis thaliana	*Saccharomyces cerevisiae*	*TPS1–TPS2*	Drought; 2-week water-deficit stress	Over expressions were glucose insensitive and also involved in carbohydrate metabolism	Miranda et al. (2007)
Nicotiana tabacum L. and *Arabidopsis thaliana* L.	*Saccharomyces cerevisiae*	*TPS1–TPS2*	Drought; 24-h water-deficit stress	Maintenance of water status and positive influence of trehalose on root development	Karim et al. (2007)
Nicotiana tabacum L. and *Arabidopsis thaliana* L.	*Glycine max* L.	*GmDREB2*	Drought; water regimes of 19 days	No growth retardation; higher proline content than wild type	Chen et al. (2007)
Festuca arundinacea L.	*Arabidopsis thaliana* L.	*AtDREB1A*	Drought; water regimes of 30 days	No growth retardation; higher proline content	Zhao et al. (2007)
Triticum aestivum L.	*Vigna aconitifolia* L.	*P5CSF129A*	Drought; water regimes of 14 days	Stress-inducible promoter resulted in high proline content	Vendruscolo et al. (2007)
Saccharum officinarum L.	*Vigna aconitifolia* L.	*P5CSF129A*	Drought; 25-day water-deficit stress	Significant increase in biomass	Molinari et al. (2007)

(continued)

Table 9.1 (continued)

Plant species	Gene source	Gene	Type and intensity of stress	Outcome	References
Oryza sativa L.	*Zea mays* L.	*OsDREB1*	Drought; water regimes of 9 days	High proline and soluble sugar content; growth retardation under non-stress condition	Ito et al. (2006)
Petunia hybrid L.	*Arabidopsis thaliana* L.	*P5CSF129A*	Drought; 2-week water-deficit stress	Transgenic petunia plants tolerated 14-day drought stress	Yamada et al. (2005)
Lycopersicon esculentum L.	*Arabidopsis thaliana* L.	*CBF1 (DREB1B)*	Drought; water regimes of 0, 7, 14, 21, 28 days	Improve drought tolerance	Hsieh et al. (2002)
Arabidopsis thaliana L.	*Arabidopsis thaliana* L.	*CBF4*	Drought; water regimes of 9 days	Improve drought tolerance	Haake et al. (2002)

regulator of drought adaptation in *Arabidopsis*. Overexpression of *CBF4* in transgenic *Arabidopsis* plant has resulted in the activation of C-repeat/dehydration-responsive element containing downstream genes that are involved in cold acclimation and drought adaptation. As a result the newly developed transgenic plants are more tolerant to freezing and drought stress. Decreased soil water potential is one of the major limiting factors for the smooth production of agricultural crops, which is basically due to the increases in global temperature and land desertification (Zhou et al. 2013). Drought stress has been reported to be induced by PLD-mediated production of phosphatidic acid (Bargmann et al. 2009) and the generation of reactive oxygen species (ROS) (Choudhury et al. 2013; Yao et al. 2013; Du et al. 2014; Hossain et al. 2014). Reactive oxygen species have been strongly linked to water-deficit-induced stomatal closure (Zhou et al. 2013), which is an important mechanism reducing transpiration-mediated water losses. Production of nitric oxide (NO) has been reported to enhance the stomatal closure in presence of water-deficit stress (Neill et al. 2008).

Transgenic *Lycopersicon esculentum* having gene *CBF1* isolated from *Arabidopsis* showed enhanced resistance to water-deficit stress (Hsieh et al. 2002). They found that in transgenic tomato plants, the stomata were closed more rapidly than the wild type after water-deficit treatment. Further, *Petunia hybrida* plants have been transformed by incorporating Δ^1-pyrroline-5-carboxylate synthetase gene (*AtP5CS*) of *Arabidopsis thaliana* L. and *OsP5CS* gene isolated from *Oryza sativa* L. (Yamada et al. 2005). The resultant transgenic *P. hybrida* plants accumulated higher amount of proline than wild types under a 14-day drought stress and provided water-deficit tolerance in the transformed plants. Zhao et al. (2007) have transferred *AtDREB1A* gene in tall fescue (*Festuca arundinacea*) plant, which was exposed to water withdrawal of 30 days. They found that the transgenic tall fescue showed increased resistance to drought and accumulated high level of proline, indicating ability of the *CBF3* gene to induce stress-related response in tall fescue. Almeida et al. (2007) have also assessed the responses of photosynthesis to water deficit of tobacco plants genetically engineered with *AtTPS1* gene obtained from *A. thaliana*. When subjected to the water-deficit stress, higher Fv/Fm, FPSII, and qP were detected for the transgenic plants which indicates a possible higher ability to withstand severe drought stress and to resist to prolonged periods without water.

Expression of *AtDREB1A* gene was found to be induced by drought stress in transgenic rice lines, which were highly tolerant to severe water-deficit stress (14-day water withdrawal) in both the vegetative and reproductive stages without affecting their morphological traits (Ravi kumar et al. 2014). *AtDREB1A* was associated with an increased accumulation of proline, maintenance of chlorophyll, increased relative water content, and decreased ion leakage under drought stress. A soybean DRE-binding transcription factor provide drought and high-salt tolerance in transgenic *Arabidopsis* and tobacco (Chen et al. 2007). The overexpression of *GmDREB2* in transgenic *Arabidopsis* resulted in enhanced tolerance to drought and high-salt stresses and did not cause growth retardation. Moreover, analysis of free proline contents in transgenic tobacco indicated that the overexpression of *GmDREB2* accumulated higher level of free proline as compared to the wild-type plants under drought condition. *TPS1* and *TPS2* genes were found to be responsible for the tolerance to drought, freezing, salt, and heat without retardation in growth of *Nicotiana* and *Arabidopsis* (Karim et al. 2007; Miranda et al. 2007). Tolerance to drought stress of the homozygous transgenic cotton (*Gossypium hirsutum* L.) plants with enhanced glycine betaine (GB) accumulation due to expression of *betA* was reported by Lv et al. (2007). In this case the GB levels in transgenic cotton were positively correlated with drought tolerance under the water-deficit stress. The results suggested that GB may not only protect the integrity of the cell membrane from drought stress damage but also be involved in osmotic adjustment in transgenic cotton plants.

Molinari et al. (2007) found that the transgenic citrus rootstock with gene *P5CSF129A* showed higher osmotic adjustment and significantly higher photosynthetic rate than the control plants at 15-day water-deficit stress. Under this condition, the leaf water potential of non-transformed plants was close to zero after 17 days, and the photosynthetic rate was almost null, 2 days later; however, in transgenic plants, similar measurements were only observed after 24 days of water-deficit stress. Kathuria et al. (2009) have shown the enhanced water-deficit stress tolerance and significant transcriptome changes in *codA*-expressing transgenic rice. They claimed that it may be an alternate possible mechanism of stress tolerance for choline oxidase expressing transgenic plants. Sakuma et al. (2006) carried out microarray analysis of transgenic *Arabidopsis* overexpressing *DREB2ACA* and found that the overexpression of *DREB2ACA* induces not only drought- and salt-responsive genes but also heat shock (HS)-related genes. They found that transient induction of the *DREB2A* occurs rapidly by HS stress and that the *sGFP-DREB2A* protein accumulates in nuclei of HS-stressed cells. *DREB2A*-upregulated genes were downregulated in DREB2A knockout mutants under stress conditions. Thermotolerance was significantly increased in plants overexpressing *DREB2ACA* and decreased in *DREB2A* knockout plants.

Heat tolerance in *Arabidopsis* by functionally characterized *GASA4* gene was found to be expressed in the presence of heat (Ko et al. 2007). The constitutive expression of *GASA4* in *Arabidopsis* led to elevated heat tolerance in transgenic lines. Stressed transgenic wheat plants showed tolerance to 15 days of water shortage as a result of overexpression of *Vigna aconitifolia* Δ^1-pyrroline-5-carboxylate synthetase (P5CS) cDNA (Vendruscolo et al. 2007). Kodama et al. (1994) have engineered transgenic tobacco plant by introduction of a chloroplast ω-3 fatty acid desaturase gene (*fad7*) isolated from *Arabidopsis thaliana* for cold tolerance. The wild-type plants exposed to 1 °C for 7 days have suppressed leaf growth, which was significantly alleviated in the transgenic plants having gene *fad7*. Moreover, chlorosis was found induced in wild types due to this cold stress, which was also managed in the plants transformed with *fad7* gene.

Cold is another important abiotic stress which significantly reduces crop yield and productivity. Plants have different mechanisms which help ameliorate the effects of cold stress. Cold acclimation correlates with a massive reprogramming of gene expression as well as the metabolome (Hara et al. 2003). Several studies suggested that exogenous NO increased cold tolerance in various plant species including *Triticum aestivum*, *Zea mays*, *Solanum lycopersicum*, etc. (Cantrel et al. 2011).

Enhanced cold tolerance and inhibition of lipid peroxidation by citrus dehydrin in transgenic tobacco have been reported by Hara et al. (2003). Malondialdehyde production was found to be enhanced by chilling stress (−4 °C), which was lowered in tobacco plants expressing citrus dehydrin than in control phenotypes. The inhibitory activity of dehydrin against liposome oxidation was stronger than that of albumin, glutathione, proline, glycine betaine, and sucrose. These results suggest that dehydrin facilitates plant cold acclimation by acting as a radical-scavenging protein to protect membrane systems under cold stress. Xiong and Yang (2003) have isolated and functionally characterized a stress-responsive mitogen-activated protein kinase (MAPK) gene (*OsMAPK5*) from rice (*Oryza sativa*). They generated and analyzed transgenic rice plants with overexpression or suppression of *OsMAPK5*, and they found that the suppression of *OsMAPK5* expression and its kinase activity resulted in the constitutive expression of pathogenesis-related genes such as *PR1* and *PR10* in dsRNAi transgenic rice plants and significantly enhanced resistance to biological pathogens, e.g., fungal (*Magnaporthe grisea*) and bacterial (*Burkholderia glumae*) pathogens; however, the same dsRNAi lines had significant reductions in drought, salt, and cold tolerance. Overexpression of lines exhibited increased OsMAPK5 kinase activity and increased tolerance to drought, salt, and cold stresses. They had concluded that *OsMAPK5* can positively regulate drought, salt, and cold tolerance and negatively modulate pathogenesis-related gene expression and broad spectrum disease resistance. Yi et al. (2004) have reported that transgenic *Arabidopsis* plants expressing *CaPF1* displayed tolerance against freezing temperatures and enhanced resistance to *Pseudomonas syringae*. They observed the disease tolerance in transgenic tobacco plants having gene *CaPF1*. The results of their study indicate that *CaPF1* is an *ERF/AP2* transcription factor in hot pepper plants that may play dual roles in response to biotic and abiotic stress in plants. The effects of expressing a heterologous tobacco mitogen-activated protein kinase (*Nicotiana PK1*), which is able to mimic H_2O_2 signaling, have been reported in maize (Shou et al. 2004). They found that low level of constitutive expression of the *Nicotiana PK1* gene enhances freezing tolerance in transgenic maize plants that are normally frost sensitive. The overexpression of *BNCBF* as reported by Savitch et al. (2005) has partially simulated cold-induced photosynthetic acclimation in *Brassica napus*. These results suggest that *BNCBF/DREB1* overexpression in *Brassica* not only resulted in increased freezing tolerance but also partially regulated chloroplast development to increase photochemical efficiency and photosynthetic capacity.

The co-expression of *N*-methyltransferase genes in *Synechococcus* caused accumulation of a significant amount of betaine and conferred salt tolerance to a freshwater cyanobacterium sufficient for it to become capable of growth in seawater (Waditee et al. 2005). Similarly *Arabidopsis* plants under 0 °C cold stress expressed *N*-methyltransferase genes which results to higher accumulation of betaine in roots, stems, leaves, and flowers that provide cold tolerance to the plants resulting in improved seed yield. Qin et al. (2004) reported that overexpression of transgenic *Arabidopsis ZmDREB1A* induced overexpression of target stress-inducible genes of *Arabidopsis DREB1A* resulting in plants with higher tolerance to drought and freezing stresses. *ZmDREB1A* is suggested to be potentially useful for producing transgenic plants that is tolerant to multiple stress drought, high salinity, and/or cold stresses.

At 450 ppm CO_2*EPF2*-overexpressing plants, with reduced stomatal density, had larger leaves and increased dry weight in comparison with controls when a collection of *Arabidopsis* epidermal patterning factor (EPF) mutants with an approximately 16-fold range of stomatal densities (approx. 20–325 % of that of control plants) were grown at three atmospheric carbon dioxide (CO_2) concentrations (200, 450, and 1000 ppm) and 30 % or 70 % soil water content (Doheny-Adams et al. 2012). The growth of these plants was also less adversely affected by reduced water availability than plants with higher stomatal density, indicating that plants with low stomatal density may be well suited to growth under predicted future atmospheric CO_2 environments and/or water-scarce environments.

9.5 Conclusions and Future Prospects

The studies cited in Table 9.1 and in the above text indicate that a cross-talk signaling is involved in stress tolerance in plants which goes across the abiotic stresses and beyond (to biotic stresses). In addition, the dehydration and oxidative stresses which are generated as secondary stress due to most of the biotic and abiotic stresses are managed by a series of metabolic reactions causing enhanced accumulation of compatible solutes and free radical-scavenging systems. All these processes involve a large number of genes to produce these tolerance mechanisms, and it is very difficult, if not impossible, to pyramid gene transfer for multiple genes. It is also not clear which component of tolerance mechanism has a major share in providing climate change-related stresses. Therefore, the regulatory genes, e.g., transcription factors, which induce a cascade of events related to signaling and biosynthesis of stress protection-related metabolites by a single gene transfer provide a hope for the future targets.

References

Almeida AM, Silva AB, Araujo SS, Cardoso LA, Santos DMJ, Torne JM, Silva JM, Paul MJ, Fevereiro PS (2007) Responses to water withdrawal of tobacco plants genetically engineered with the AtTPS1 gene: a special reference to photosynthetic parameters. Euphytica 154:113–126

Bakht J, Shafi M, Jan MT, Shah Z (2009) Influence of crop residue management, cropping system and N fertilizers on soil N and C dynamics and sustainable wheat (*Triticum aestivum* L.) production. Soil Till Res 104:233–240

Bargmann BOR, Laxalt AM, Riet B, van Schooten B, Merquiol E, Testerink C, Haring MA, Bartels D, Munnik T (2009) Multiple PLDs required for high salinity and water deficit tolerance in plants. Plant Cell Physiol 50:78–89

Bernard F, Brulle F, Dumez S, Lemiere S, Platel A, Nesslany F, Cuny D, Deram A, Vandenbulcke F (2014) Antioxidant responses of Annelids, *Brassicaceae* and *Fabaceae* to pollutants: a review. Ecotoxicol Environ Safety. 114:273–303. doi:10.1016/j.ecoenv.2014.04.024

Cantrel C, Vazquez T, Puyaubert J, Reze N, Lesch M, Kaiser WM et al (2011) Nitric oxide participates in cold-responsive phosphosphingolipid formation and gene expression in Arabidopsis thaliana. New Phytol 189:415–427

Chandna P, Khurana ML, Ladha JK, Punia M, Mehla RS, Gupta R (2011) Spatial and seasonal distribution of nitrate- N in groundwater beneath the rice-wheat cropping system of India: a geospatial analysis. Environ Monit Assess 178:545–562

Chen M, Wang QY, Cheng XG, Xu ZS, Li LC, Ye XG, Xia LQ, Ma YZ (2007) *GmDREB2*, a soybean DRE-binding transcription factor, conferred drought and high-salt tolerance in transgenic plants. Biochem Bioph Res Co 353:299–305

Choudhury S, Panda P, Sahoo L, Panda SK (2013) Reactive oxygen species signaling in plants under abiotic stress. Plant Signal Behav 8:4, e23681. doi:10.4161/psb.23681

Cui F, Yan G, Zhou Z, Zheng X, Deng J (2012) Annual emission of nitrous oxide and nitric oxide from a wheat- maize cropping system on a silt loam calcareous soil in the North china plain. Soil Biol Biochem 48:10–19

Danquah A, de Zelicourt A, Colcombet J, Hirt H (2014) The role of ABA and MAPK signaling pathways in plant abiotic stress responses. Biotechnol Adv 32:40–52

de Paiva Rolla AA, de Fa'tima Corrêa Carvalho J, Fuganti-Pagliarini R et al (2013) Phenotyping soybean plants transformed with rd29A: AtDREB1A for drought tolerance in the greenhouse and field. Transgenic Res 23:75–87. doi:10.1007/s11248-013-9723-6

Doheny-Adams T, Hunt L, Franks PJ, Beerling DJ, Gray JE (2012) Genetic manipulation of stomatal density influences stomatal size, plant growth and tolerance to restricted water supply across a growth carbon dioxide gradient. Philos Trans R Soc B Biol Sci 19 367(1588):547–555

Du F, Shi H, Zhang X, Xu X (2014) Responses of reactive oxygen scavenging enzymes, proline and malondialdehyde to water deficits among six secondary successional cereal species in Loess Plateau. PLoS One 9(6):e98872. doi:10.1371/journal.pone.0098872

Evrard A, Kumar M, Lecourieux D, Lucks J, von Koskull-Doring P, Hirt H (2013) Regulation of the heat stress response in Arabidopsis by MPK6-targeted phosphorylation of the heat stress factor HsfA2. Peer J 1:e59

Ewert F, Rodriguez D, Jamieson P, Semenov MA, Mitchell RAC, Goudriaan J, Porter JR, Kimball BA, Pinter PJ, Manderscheid R, Weigel HJ, Fangmeier A, Fereres E, Villalobos F (2002) Effects of elevated CO_2 and drought on wheat: testing crop simulation models for different experimental and climatic conditions. Agri Eco Environ 93:249–266

Ferrandino A, Lovisolo C (2014) Abiotic stress effects on grapevine (*Vitis vinifera* L.): Focus on abscisic acid-mediated consequences on secondary metabolism and berry quality. Environ Exp Bot 103:138–147

Foulkes MJ, Hawkesford MJ, Barraclough PB, Holdsworth MJ, Kerr S, Kightley S, Shewry PR (2009) Identifying traits to improve the nitrogen economy of wheat: recent advances and future prospects. Field Crop Res 114:329–324

Galloway JN, Townsend AR, Erisman JW, Bekunda M, Cai Z, Freney JR, Martinelli SP, Seitzinger SP, Sutton MA (2008) Transformation of nitrogen cycle: recent trends, questions, and potential solutions. Science 320:889–892

Gill SS, Anjum NA, Hasanuzzaman M, Gill R, Trivedi DK, Ahmad I, Pereira E, Tuteja N (2013) Glutathione and glutathione reductase: a boon in disguise for plant abiotic stress defense operations. Plant Physiol Biochem 70:204–212

Gogoi B, Baruah KK (2012) Nitrous oxide emission from fields with different wheat and rice varieties. Pedosphere 22:112–121

Haake V, Cook D, Riechmann JL, Pineda O, Thomashow MF, Zhang JZ (2002) Transcription factor CBF4 is a regulator of drought adaptation in Arabidopsis. Plant Physiol 130:639–648

Hanumappa M, Nguyen HT (2010) Genetic approaches toward improving heat tolerance in plants. In: Jenks MA, Wood AJ (eds) Genes for plant abiotic stress. Blackwell Publishing, Ames, pp 221–260

Hara M, Terashima S, Fukaya T, Kuboi T (2003) Enhancement of cold tolerance and inhibition of lipid peroxidation by citrus dehydrin in transgenic tobacco. Planta 217:290–298

Hong B, Ma C, Yang Y, Wang T, Yamaguchi-Shinozaki K, Gao J (2009) Over-expression of AtDREB1A in chrysanthemum enhances tolerance to heat stress. Plant Mol Biol 70:231–240

Hossain MA, Mostofa MG, Burritt DJ, Fujita M (2014) Modulation of reactive oxygen species and methylglyoxal detoxification systems by exogenous glycinebetaine and proline improves drought tolerance in mustard (*Brassica juncea* L.). Int J Plant Biol Res 2:1–14

Howarth CJ (2005) Genetic improvements of tolerance to high temperature. In: Ashraf M, Harris PJC (eds) Abiotic stresses: plant resistance through breeding and molecular approaches. Howarth Press Inc, New York

Hsieh TH, Lee JT, Charng YY, Chan MT (2002) Tomato plants ectopically expressing Arabidopsis CBF1 show enhanced resistance to water deficit stress. Plant Physiol 130:618–626

Ito Y, Katsura K, Maruyama K, Taji T, Kobayashi M, Seki M, Shinozaki K, Yamaguchi-Shinozaki K (2006) Functional analysis of rice DREB1/CBF-type transcription factors involved in cold-responsive gene expression in transgenic rice. Plant Cell Physiol 47:141–153

Karim S, Aronsson H, Ericson H, Pirhonen M, Leyman B, Welin B, Mantyla E, Palva ET, van Dijck P, Holstrom KO (2007) Improved drought tolerance without undesired side effects in transgenic plants producing trehalose. Plant Mol Biol 64:371–386

Kathuria H, Giri J, Tyagi H, Tyagi AK (2009) Advances in transgenic rice biotechnology. Crit Rev Plant Sci 26:65–103

Kishitani S, Takanami T, Suzuki M, Oikawa M, Yokoi S, Ishitani M, Alvarez-Nakase AM, Takabe T, Takabe T (2000) Compatibility of glycinebetaine in rice plants: evaluation using transgenic rice plants with a gene for peroxisomal betaine aldehyde dehydrogenase from barley. Plant Cell Environ 23:107–114

Knight H (2000) Calcium signaling during abiotic stress in plants. Int Rev Cytol 195:269–324

Ko C, Woo Y, Lee DJ, Lee M, Kim CS (2007) Enhanced tolerance to heat stress in transgenic plants expressing the GASA4 gene. Plant Physiol Biochem 45:722–728

Kodama H, Hamada T, Horiguchi G, Nishimura M, Iba K (1994) Genetic enhancement of cold tolerance by expression of a gene for chloroplast [omega]-3 fatty acid desaturase in transgenic tobacco. Plant Physiol 105:601–605

Larkindale J, Knight MR (2002) Protection against heat stress-induced oxidative damage in Arabidopsis involves calcium, abscisic acid, ethylene, and salicylic acid. Plant Physiol 128:628–695

Liu F, Andersen MN, Jensen CR (2003) Loss of pod set caused by drought stress is associated with water status and ABA content of reproductive structures in soybean. Func Plant Biol 30:271–280

Lv S, Yang A, Zhang K, Wang L, Zhang J (2007) Increase of glycine betaine synthesis improves drought tolerance in cotton. Mol Breed 20:233–248

Minocha R, Majumdar R, Minocha SC (2014) Polyamines and abiotic stress in plants: a complex relationship. Front Plant Sci 5:1–17

Miranda JA, Avonce N, Suarez R, Thevelein JM, Dijck PV, Iturriaga G (2007) A bifunctional TPS–TPP enzyme from yeast confers tolerance to multiple and extreme abiotic conditions in transgenic Arabidopsis. Planta 226:1411–1421

Mittler R, Vanderauwera S, Gollery M, Van Breusegem F (2004) Reactive oxygen gene network of plants. Trends Plant Sci 9:490–498

Miura K, Okamoto H, Okuma E, Shiba H, Kamada H, Hasegawa PM et al (2013) *SIZ1* deficiency causes reduced stomatal aperture and enhanced drought tolerance via controlling salicylic acid-induced accumulation of reactive oxygen species in Arabidopsis. Plant J 73:91–104

Molinari HBC, Marur CJ, Filho JCB, Kobayashi AK, Pileggi M, Junior RPL, Pereira LFP, Vieira LGE (2007) Osmotic adjustment in transgenic citrus rootstock Carrizo citrange (*Citrus sinensis* Osb. 6 *Poncirus trifoliata* L. Raf.) overproducing proline. Plant Sci 167:1375–1381

Morison JIL, Gifford RM (1983) Stomatal sensitivity to carbon dioxide and humidity. Plant Physiol 71:789–796

Neill S, Barros R, Bright J, Desikan R, Hancock J, Harrison J, Morris P, Ribeiro D, Wilson I (2008) Nitric oxide, stomatal closure, and abiotic stress. J Exp Bot 59(2):165–167

Prior SA, Runion GB (2011) A review of elevated atmospheric CO2 effects on plant growth and water relations: implications for horticulture. Hortscience 46:158–162

Prior SA, Torbert HA, Runion GB, Rogers HH (2003) Implications of elevated CO_2 induced changes in agroecosystem productivity. J Crop Prod 8:217–244

Qin F, Sakamura Y, Li J, Liu Q, Li YQ, Shinozaki K, Yamaguchi-Shinozaki K (2004) Cloning and functional analysis of a novel DREB1/CBF transcription factor involved in cold-responsive gene expression in *Zea mays* L. Plant Cell Physiol 45:1042–1052

Qin Q, Liu J, Zhang Z, Peng R, Xiong A, Yao Q, Chen J (2007) Isolation, optimization, and functional analysis of the cDNA encoding transcription factor OsDREB1B in *Oryza Sativa* L. Mol Breed 19:329–340

Ratnakumar P, Vadez V, Nigam SN, Krishnamurthy L (2009) Assessment of transpiration efficiency in peanut (*Arachis hypogaea* L.) under drought using a lysimeter system. Plant Biol 11:124–130

Ravikumar G, Manimaran P, Voleti SR, Subrahmanyam D, Sundaram RM, Bansal KC, Viraktamath BC, Balachandran SM (2014) Stress-inducible expression of *AtDREB1A* transcription factor greatly improves drought stress tolerance in transgenic *Indica* rice. Transgenic Res 23:421–439. doi:10.1007/s11248-013-9776-6

Rawat SK, Singh RK, Singh RP (2010) Seasonal variation of nitrate level in ground and surface waters of Lucknow and its remediation using certain aquatic macrophytes. Int J Lake Riv 3(1):25–35

Rawat SK, Singh RK, Singh RP (2012) Remediation of nitrite in ground and surface waters using aquatic macrophytes. J Environ Biol 33:51–56

Sakuma Y, Maruyama K, Qin F, Osakabe Y, Shinozaki K, Yamaguchi-Shinozaki K (2006) Dual function of an Arabidopsis transcription factor *DREB2A* in water-stress-responsive and heat-stress-responsive gene expression. Proc Natl Acad Sci U S A 103:18822–18827

Savitch LV, Allard G, Seki M, Robert LS, Tinker NA, Huner NPA, Shinozaki K, Singh J (2005) The effect of over-expression of two *Brassica* CBF/DREB1-like transcription factors on photosynthetic capacity and frost tolerance in *Brassica napus*. Plant Cell Physiol 46:1525–1539

Shabala S, Pottosin I (2014) Regulation of potassium transport in plants under hostile conditions: implications for abiotic and biotic stress tolerance. Physiol Plantarum 151:257–279

Shi H, Chan Z (2014) Improvement of plant abiotic stress tolerance through modulation of the polyamine pathway. J Integr Plant Biol 56:114–121

Shou H, Bordallo P, Fan J, Yeakley JM, Bibikova M, Sheen J, Wang K (2004) Expression of an active tobacco mitogen-activated protein kinase enhances freezing tolerance in transgenic maize. Proc Natl Acad Sci U S A 101:3298–3303

Srivastava HS, Singh RP (1987) Role and regulation of L-glutamate dehydrogenase activity in higher plants. Phytochemistry 26(3):597–610

Su J, Wu R (2004) Stress-inducible synthesis of proline in transgenic rice confers faster growth under stress conditions than that with constitutive synthesis. Plant Sci 166:941–948

Sung D-Y, Kaplan F, Lee K-J, Guy CL (2003a) Acquired tolerance to temperature extremes. Trends Plant Sci 8:179–187

Sung DY, Kaplan F, Lee KJ, Guy CL (2003b) Acquired tolerance to temperature extremes. Trends Plant Sci 8:179–187

Theocharis A, Clement C, Barka EA (2012) Physiological and molecular changes in plants grown at low temperatures. Planta 235:1091–1105

Tripathi P, Rabara RC, Rushton PJ (2014) A systems biology perspective on the role of WRKY transcription factors in drought responses in plants. Planta 239:255–266

Valifard M, Mohsenzadeh S, Kholdebarin B, Rowshan V (2014) Effects of salt stress on volatile compounds, total phenolic content and antioxidant activities of *Salvia mirzayanii*. South Afr J Bot 93:92–97

Vendruscolo ECG, Schuster I, Pileggi M, Scapim CA, Molinari HBC, Marur CJ, Vieira LGE (2007) Stress-induced synthesis of proline confers tolerance to water deficit in transgenic wheat. J Plant Physiol 164:1367–1376

Waditee R, Bhuiyan MNH, Rai V, Aoki K, Tanaka Y, Hibino T, Suzuki S, Takano J, Jagenodorf AT, Takabe T, Takabe T (2005) Genes for direct methylation of glycine provide high levels of glycinebetaine and abiotic-stress tolerance in *Synechococcus* and Arabidopsis. Proc Natl Acad Sci U S A 102:1318–1323

Wahid A, Gelani S, Ashraf M, Foolad MR (2007) Heat tolerance in plants: an overview. Environ Exp Bot 61:199–223

Wang Q, Li F, Zhao L, Zhang E, Shi S, Zhao W, Song W, Vance MM (2010) Effect of irrigation and nitrogen application rates on nitrate nitrogen distribution and fertilizer nitrogen loss, wheat yield and nitrogen uptake on a recently reclaimed sandy farmland. Plant soil 337:325–339

Wang X, Cai J, Liu F, Dai T, Cao W, Wollenweber B, Jiang D (2014) Multiple heat priming enhances thermo-tolerance to a later high temperature stress via improving subcellular antioxidant activities in wheat seedlings. Plant Physiol Biochem 74:185–192

Webb NP, Stokes CJ, Scanlan JC (2012) Interacting effects of vegetation, soils and management on the sensitivity of savanna rangelands to climate change. Clim Change 112:925–943

Xiong L, Yang Y (2003) Disease resistance and abiotic stress tolerance in rice are inversely modulated by an abscisic acid–inducible mitogen-activated protein kinase. Plant Cell 15:745–759

Xu-rong Mei, Xiu-li Zhong, Vincent V, Xiao-ying, L (2013) Improving water use efficiency of wheat crop varieties in the North China plain: review and analysis. J Integ Agric 12:1243–1250

Yamada M, Morishita H, Urano K, Shiozaki N, Yamaguchi-Shinozaki K, Shinozaki K, Yoshiba Y (2005) Effects of free proline accumulation in petunias under drought stress. J Exp Bot 56:1975–1981

Yamamoto K, Sakamoto H, Momonoki YS (2011) Maize acetylcholinesterase is a positive regulator of heat tolerance in plants. J Plant Physiol 168:1987–1992

Yang YC, Zhang M, Zheng L, Cheng DD, Liu M, Geng YQ (2011) Controlled release urea improved nitrogen use efficiency, yield and quality of wheat. Agron J 103:479–485

Yao Y, Liu X, Li Z, Ma X, Rennenberg H, Wang X et al (2013) Drought-induced HO accumulation in subsidiary cells is involved in regulatory signaling of stomatal closure in maize leaves. Planta 238:217–227

Yi SY, Kim JH, Joung YH, Lee S, Kim WT, Yu SH, Choi D (2004) The pepper transcription factor CaPF1 confers pathogen and freezing tolerance in Arabidopsis. Plant Physiol 136:2862–2874

Zaman M, Nguyen ML, Blennerhasssett D (2010) The effect of different rates of urea with or without urease inhibitor (NBPT) on wheat yield and quality. Agric J 5:309–312

Zhang ZB, Xu P, Shao HB, Liu MJ, Fu ZY, Chu LY (2011) Advances and prospects: biotechnologically improving crop water use efficiency. Crit Rev Biotechnol 31:281–293

Zhao J, Ren W, Zhi D, Wang L, Xia G (2007) Arabidopsis DREB1A/CBF3 bestowed transgenic tall fescue increased tolerance to drought stress. Plant Cell Rep 26:1521–1528

Zhou XF, Jin YH, Yoo CY, Lin XL, Kim WY, Yun DJ et al (2013) *CYCLINH;1* regulates drought stress responses and blue light-induced stomatal opening by inhibiting ROS accumulation in Arabidopsis. Plant Physiol 162:1030–1041

Zhu X, Li X, Chen W, Lu W, Mao J, Liu T (2013) Molecular cloning, characterization and expression analysis of CpCBF2 gene in harvested papaya fruit under temperature stresses. Electron J Biotechnol. doi:10.2225/vol16-issue4-fulltext-1

Impacts of Anthropogenic Carbon Dioxide Emissions on Plant-Insect Interactions

10

Jorge A. Zavala and Linus Gog

Abstract

Human industrialization has steadily raised atmospheric levels of CO_2 from 280 ppm prior to industrialization to current levels at 400 ppm and by 2050 are expected levels of 550 ppm. Climate change has important impacts on plant-insect interactions, and gaps in current understanding of plant responses to herbivory exist. Lately new empirical data has started to illuminate the mechanisms of the effects of elevated CO_2 in plant-insect interactions. Research has shown that the resource allocation to allelochemicals is interconnected among photosynthesis, genetic regulation, and hormonal signaling. Recent molecular approaches have revealed that insect damage is perceived by plants, and the signal is amplified by the participation of regulatory elements modulated by JA and ET, which induce plant responses to increase chemical defenses against herbivores. Elevated CO_2 inhibits JA and ET pathways and increases susceptibility of plants to herbivore attack by decreasing both constitutive and inducible chemical defenses against certain insects. Conversely, enriched atmospheric CO_2 increases SA, which increases other chemical defense pathways that are not regulated by JA. Identifying how atmospheres with high CO_2 levels moderate resource allocation to secondary metabolism would help to avoid any interference in natural plant defenses. In this chapter, we discuss current understanding of the mechanisms controlling insect herbivory pertaining to the global rise in atmospheric CO_2 concentrations.

Keywords

Elevated CO_2 • Global change • Plant insect-interactions • Plant defenses • Secondary metabolites

J.A. Zavala (✉)
Cátedra de Bioquímica/Instituto de Investigación en Biociencias Agrícolas y Ambientales, Facultad de Agronomía, University of Buenos Aires-Consejo Nacional de Investigaciones Científicas y Técnicas de Argentina, C1417DSE Buenos Aires, Argentina
e-mail: zavala@agro.uba.ar

L. Gog
Department of Plant Biology, University of Illinois at Urbana-Champaign, Champaign, IL, USA

10.1 Introduction

At the beginning of the twenty-first century, the scientific literature outlines global boundary limits to agricultural production. Estimates of land use, for instance, hold that croplands, pastures, and planted forests occupy between 4000 and 5000 million hectares of land on Earth, with a

P.K. Jaiwal et al. (eds.), *Genetic Manipulation in Plants for Mitigation of Climate Change*,
DOI 10.1007/978-81-322-2662-8_10

predicted addition of 120–280 million hectares by 2030 (Lambin and Meyfroidt 2011). The ceiling on energy efficiency of plants is known as well: Zhu et al. (2008) calculate that at contemporary CO_2 concentrations, C3 plants are up to 4.6 % efficient at capturing solar energy as chemical bonds, while C4 plants are 6 %. Before that energy can be appropriated for anthropic purposes, however, herbivory by insects annually deducts up to 13 % from global agricultural yield, at an estimated economic cost of 400 billion dollars (Pimental 2004).

Responses of insect populations to rising anthropogenic CO_2 emissions are a topic of uncertainty and important to agriculture. Human industrialization has steadily raised atmospheric levels of CO_2 from 280 ppm prior to industrialization to current levels at 400 ppm and by 2050 are expected levels of 550 ppm (IPCC 2007). Atmospheric CO_2 concentration modulates the overall physiology of C3 plants (Leakey et al. 2009) and changes the quality of foliage, which makes herbivore insects sensitive to the rise in atmospheric CO_2 (DeLucia et al. 2008). Since climate change influences insect populations through their interactions with host plants, predicting how insect populations will respond to climate change centers on understanding of plant physiology responses to herbivory. Especially, the physiological mechanisms controlling allocation of resources to defenses against herbivores in plants grown under elevated CO_2 are most uncertain (Zavala et al. 2013).

Understanding the physiological mechanisms that lie beneath plant-insect interactions in elevated CO_2 environments could help to predict the impacts of climate change on insect populations. Yet adjusting plant chemical defenses to changing environment could be an option for managing the stability of agroecosystems. Additionally, understanding how plants allocate resources to secondary metabolism under elevated CO_2 environment would help to prevent any interference on the innate defenses in genetically modified plants. In this chapter, we discuss current understanding of the mechanisms controlling insect herbivory pertaining to the global rise in atmospheric CO_2 concentrations.

10.2 Role of Plant Ecophysiology on Quality of Foliage

10.2.1 Photosynthesis and Stomata Closure

Photosynthesis can be imagined as the initial stage of a process that distributes carbon and energy through interactions between plants and herbivore insects to culminate in considerable portion of the terrestrial ecosystem (Weisser and Siemann 2008). The carbon fixation enzyme ribulose bisphosphate carboxylase/oxidase (rubisco) is responsible for fixing atmospheric CO_2, and it is situated at the beginning of this process. Competitive substrate binding between CO_2 and O_2 at the active site of rubisco accounts for the sensitivity of photosynthesis to changes in levels of atmospheric CO_2. Although some grasses and arid succulents (C4 and CAM plants) fix carbon independent of atmospheric CO_2 concentrations, the majority of plants (those which possess C3 metabolism) photosynthesize with increasing efficiency as their supply of CO_2 becomes more concentrated (Leakey et al. 2009).

The stimulation of photosynthesis by rising CO_2 levels has proliferant consequences on the general physiology of plants. For instance, stomatal apertures constrict as photosynthetic capacity increases, which leads to reduced transpiration rates, resulting in increased foliar temperature (Ainsworth and Rogers 2007; Bernacchi et al. 2007). The substrate saturation of rubisco by CO_2 assimilation not only influences plant physiology by increasing starch accumulation and sucrose signaling but also extends into the ecological dynamic between plants and insect herbivores.

10.2.2 The Impact of Leaf Quality on Folivores

The main source of N for insects comes from rubisco, the dominant protein in the leaves of most plants and also the main source of valuable amino acids. Leaves contain up to nearly 40 % of amino acids and this is a potentially greater source of energy than a few micrograms of starch

(Farmer 2014). The ratio of carbon (C) to nitrogen (N) in plant tissue is higher than that of insects and converts N a limiting factor to the growth of insect herbivores (Mattson 1980). Although insects in general tend to compensate the lack of N in the diet by consuming more foliage, some insects are phagostimulated by high levels of carbohydrates (Hamilton et al. 2005). In one study, the CO_2-driven increase of carbohydrates in soybean increased fecundity of Japanese beetles (O'Neill et al. 2008).

Although insects want nutritional resources from leaves, plants need to conserve these same resources for their own growth and reproduction. While the plant nutritional resources support insect growth, many plant secondary metabolites specifically reduce herbivore feeding efficiency (Felton 2005). Optimal defense theory predicts that the products of primary metabolism are balanced against the products of secondary metabolism in plant tissue, such that plant fitness is maximized under challenge by insect herbivores (McKey 1974). When considering the effects of elevated CO_2 on insect populations, both the nutritional content and the titer of chemical defenses must be taken into account.

10.2.3 Consequences on Plant Growth Rate

Increasing atmospheric CO_2 accelerates plant growth. Elevated CO_2 environment spurns plants to assimilate carbon at a greater rate, leading to increased carbohydrate content as well as a shift to lower specific leaf areas (Stiling and Cornelissen 2007). The acceleration of plant growth rate can result a form of tolerance to insect herbivores by decreasing the relative foliar N content (Fornoni 2011). At the same time, the reduction in ratio of N to C obliges insect herbivores to compensate by consuming more plant tissue to meet their basic dietary N requirements (DeLucia et al. 2008).

The phenology of plants also can be affected by the environmental concentration of CO_2; the timing and extent of developmental stages of plants are typically delayed with rising CO_2 atmospheres. Soybeans grown under elevated CO_2, for instance, begin to flower later than their control counterparts grown in ambient CO_2 atmosphere (Castro et al. 2009). Likewise, senescence is both delayed in onset and prolonged in duration for plants exposed to elevated CO_2 levels (Taylor et al. 2008). This shift in phenological timing can lead to developmental mismatches between plants and their insect counterparts (DeLucia et al. 2012). Stages in the life cycles of insect pollinators, in particular, often are synchronized with the flowering times of plants they frequent and can drift out of temporal overlap under elevated CO_2 (DeLucia et al. 2012).

10.2.4 Consequences of Ecophysiology Changes on Secondary Metabolism

10.2.4.1 C:N Balance and Plant-Insect Interactions

The carbon-nutrient balance hypothesis postulates that the nutrition status (C:N) of plants directly controls allocation of secondary metabolites (Bryant et al. 1983). Similarly, the resource availability hypothesis (Coley et al. 1985) predicated the physiological role of resource availability as a control point connecting the plant's primary growth with its chemical defenses against insect herbivores (Fig. 10.1). Based on these hypotheses, one can predict that plants grown under elevated CO_2 environment produce more carbon-based secondary metabolites because the higher CO_2 assimilated is diverted to the production of these types of metabolites. This prediction, however, has been disputed by many empirical observations on the production of carbon-based secondary metabolites by plants grown under conditions of enriched CO_2 (Lindroth 2010). Continuous progresses in understanding patterns of synthesis of secondary metabolites have shown that resource allocation is interwoven among at least three physiological operators: photosynthesis, genetic regulation, and hormonal signaling (Kerchev et al. 2012; Fig. 10.2).

While the earlier models of plant responses to elevated CO_2 clearly conceptualize the effects of plant growth habit and nutrition on insect populations, they unsuccessfully predict the

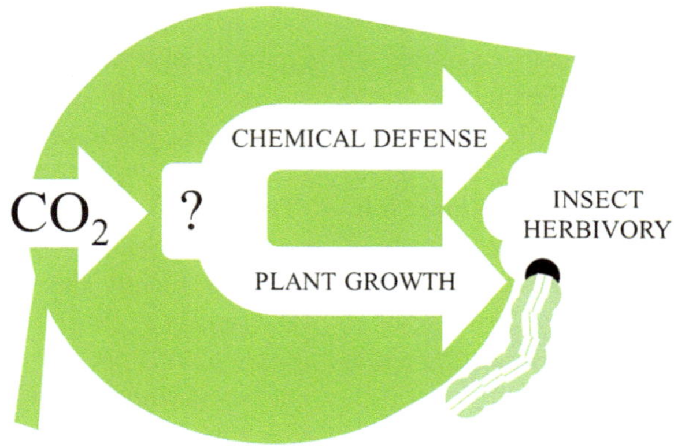

Fig. 10.1 Early conceptual models of resource allocation in plants, such as the carbon-nutrient balance hypothesis, based the partitioning of resources to growth and chemical defenses on only a few physiological operators, such as C:N balance. However, the variable influence of elevated CO_2 on the production of plant secondary metabolites against insect herbivory suggests that the task of resource allocation in plants is more complex than previously supposed

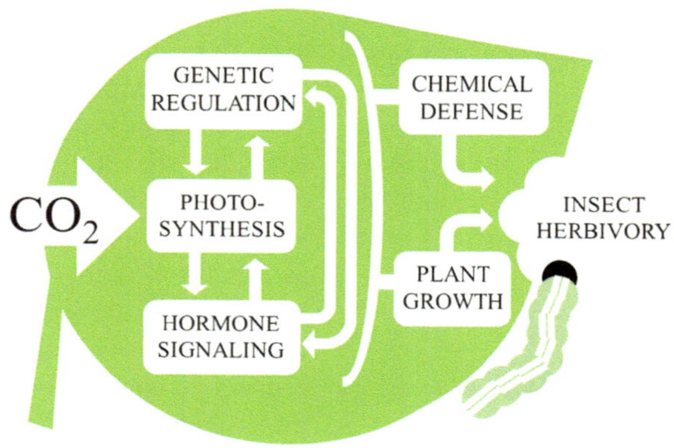

Fig. 10.2 The physiological function of resource allocation is now understood to be interwoven among at least three operators: photosynthesis, genetic regulation, and hormone signaling. The uncertainty in predicting how CO_2 influences populations of insect herbivores lies in the way these components interact with one another to yield plant growth and defense

production of secondary metabolites in defense against insect herbivory (Zavala et al. 2013; Hamilton et al. 2001). Existing mechanistic models of plant secondary metabolism regulation by environmental CO_2 have yet to discern underlying patterns among the empirical data available. Evidence suggests that photosynthesis plays a direct regulatory role in the mobilization of plant chemical defenses against herbivores. Under biotic stress, for instance, plants universally downregulate genes related to photosynthesis (Bilgin et al. 2010). The redox state within the chloroplast is thought to link photosynthesis with secondary metabolism; under attack from herbivore insects, plants interrupt the electron transport chain in photosynthesis, leading to a rapid accumulation of reactive oxygen species within the chloroplast (Kerchev et al. 2012). In turn, metabolic pathways responsible for the synthesis of several defense hormones are sensitive to the redox state of the cell (Kerchev et al. 2012).

10.2.4.2 Unpredictability of CO_2 Effects on Secondary Metabolism

Secondary metabolites are both diverse and idiosyncratic in that their distribution does not always follow phylogenetic relationships among the plants that produce them (Wink 2003). This characteristic unpredictability remains consistent in response to rising atmospheric CO_2; the influence of CO_2 on plant chemical defenses results in a range of responses without any apparent pattern (Zavala et al. 2013). Of the classes of compounds studied with regard to CO_2, phenolics and terpenoids represent the majority (Lindroth 2010). Phenolic compounds appear largely unaffected by carbon dioxide, with the exception of condensed tannins which vary widely in response across species studied. Likewise, the response of terpenoids to rising CO_2 concentration varies without pattern by species of plant (Lindroth 2010). The unpredictability of secondary compound production in plants grown under elevated CO_2 remains much the same for other studied classes of secondary compounds.

In addition to inducing direct chemical defenses against insects, herbivore attack on plants elicits the release of airborne compounds that interact with the surrounding ecosystem and function as a potential indirect defense (Karban 2008). The "bouquet" of volatiles can be composed of terpenoids, lipids, and methyl-esterated defense hormones typically released from herbivore-induced wounds in plant tissue (Karban 2008). Some of the potential ecological effects of these biogenic volatile organic compounds (BVOCs) are from activating chemical defenses in neighboring plants (Karban et al. 2010), to recruit natural predators of herbivores challenging the plant (Heil 2008). BVOCs emission can reduce the herbivore attack of the plant by more than 90 % because it attracts predators and reduces herbivore oviposition rates (Kessler and Baldwin 2001). Although some studies are available regarding to the impacts of atmospheric CO_2 on BVOCs production (Vuorinen et al. 2004a, b; Klaiber et al. 2013), more studies are required to understand biological impacts of elevated CO_2 environments on BVOC emissions.

10.2.5 Consequences of Stomata Closure

10.2.5.1 Transpiration and Leaf Temperature

In addition to altering the nutritive and chemical properties of leaves to prospective herbivore insects, elevated CO_2 also influences the hydraulic and thermal characteristics of leaves by reducing transpiration rates through closure of stomata (Bernacchi et al. 2007). As phloem feeders, aphid populations are especially sensitive to changes in the rate of nutrient translocation within their host plants (Hullé et al. 2010). Lowering transpiration rates in the plant could reduce the growth of insect populations with piercing/sucking mouthparts.

As an effect of reduced evapotranspiration due to lowered stomatal conductance, leaf temperatures rise under elevated CO_2. As ectotherms, insects in close contact with leaf surfaces experience increased metabolic rates as the temperature of the leaf rises (Pincebourde and Woods 2012). This principle especially applies to aphids, whose feeding rates and population growth can be expected to steepen as leaf temperatures increase (O'Neill et al. 2011).

10.3 Solving the Ambiguity of Induced Chemical Defenses to Insect Damage

In the last years, many new studies have been published suggesting that elevated CO_2 modifies plant chemical defenses against insects and alters their performance (Zavala et al. 2013). Insect damage is perceived by plants and induces signals through different pathways, which further produce biochemical and physiological changes that can be affected or modulated by variations in environmental CO_2 concentrations (Fig. 10.2). These changes in damaged leaves induce the production of plant secondary metabolites and decrease insect performance. However, it is not clear how high environmental CO_2 levels can interfere with the production of plant chemical defenses and its effects on plant-insect interactions. An earlier study on field-grown soybean demonstrated that changes in plant defenses against herbivore attack in plants grown in elevated CO_2 atmosphere are regulated by phytohormones (Zavala et al. 2008). Lately, new progress has been done in revealing the mechanisms of the influences of atmosphere with high CO_2 levels on plant defense elicitation by herbivore damage, which has been recently reviewed (DeLucia et al. 2012; Zavala et al. 2013). Although our understanding of the plant-herbivore interphase is limited, recent molecular approaches have revealed the participation of regulatory elements, such as Ca^{2+} ion fluxes, mitogen-activated protein kinases (MAPKs), jasmonic acid (JA), ethylene (ET), and reactive oxygen species (ROS), which detect insect damage, amplify the signal, and induce plant responses to increase chemical defenses against herbivores (Fig. 10.3).

10.3.1 Early Responses to Herbivory

Since plant defenses against insect attack are costly for plants, they must avoid wasting defensive resources and differentiate insect feeding from simple mechanical damage (Zavala et al. 2004; Zavala and Baldwin 2004). Herbivore attack is commonly associated with wounding of plant tissues and direct contact of insect oral secretions (OS) with putative cell receptors, which transduce the alarm signal and induce the accumulation of defensive metabolites. Volicitin [N-(17-hydroxylinolenoyl)-L-glutamine], the first identified fatty acid-amino acid conjugates (FACs) and herbivore-derived elicitor, was obtained from *Spodoptera exigua* OS (Alborn et al. 1997). FACs are not only present in OS of

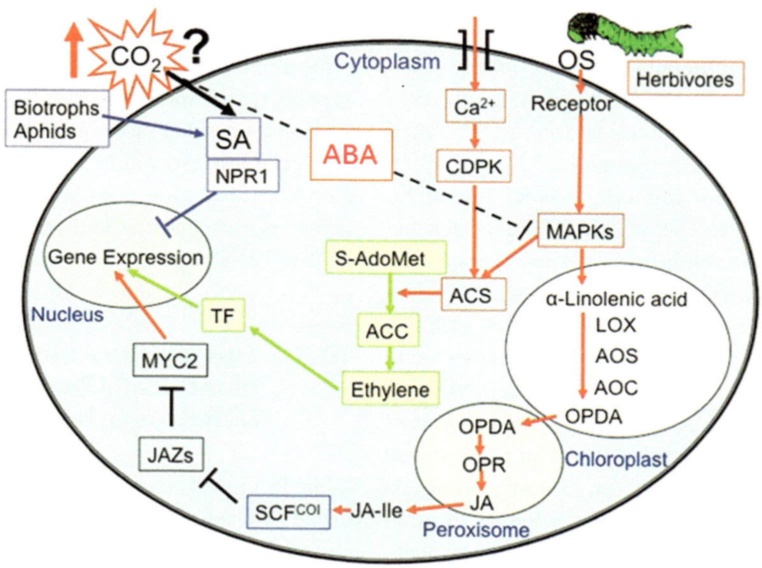

Fig. 10.3 A simplified model of the effects of elevated CO_2 on signaling events in a cell as a response of herbivore attack (Modified from Zavala et al. 2013). In ambient CO_2 herbivore oral secretions are perceived by unidentified receptors and trigger the activation of Ca^{2+} channels, resulting in Ca^{2+} influxes. Ca^{2+} binds to calmodulins and calcium-dependent protein kinases (*CDPKs*). Mitogen-activated protein kinases (*MAPKs*) are also rapidly activated and trigger the biosynthesis of jasmonic acid (*JA*) and thus JA-Ile. JA-Ile binds to the COI1 receptor that leads to the degradation of JAZ proteins, resulting in the release of their inhibitory effect on MYC2, which induces defense genes in the nucleus. MAPK and CDPK phosphorylate ACS proteins and increase ethylene production, which leads to the increased activity of ethylene-responsive transcription factors, inducing defense genes. However, elevated CO_2 induces salicylic acid (*SA*) accumulation and NPR1 activation by changing redox status in the cytosol by inducing thioredoxins and glutathione S-transferase (*black arrow*). The activated NPR1 functions as a transcription factor (*TF*) in the nucleus and inhibits JA-induced defense gene expression. In addition, we hypothesize that elevated CO_2 increases ABA concentration, which decrease ethylene emission. In addition elevated CO_2 decreases MPK4 activity and activates downstream genes of SA signaling (*dashed black arrow*). The inhibition of early signaling events by elevated CO_2 decreases the accumulation of metabolites that function as defense against herbivores.

some lepidopteran species (Halitschke et al. 2001; Pohnert et al. 1999; Spiteller and Boland 2003; Spiteller et al. 2004) but also in OS of fruit flies (*Drosophila melanogaster*) and crickets (*Teleogryllus taiwanemma*) (Yoshinaga et al. 2007). A new study has demonstrated that plants can differentiate specialists from generalist insects by the amount of FAC present in the OS; this identification of the attacker allows plants to upregulate specific defenses accordingly (Hettenhausen et al. 2013).

Membrane depolarization is one of the earliest cellular responses to damage by herbivory. Bites of *Spodoptera littoralis* larvae on leaves of *Phaseolus lunatus* (lima bean) produced large membrane depolarization (Maffei et al. 2004). Since Ca^{+2} fluxes across the cell membranes' change membrane potentials, this ion has been implicated as a second messenger in many plant signaling pathways, including responses to herbivory (Maffei et al. 2007). Larva FACs contribute to the Ca^{+2} influxes by forming ion channels in planar lipid bilayer membranes (Maischak et al. 2007). It has been suggested that calcium-dependent protein kinases are part of plant defense system against herbivores because these Ca^{+2}-dependent protein kinases regulate the production of reactive oxygen species (ROS) in potato, which in turn upregulate plant defenses (Kobayashi et al. 2007).

Activation of mitogen-activated protein kinase (MAPK) cascades is a common response of plants to abiotic and biotic stimuli, including pathogens and herbivory (Romeis et al. 2001; Zhang and Klessig 2001; Wu et al. 2007). Herbivory and application of *Manduca sexta* OS to wounded leaves of *Nicotiana attenuata* and *Solanum lycopersicum* (tomato) highly elicit both salicylic acid-induced protein kinase (SIPK) and wound-induced protein kinase (WIPK), which induced transcriptional regulation of many defense-related genes (Wu et al. 2007; Kandoth et al. 2007). In potato the homologous SIPK and WIPK are involved in Mi-1-mediated resistance to aphids (Li et al. 2006). The *Arabidopsis* orthologous MPK3 and MPK6 to tobacco (*Nicotiana*) WIPK and SIPK, respectively (Zhang and Klessig 2001; Ichimura et al. 2002; Ren et al. 2002), phosphorylate the transcription factor WRKY33 and induce important plant defenses (Mao et al. 2011). In addition, SIPK regulates the activity of chloroplastic GLA1 phospholipase and releases polyunsaturated fatty acids, such as linolenic acid, from the plastidial membranes (Kallenbach et al. 2010). Linolenic acid can be used to the synthesis of the defense hormone JA, the most important phytohormone that controls plant defenses against herbivores (Wasternack 2007).

10.3.2 Jamonates Regulation as Key of Inducing Defenses

Variations of atmospheric CO_2 levels can modify jasmonic acid (JA) accumulation in plant tissues and alter plant chemical defenses against herbivore insects (e.g., Zavala et al. 2008; Sun et al. 2010, 2013; Casteel et al. 2012). Leaf damage inflicted by chewing insects or mechanical damage increases (<30 min) accumulation of JA at the site of wounding; this key cellular signal is involved in the activation of the immune responses to most insect herbivores and necrotrophic microorganisms (Farmer 2014). JA is synthesized via the octadecanoid pathway, from which nearly all jasmonate biosynthetic enzymes have been identified in *Arabidopsis* and characterized in several species (Schaller et al. 2005; Halitschke and Baldwin 2003; Fig. 10.3). The linolenic acid liberated by phospholipases from lipids of chloroplast membranes is transformed to a series of reactions to 12-oxo-phytodienoic acid (OPDA) by the enzymes, lipoxygenase (LOX), allene oxide synthase (AOS), and allene oxide cyclase (AOC) (Fig. 10.3). Then in the peroxisomes, after three steps of β-oxidation, OPDA is converted to JA (Wasternack 2007).

Studies using microarray analysis and RT-PCR of field-grown soybean (*Glycine max*) showed that constitutive and inducible levels of key transcripts associated with JA pathway, LOX7 and LOX8, AOS, and AOC, were downregulated under elevated CO_2 (Casteel et al. 2008; Zavala et al. 2008). Downregulation of JA pathway resulted in lower level of chemical defenses in soybean foliage after 24 h of Japanese beetle (*Popillia japonica*) herbivory (Casteel 2010). Similar results were found in tomato and *Arabidopsis*; elevated CO_2 levels reduced the JA pathway and accumulation in the wild-type plants and in a genotype in which the JA pathway is constitutively upregulated (Sun et al. 2011, 2013).

The new insights into the mechanisms of JA bioactive perception and signal transduction into the activation of defense transcriptional responses have been discussed in several review papers and suggest that the core signal transduction chain is composed for relative few links (Ballaré 2011; Wu and Baldwin 2010; Wasternack and Hause 2013). The first step is conjugation of JA to isoleucine (Ile), which is required for direct defense response against herbivores, such as caterpillar feeding (Kang et al. 2006). The ubiquitin ligase SCFCOI1 complex perceives the jasmonoyl-isoleucine (JA-Ile), and the F-box protein CORONATINE-INSENSITIVE1 (COI1) recognizes JA-Ile, which triggers the ubiquitination and subsequent proteosomal degradation of JASMONATE ZIM DOMAIN (JAZ) proteins. Degradation of the JAZ repressors releases the expression of MYC2 transcription factor (TF). Therefore, MYC2 TF induces downstream defense responses related to JA burst. Interestingly, increases of JA synthesis (possibly via MYC2) induce the transcription of JAZ repressors (Thines et al. 2007; Chico et al. 2008), controlling the levels of signal transduction and production of chemical defenses (Fig. 10.3). Since rapid resynthesis of JAZ repressors restrains the expression of genes related with defenses, this negative feedback presumably avoids the synthesis of energetically costly compounds once insect attack is finished and allows plants to save energy.

10.3.3 Cross Talk among Defense Pathways

Plants grown in elevated CO_2 atmosphere presented low JA accumulation and emission of ET, which reduced antiherbivore defenses and increased insect attack (Zavala et al. 2008, 2009; Sun et al. 2011, 2013). ET and JA have synergistic effects on upregulating plant defenses against herbivores, such as the upregulation of protease inhibitors (PIs) in tomato (O'Donnell et al. 1996). It has been demonstrated that both ET and JA can activate common TF in *Arabidopsis* and sunflower (*Helianthus annuus*) (Manavella et al. 2008; Lorenzo et al. 2003; Pre et al. 2008), explaining the synergistic effect. The synergistic cross talk between JA and ET is known to occur preferentially for the response to necrotrophic pathogens (Pieterse et al. 2012). Two central TFs of ET signaling, ETHYLENE-INSENSITIVE3 (EIN3) and EIN3-like (EIL1), bind JAZ1, JAZ3, and JAZ9 via the Jas domain of JAZs, resulting in the suppression of EIN3/EIL1 activity (Zhu et al. 2011).

ET pathway is initiated by the synthesis of S-adenosylmethionine from methionine, which after being oxidized by 1-aminocyclopropane-1-carboxylic acid (ACC) oxidases is converted to the ET precursor ACC by ACC synthase (ACCs) (Fig. 10.3). Since ACC forms ET, ACC synthesis is considered to be rate limiting of ET biosynthesis, which can be inhibited in plants grown under elevated CO_2, increasing herbivory susceptibility. Both constitutive and inducible expression levels of *acc* are diminished in foliage of field-grown soybeans under elevated CO_2 atmosphere (Casteel et al. 2008; Zavala et al. 2008). In addition, elevated CO_2 environment decreased both the accumulation of JA and the emission of ET in *Arabidopsis* (Sun et al. 2013). While JA and ET are important modulator for chewing insects, salicylic acid (SA) pathway is activated by plants in response to attack by phloem feeding insects, such as aphids and silver leaf whiteflies and biotrophic pathogens (Walling 2000).

One of the most studied cross talk is the antagonistic interaction between the JA and SA pathways (Pieterse et al. 2009; Kunkel and Brooks 2002). Upregulation of JA pathway can repress SA defense responses (Brooks et al. 2005; Uppalapati et al. 2007), and conversely, induction of SA by biotrophic pathogens inhibits defenses regulated by JA (Felton and Korth 2000; Spoel et al. 2007). Similar antagonistic interaction between JA and SA was observed in different plant species grown under enriched CO_2 atmosphere, increasing the allocation of resources to SA-dependent over JA-dependent defenses. Elevated CO_2 increased the induce defenses based on SA pathway in tomato, such as the pathogenesis-related protein (PR), and decreased JA-regulated defenses, such as PIs (Sun et al. 2011; Huang et al. 2012). Similar results were found in *Arabidopsis* growing in open chambers with elevated CO_2 atmosphere (Sun et al. 2013). Field-grown soybeans under elevated CO_2 amplified SA-regulated defenses and pathway and diminished gene expression of JA pathway and JA accumulation compared to ambient-grown plants (Casteel et al. 2008, 2012; Zavala et al. 2008).

Synthesis of defense compounds against plant pathogens is regulated by SA signaling pathway and can be modulated by CO_2 (Glazebrook 2005; Zarate et al. 2007). Recently, it has been demonstrated that some soybean pathogens are reduced in elevated CO_2 treatments at SoyFACE (Eastburn et al. 2010). In addition, elevated CO_2 increased SA and decreased JA accumulation in tomato, enhancing resistant to tomato yellow leaf curl virus (Huang et al. 2012). The mechanism by which elevated CO_2 alters the hormonal response to herbivory is not known, but the antagonism between SA and JA responses is discussed in several recent reviews (Pieterse et al. 2009, 2012).

SA can decrease JA biosynthesis and sensitivity through a negative interaction (Spoel et al. 2003). NPR1 (NONEXPRESSOR OF PATHOGENESIS-RELATED GENES1) is an important protein component of SA pathway and activated by SA (Fig. 10.3). Some NPR1 multimers monomerize by SA-induced changes of the redox state via thioredoxin followed by the transport of the monomeric forms into the nucleus. Here, they bind as activators to TGA TFs specific for SA-inducible genes and phosphorylated them and induce transcription (Fu et al. 2012). Elevated CO_2 alters the transcripts regulating the redox status of soybeans by inducing thioredoxins and glutathione S-transferase (Casteel et al. 2008). Moreover, elevated CO_2 alters ascorbate or glutathione, albeit with some degree of plant specificity (Gillespie et al. 2011; Perez-Lopez et al. 2009), and may allow TFs to interact with reduced NPR1 and facilitate efficient DNA binding for induction of immune signaling (Spoel and Loake 2011).

Assuming that NPR1 may be responsible for downregulating JA-related defenses and increasing herbivory susceptibility in plants grown in enriched CO_2 environment, what is the early signal perceived by plants that upregulate the SA pathway? Recent experiments demonstrated that SA and ROS accumulation and PR expression were increased in MPK4-silenced soybeans and *Arabidopsis* (Liu et al. 2011; Petersen et al. 2000). Interestingly, expression of WRKY33 increased 16-fold in MPK4-silenced soybeans, suggesting that MPK4 negatively control WRKY33 at both the posttranslational and transcriptional level (Liu et al. 2011). MPK4 functions to sequester WRKY33 in the nucleus and prevent it to activating downstream genes of SA signaling (Qiu et al. 2008). Conversely, the cross talk between JA with other hormones can also interfere with either JA or ET pathway.

Atmosphere of elevated CO_2 concentration increases ABA levels during leaf development and promotes stomatal closure, reducing stomatal conductance by decreasing pore size in developing leaves (Ainsworth and Rogers 2007; Herrick et al. 2004). ABA and SA signaling required for stomatal closure (Pieterse et al. 2012) are modulated by MAPK4 (Hettenhausen et al. 2013). Upon wounding or herbivory, ABA production increases and antagonizes the ERF pathway, compromising ET signaling and perception (Abe et al. 2003; Anderson et al. 2004; Fig. 10.3). ET can play a critical role during SA-JA interaction, and the final outcome of this signal cross talk can be shaped by levels of ET. When ET signal is high, JA-SA antagonisms are diluted, while with low ET levels in plant tissue, the JA-SA antagonistic relationship becomes greater (Pieterse et al. 2012).

In summary, we suggest that elevated CO_2 atmosphere induces changes in gene expression and activity of MAPK4 and the production of ABA in leaves, which induces SA signaling and inhibits ET emission and pathway that increases the JA-SA antagonistic relationship. Thus, in plants grown in elevated CO_2 environment, levels of SA increase and production of JA and ET decrease compared to ambient-grown plants (Casteel et al. 2007; Zavala et al. 2008; Casteel et al. 2008, 2012; Sun et al. 2011; Huang et al. 2012). Although the mechanism of the impact of elevated CO_2 on the early signaling steps involved in the perception of insect damage and the activation of JA biosynthesis remain to be elucidated, some advances have been done to explain CO_2 regulation on secondary metabolite production. Elevated CO_2 alters hormonal responses that are transduced into the activation of transcriptional responses and affect plant-insect interactions.

10.3.4 CO_2 Regulates Chemical Defenses through Phytohormones

Many empirical studies about the consequences of elevated CO_2 on plant chemical responses to insect damage and their effects on herbivore performance seem to be contradictory. Whereas some studies report that plants grown under elevated CO_2 environment are more defensed against herbivores (e.g., Lindroth et al. 1993; Holton et al. 2003), field studies demonstrated that elevated CO_2 atmosphere diminished plant defenses against some herbivore insects (e.g., Zavala et al. 2008; 2009; Hamilton et al. 2005). Recently new research has started to study the mechanisms of the consequences of elevated CO_2 on chemical defenses and its relationships with hormonal regulation rather than just study the effects of defenses on herbivore performance (Zavala et al. 2013; DeLucia et al. 2012). Although elevated CO_2 increases SA in plant tissue and increases chemical defense pathways that are not regulated by JA (e.g., Sun et al. 2011; Ghasemzadeh et al. 2010; Casteel et al. 2012), plants grown in enriched atmospheric CO_2 downregulate JA and ET pathways and increase susceptibility to herbivore attack by disrupting both constitutive and inducible important chemical defenses (Zavala et al. 2008, 2009); Interestingly, a new study has proposed that an unknown JA-independent defense pathway may defend *Nicotiana attenuata* against larvae of *M. sexta* (Hettenhausen et al. 2013).

Revealing the regulation of plant chemical defenses in response to herbivory will allow us to determine the adaptive function of anti-herbivore defenses (Karban and Baldwin 1997) that are modified by elevated CO_2 and these modifications should be investigated in concert with predictable changes in CO_2-driven plant traits.

Phytohormones differentiate among phenolic pathways and synthesize both isoflavonoids and flavonols compounds in accordance to environmental cues (Ferrer et al. 2008). Although *Nicotiana attenuata* impaired in JA biosynthesis was unable to produce phenolic conjugates after damage produced by herbivores, these plants were able to produce rutin (a flavonoid) and chlorogenic acid (Demkura et al. 2010; Hoffman-Campo et al. 2001). The production of carbon-based defenses is modified by hormones affected by elevated CO_2. Elevated CO_2 levels increased C:N ratio and concomitantly the production of SA-regulated flavonoids in plants, such as quercetin, kaempferol, and fisetin (Ghasemzadeh et al. 2010), but decreased the concentration of JA-regulated isoflavonoids such as genistein (O'Neill et al. 2010), an important antiherbivore defense (Piubelli et al. 2005; Figs 10.1 and 10.2). In addition, elevated CO_2 decreased the JA-regulated triterpenoid cardenolides in four different genotypes of milkweed (*Asclepias syriaca*) another important chemical defense against herbivores (Vanette and Hunter 2011; Rasmann et al. 2009).

One of the first defenses against herbivores studied is protein proteases inhibitors (PIs) in tomato, which are regulated systemically and locally by JA (Ryan 1990). PIs damage the ability of insects to digest proteins and decrease amino acid assimilation (Birk 2003; Zavala et al. 2004). The accumulation of these nitrogen-based defenses also can change C:N ratio in plant tissue (Howe and Jander 2008; Figs. 10.1 and 10.2). Foliage of field-grown soybean had low constitutive and induced expression and activity levels of cysteine proteinase inhibitors (CystPIs) in an atmosphere with high CO_2 levels, the principal antiherbivore defenses against coleopteran herbivores (Zavala et al. 2008, 2009; Sun et al. 2011). CystPIs are typically regulated by the defense hormones JA and ET that are downregulated by elevated CO_2 (Zavala et al. 2008; Sun et al. 2013). Glucosinolates, another nitrogen-based defense and regulated by JA and SA signaling pathways, are differentially regulated by concentration levels of CO_2 in the atmosphere. Plants grown in atmosphere with high CO_2 levels had incremented constitutive glucosinolates regulated by SA, such as aliphatic glucosinolates and methyl sulfinylalkyl glucosinolates glucoraphanin and glucoiberin in broccoli (*Brassica oleracea*), and these metabolites were induced in *Arabidopsis* by diamondback moths (*Plutella xylostella*) herbivory (Schonhof et al. 2007; Bidart-Bouzat et al. 2005; Doughty et al. 1991). Conversely, JA-regulated indole glucosinolates decreased in plants grown under elevated CO_2, predominantly because of a reduction of glucobrassicin and 4-methoxy-glucobrassicin contents, which are regulated by JA (Schonhof et al. 2007; Brader et al. 2001).

Analogously, levels of plant BVOC emission depend on whether they are regulated by JA or SA. Downregulation of JA levels by elevated CO_2 environment decreased the emission of terpene volatile compounds in cabbage (*Brassica oleracea*; Vuorinen et al. 2004a, b). Plants exposed to elevated CO_2 exhibited a great reduction of terpene emission, decreasing aphid (*Brevicoryne brassicae*) colonization and attraction of herbivores' natural enemies (Vuorinen et al. 2004a; Klaiber et al. 2013). However, emission of GLV following herbivory was enhanced in plants grown under elevated CO_2 environment (Vuorinen et al. 2004b), and it appears that the regulation of these compounds does not rely on JA signaling (Halitschke and Baldwin 2003; Allmann et al. 2010). Based on the empirical evidence presented here, we suggest that JA/ET and SA may provide new insights into how elevated CO_2 modulates plant chemical responses to herbivory.

10.4 Impact of Atmosphere with High CO_2 Levels on Agriculture

By 2050 crops will grow in an atmosphere with CO_2 levels 50 % higher than today (Prather et al. 2001) and by that date the projected world population will be about 9 billion (Ray et al. 2013). Although it has been suggested that global crop production needs to increase twice the present rate to meet global food demand, climate change is likely to worsen the situation by increasing infestation of diseases and insects on crops (Khoury et al. 2014). As a consequence of anthropogenic global change, elevated levels of atmospheric CO_2 can greatly modify the interactions between crops and insect pests and may promote the rapid establishment of invasive species (Zavala et al. 2008; 2013).

Whereas higher atmospheric CO_2 concentrations stimulate photosynthetic activity in C3 plants (such as wheat and soybean), elevated CO_2 increases water-use efficiency in both C3 and C4 plants (such as maize and millet), suggesting a potential increment in yield of crops (Ainsworth and Long 2005). However, field experiments demonstrated that the projections of increasing agricultural production under climate change cannot be reached, probably because of insect herbivory, an aspect that was not considered in open chambers (Muller 2013; Long et al. 2006). New studies have suggested that climate change may affect plant natural defenses against insects, especially when they are grown under elevated atmospheric CO_2 concentrations (DaMatta et al. 2010; Taub et al. 2008; Zavala et al. 2013). Downregulation of natural plant defenses may also affect pesticide use, to compensate the altered chemical composition changes that increase susceptibility of crops to insect damage (Dermody et al. 2008; Zavala et al. 2008).

One of the main components of integrated pest management programs is the utilization of natural plant defenses against insect pest, which together with the expression of foreign genes with inducible promoters has been suggested as an important tool to protect crops against insect herbivores. However, the ability of insect to adapt to plant defenses may limit the extent of those programs (Bolter and Jongsma 1995). In addition, high CO_2 levels in the atmosphere in the future may restrain inducible defenses of crops, such as CystPI in soybean, which will be downregulated and may compromise defenses against insects (Zavala et al. 2008, 2013). Elevated CO_2 may affect the expression of JA-regulated defenses of soybean, which include not only CystPIs but also serine proteinase inhibitors, isoflavonoid content, or polyphenol oxidase activity, potentially making plants more vulnerable to herbivores other than beetles and leading to even greater losses. *Nicotiana attenuata* unable to produce JA planted into native habitats was more vulnerable to adapted herbivores and also was more colonized at a higher rate by novel herbivore species, which fed and reproduced successfully (Kessler et al. 2004). On the other side, insect herbivores preferred to feed on younger leaves of plants grown under elevated CO_2 in FACE experiments (Zavala et al. 2009), which is a potential mechanism to explain the offset of the predicted increases in agricultural productivity associated with greater levels of CO_2 in the atmosphere (Ainsworth and Long 2005; Long et al. 2006). Furthermore, it is likely that the impacts of elevated CO_2 atmosphere on plant chemical defenses reported here also affect communities of natural enemies and the entire trophic structure of agroecosystems (Richards et al. 2015), potentially exacerbating pest problems by multiple mechanisms. Additional scientific attention is necessary to understand and evaluate the impact of CO_2 fertilization on agricultural productivity and crop quality (Muller 2013).

10.5 Concluding Remarks

Recent evidence has demonstrated that anthropogenic elevation in CO_2 concentration alters plant-insect interactions by altering hormonal regulation that in turn change chemical defenses of plants against herbivore insects (e.g., Zavala et al. 2008, DeLucia et al. 2008). However, many aspects of the mechanisms that regulate the impact of elevated CO_2 atmosphere on defense metabolisms are still to be elucidated. As recent review articles have remarked, the task of predicting how insect populations respond to changes in host plant growth habits and nutritive composition would be easier if the basic regulatory connections between plant primary and secondary metabolism were better understood.

References

Abe H, Urao T, Ito T, Seki M, Shinozaki K, Yamaguchi-Shinozaki K (2003) Arabidopsis AtMYC2 (bHLH) and AtMYB2 (MYB) function as transcriptional activators in abscisic acid signaling. Plant Cell 15:63–78

Ainsworth EA, Long SP (2005) What have we learned from 15 years of free-airCO_2 enrichment (FACE)? A meta-analytic review of the responses of photosynthesis, canopy properties and plant production to rising CO_2. New Phytol 165:351–71

Ainsworth EA, Rogers A (2007) The response of photosynthesis and stomatal conductance to rising [CO_2]: mechanisms and environmental interactions. Plant Cell Environ 30:258–270

Alborn HT, Turlings TCJ, Jones TH, Stenhagen G, Loughrin JH, Tumlinson JH (1997) An elicitor of plant volatiles from beet armyworm oral secretion. Science 276:945–49

Allmann S, Halitschke R, Schuurink RC, Baldwin IT (2010) Oxylipin channelling in *Nicotiana attenuata*: lipoxygenase 2 supplies substrates for green leaf volatile production. Plant Cell Environ 33:2028–2040

Anderson JP, Badruzsaufari E, Schenk PM, Manners JM, Desmond OJ et al (2004) Antagonistic interaction between abscisic acid and jasmonate-ethylene signaling pathways modulates defense gene expression and disease resistance in Arabidopsis. Plant Cell 16:3460–79

Ballaré CL (2011) Jasmonate-induced defenses: a tale of intelligence, collaborators and rascals. Trends Plant Sci 16:249–257

Bernacchi CJ, Kimball BA, Quarles DR, Long SP, Ort DR (2007) Decreases in stomatal conductance of soybean under open-air elevation of [CO_2] are closely coupled with decreases in ecosystem evapotranspiration. Plant Physiol 143:134–144

Bernays EA (1998) Evolution of feeding behavior in insect herbivores. BioScience 48:35–44

Bidart-Bouzat MG, Mithen R, Berenbaum MR (2005) Elevated CO_2 influences herbivory-induced defense responses of *Arabidopsis thaliana*. Oecologia 145:415–424

Bilgin DD, Zavala JA, Zhu JIN, Clough SJ, Ort DR, DeLucia EH (2010) Biotic stress globally downregulates photosynthesis genes. Plant Cell Environ 33:1597–1613

Birk Y (2003) Plant protease inhibitors: significance in nutrition, plant protection, cancer prevention and genetic engineering. Springer, Berlin

Bolter C, Jongsma MA (1995) Colorado potato beetles (*Leptinotarsa decemlineata*) adapt to proteinase inhibitors induced in potato leaves by methyl jasmonate. J Insect Physiol 41:1071–1078

Brader G, Tas E, Palva ET (2001) Jasmonate-dependent induction of indole glucosinolates in *Arabidopsis* by culture filtrates of the nonspecific pathogen *Erwinia carotovora*. Plant Physiol 126:849–860

Brooks DM, Bender C, Kunkel BN (2005) The *Pseudomonas syringae* phytotoxin coronatine promotes virulence by overcoming salicylic acid dependent defences in *Arabidopsis thaliana*. Mol Plant Pathol 6:629–639

Bryant JP, Chapin FS III, Klein DR (1983) Carbon/nutrient balance of boreal plants in relation to vertebrate herbivory. Oikos 40:357–368

Casteel CL (2010) Impacts of climate change on herbivore induced plant signaling and defenses. Doctoral dissertation, University of Illinois, Urbana-Champaign, USA, p 155

Casteel CL, O'Neill BF, Zavala JA, Bilgin DD, Berenbaum MR, DeLucia EH (2008) Transcriptional profiling reveals elevated CO_2 and elevated O_3 alter resistance of soybean (*Glycine max*) to Japanese beetles (*Popillia japonica*). Plant Cell Environ 31:419–434

Casteel CL, Segal LM, Niziolek O, Berenbaum MR, DeLucia EH (2012) Elevated carbon dioxide increases salicylic acid in *Glycine max*. Environ Entomol 41:1435–1442

Castro JC, Dohleman FG, Bernacchi CJ, Long SP (2009) Elevated CO_2 significantly delays reproductive development of soybean under Free-Air Concentration Enrichment (FACE). J Exp Bot 60:2945–2951

Chico JM, Chini A, Fonseca S, Solano R (2008) JAZ repressors set the rhythm in jasmonate signaling. Curr Opin Plant Biol 11:486–494

Coley PD, Bryant JP, Chapin FS III (1985) Resource availability and plant antiherbivore defense. Science 230:895–899

Curzi MJ, Zavala JA, Spencer JL, Seufferheld MJ (2012) Abnormally high digestive enzyme activity and gene expression explain the contemporary evolution of a *Diabrotica* biotype able to feed on soybeans. Ecol Evol 2:2005–2017

DaMatta FM, Grandis A, Arenque BC, Buckeridge MS (2010) Impacts of climate changes on crop physiology and food quality. Food Res Int 43:1814–23

DeLucia EH, Casteel CL, Nabity PD, O'Neill BF (2008) Insects take a bigger bite out of plants in a warmer, higher carbon dioxide world. Proc Natl Acad Sci U S A 105:1781–1782

DeLucia EH, Nabity PD, Zavala JA, Berenbaum MR (2012) Climate change: resetting plant-insect interactions. Plant Physiol 160:1677–1685

Demkura PV, Abdala G, Baldwin IT, Ballaré CL (2010) Jasmonate-dependent and -independent pathways mediate specific effects of solar ultraviolet B radiation on leaf phenolics and antiherbivore defense. Plant Physiol 152:1084–95

Dermody O, O'Neill BF, Zangerl AR, Berenbaum MR, DeLucia EH (2008) Effects of elevated CO_2 and O_3 on leaf damage and insect abundance in a soybean agroecosystem. Arthropod Plant Interact 2:125–35

Doughty KJ, Porter AJR, Morton AM, Kiddle G, Bock CH, Wallsgrove R (1991) Variation in the glucosinolate content of oilseed rape (*Brassica napus* L.) leaves. II. Response to infection by *Alternaria brassicae* (Berk.) Sacc. Ann Appl Biol 118:469–77

Eastburn DM, Degennaro MM, DeLucia EH, Dermody O, Mcelrone AJ (2010) Elevated atmospheric carbon dioxide and ozone alter soybean diseases at SoyFACE. Glob Chang Biol 16:320–330

Farmer EE (2014) Leaf defence. Oxford University Press, Oxford, p 216

Felton GW (2005) Indigestion is a plant's best defense. Proc Natl Acad Sci U S A 102:18771–18772

Felton GW, Korth KL (2000) Trade-offs between pathogen and herbivore resistance. Curr Opin Plant Biol 3:309–314

Ferrer JL, Austin MB, Stewart C, Noe JP (2008) Structure and function of enzymes involved in the biosynthesis of phenylpropanoids. Plant Physiol Biochem 46:356–70

Fornoni J (2011) Ecological and evolutionary implications of plant tolerance to herbivory. Funct Ecol 25:399–407

Fu ZQ, Yan S, Saleh A et al (2012) NPR3 and NPR4 are receptors for the immune signal salicylic acid in plants. Nature 486:228–232

Ghasemzadeh A, Jaafar H, Rahmat A (2010) Elevated carbon dioxide increases contents of flavonoids and phenolic compounds, and antioxidant activities in Malaysian young ginger (*Zingiber officinale* Roscoe.) varieties. Molecules 15:7907–7922

Gillespie KM, Rogers A, Ainsworth EA (2011) Growth at elevated ozone or elevated carbon dioxide concentration alters antioxidant capacity and response to acute oxidative stress in soybean (*Glycine max*). J Exp Bot 62:2667–2678

Glazebrook J (2005) Contrasting mechanisms of defense against biotrophic and necrotrophic pathogens. Annu Rev Phytopathol 43:205–27

Halitschke R, Baldwin IT (2003) Antisense LOX expression increases herbivore performance by decreasing defense responses and inhibiting growth-related transcriptional reorganization in *Nicotiana attenuata*. Plant J 36:794–807

Halitschke R, Schittko U, Pohnert G, Boland W, Baldwin IT (2001) Molecular interactions between the specialist herbivore *Manduca sexta* (Lepidoptera, Sphingidae) and its natural host *Nicotiana attenuata*. III. Fatty acid-amino acid conjugates in herbivore oral secretions are necessary and sufficient for herbivore-specific plant responses. Plant Physiol 125:711–17

Hamilton JG, Zangrerl AR, DeLucia EH, Berenbaum MR (2001) The carbon-nutrient balance hypothesis: its rise and fall. Ecol Lett 4:86–95

Hamilton JG, Dermody O, Aldea M, Zangerl AR, Rogers A, Berenbaum MR, DeLucia EH (2005) Anthropogenic changes in tropospheric composition increase susceptibility of soybean to insect herbivory. Environ Entomol 34:479–485

Heil M (2008) Indirect defence via tritrophic interactions. New Phytol 178:41–61

Herrick J, Maherali H, Thomas R (2004) Reduced stomatal conductance in sweetgum (*Liquidambar styraciflua*) sustained over long-term CO_2 enrichment. New Phytol 162:387–96

Hettenhausen C, Baldwin IT, Wu J (2013) *Nicotiana attenuata* MPK4 suppresses a novel jasmonic acid (JA) signaling-independent defense pathway against the specialist insect *Manduca sexta*, but is not required for the resistance to the generalist *Spodoptera littoralis*. New Phytol 199:787–799

Hoffman-Campo CB, Harborne JB, McCaffery AR (2001) Pre-ingestive and post-ingestive effects of soya bean extracts and rutin on *Trichoplusiani* growth. Entomol Exp Appl 98:181–94

Holton MK, Lindroth RL, Nordheim EV (2003) Foliar quality influences tree-herbivore-parasitoid interactions: effects of elevated CO_2, O_3, and plant genotype. Oecologia 137:233–244

Howe GA, Jander G (2008) Plant immunity to insect herbivores. Annu Rev Plant Biol 59:41–66

Huang L, Ren Q, Sun Y, Ye L, Cao H, Ge F (2012) Lower incidence and severity of tomato virus in elevated CO_2 is accompanied by modulated plant induced defence in tomato. Plant Biol 14:905–913

Hullé M, Cœurd'Acier A, Bankhead-Dronnet S, Harrington R (2010) Aphids in the face of global changes. C R Biol 333:497–503

Ichimura K et al (2002) Mitogen-activated protein kinase cascades in plants: A new nomenclature. Trends Plant Sci 7:301–308

IPCC (2007) Contribution of working group I to the fourth assessment report of the intergovernmental panel on climate change. In: Solomon S, Qin D, Manning M, Chen Z, Marquis M, Averyt KB, Tignor M, Miller HL (eds) Climate change 2007: the physical science basis. Cambridge University Press, Cambridge

Kallenbach M, Alagna F, Baldwin IT, Bonaventure G (2010) *Nicotiana attenuata* SIPK, WIPK, NPR1, and fatty acid-amino acid conjugates participate in the induction of jasmonic acid biosynthesis by affecting early enzymatic steps in the pathway. Plant Physiol 152:96–106

Kandoth PK, Ranf S, Pancholi SS, Jayanty S, Walla MD, Miller W, Howe GA, Lincoln DE, Stratmann JW (2007) Tomato MAPKs LeMPK1, LeMPK2, and LeMPK3 function in the systemin-mediated defense response against herbivorous insects. Proc Natl Acad Sci U S A 104:12205–10

Kang JH, Wang L, Giri A, Baldwin IT (2006) Silencing threonine deaminase and JAR4 in *Nicotiana attenuata* impairs jasmonic acid-isoleucine-mediated defenses against *Manduca sexta*. Plant Cell 18:3303–3320

Karban R (2008) Plant behaviour and communication. Ecol Lett 11:727–739

Karban R, Baldwin IT (1997) Induced responses to herbivory. The University of Chicago Press, Chicago, p 319

Karban R, Shiojiri K, Ishizaki S (2010) An air transfer experiment confirms the role of volatile cues in communication between plants. Am Nat 176:381–384

Kerchev PI, Fenton B, Foyer CH, Hancock RD (2012) Plant responses to insect herbivory: interactions between photosynthesis, reactive oxygen species and hormonal signaling pathways. Plant Cell Environ 35:441–453

Kessler A, Baldwin IT (2001) Defensive function of herbivore-induced plant volatile emissions in nature. Science 291:2141–2144

Kessler A, Halitschke R, Baldwin IT (2004) Silencing the jasmonate cascade: induced plant defenses and insect populations. Science 305:665–668

Khoury CK, Bjorkman AD, Dempewolf H, Ramirez-villegas J, Guarino L, Jarvis A, Rieseberg LH, Struik PC (2014) Increasing homogeneity in global food supplies and the implications for food security. Proc Natl Acad Sci U S A 111:4001–4006

Klaiber J, Najar-Rodriguez AJ, Piskorski R, Dorn S (2013) Plant acclimation to elevated CO_2 affects important plant functional traits, and concomitantly reduces plant colonization rates by an herbivorous insect. Planta 237:29–42

Kobayashi M, Ohura I, Kawakita K, Yokota N, Fujiwara M et al (2007) Calcium-dependent protein kinases regulate the production of reactive oxygen species by potato ADPH oxidase. Plant Cell 19:1065–80

Kunkel BN, Brooks DM (2002) Cross talk between signaling pathways in pathogen defense. Curr Opin Plant Biol 5:325–331

Lambin EF, Meyfroidt P (2011) Global land use change, economic globalization, and the looming land scarcity. Proc Natl Acad Sci U S A 108:3465–3472

Leakey AD, Ainsworth EA, Bernacchi CJ, Rogers A, Long SP, Ort DR (2009) Elevated CO_2 effects on plant carbon, nitrogen, and water relations: six important lessons from FACE. J Exp Bot 60:2859–2876

Li Q, Xie QG, Smith-Becker J, Navarre DA, Kaloshian I (2006) Mi-1-mediated aphid resistance involves salicylic acid and mitogen-activated protein kinase signaling cascades. Mol Plant Microbe Interact 19:655–64

Lindroth RL (2010) Impacts of elevated atmospheric CO_2 and O_3 on forests: phytochemistry, trophic Interactions, and ecosystem dynamics. J Chem Ecol 36:2–21

Lindroth RL, Kinney KK, Platz CL (1993) Responses of deciduous trees to elevated atmospheric CO_2: productivity, phytochemistry, and insect performance. Ecology 74:763–777

Liu Z, Horstman HD, Braun E, Graham MA, Zhang C, Navarre D, Qiu W, Lee Y, Nettleton D, Hill JH, Whitham SA (2011) Soybean homologs of MPK4 negatively regulate defense responses and positively regulate growth and development. Plant Physiol 157:1363–1378

Long SP, Ainsworth EA, Leakey ADB, Nosberger J, Ort DR (2006) Food for thought: lower-than expected crop yield stimulation with rising CO_2 concentrations. Science 312:1918–21

Lorenzo O, PiquerasR S-SJJ, Solano R (2003) ETHYLENE RESPONSE FACTOR1 integrates signals from ethylene and jasmonate pathways in plant defense. Plant Cell 15:165–178

Maffei ME, Bossi S, Spiteller D, Mithofer A, Boland W (2004) Effects of feeding *Spodoptera littoralis* on lima bean leaves. I. Membrane potentials, intracellular calcium variations, oral secretions, and regurgitate components. Plant Physiol 134:1752–62

Maffei ME, Mithofer A, Boland W (2007) Before gene expression: early events in plant-insect interaction. Trends Plant Sci 12:310–16

Maischak H, Grigoriev PA, Vogel H, Boland W, Mithofer A (2007) Oral secretions from herbivorous lepidopteran larvae exhibit ion channel–forming activities. FEBS Lett 581:898–904

Manavella PA, Dezar CA, Bonaventure G, Baldwin IT, Chan RL (2008) HAHB4, a sunflower HD-Zip protein, integrates signals from the jasmonic acid and ethylene pathways during wounding and biotic stress responses. Plant J 56:376–388

Mao G, Meng X, Liu Y, Zheng Z, Chen Z, Zhang S (2011) Phosphorylation of WRKY transcription factor by two pathogen-responsive MAPKs drives phytoalexin biosynthesis in Arabidopsis. Plant Cell 23:1639–1653

Mattson WJ (1980) Herbivory in relation to plant nitrogen content. Annu Rev Ecol Syst 11:119–161

McKey D (1974) Adaptive patterns in alkaloid physiology. Am Nat 108:305–320

Muller C (2013) African lessons on climate change risks for agriculture. Annu Rev Nutr 33:395–411

O'Donnell PJ, Calvert C, Atzorn R, Wasternack C, Leyser HMO, Bowles DJ (1996) Ethylene as a signal mediating the wound response of tomato plants. Science 274:1914–17

O'Neill BF, Zangerl AR, Dermondy O, Bilgin DD, Casteel CL, Zavala JA, DeLucia EH, Berenbaum MR (2010) Impact of elevated levels of atmospheric CO_2 and herbivory on flavonoids of soybean (*Glycine max* L.). J Chem Ecol 36:35–45

O'Neill BF, Zangerl AR, DeLucia EH, Berenbaum MR (2008) Longevity and fecundity of Japanese beetle (*Popillia japonica*) on foliage grown under elevated carbon dioxide. Environ Entomol 37:601–607

O'Neill BF, Zangerl AR, DeLucia EH, Casteel C, Zavala JA, Berenbaum MR (2011) Leaf Temperature of Soybean grown under elevated CO_2 increases *Aphis glycines* (Hemiptera: Aphididae) population growth. Insect Sci 18:419–425

Perez-Lopez U, Robredo A, Lacuesta M, Sgherri C, Munoz-Rueda A, Navari-Izzo F, Mena-Petite A (2009) The oxidative stress caused by salinity in two barley cultivars is mitigated by elevated CO_2. Physiol Plant 135:29–42

Petersen M, Brodersen P, Naested H, Andreasson E, Lindhart U, Johansen B, Nielsen HB, Lacy M, Austin MJ, Parker JE, Sharma SB, Klessig DF, Martienssen R, Mattsson O, Jensen AB, Mundy J (2000) Arabidopsis map kinase 4 negatively regulates systemic acquired resistance. Cell 103:1111–1120

Pieterse CMJ, Leon-Reyes A, Van der Ent S, Van Wees SCM (2009) Networking by small-molecule hormones in plant immunity. Nat Chem Biol 5:308–316

Pieterse CMJ, Van der Does D, Zamioudis C, Leon-Reyes A, VanWees SCM (2012) Hormonal modulation of plant immunity. Annu Rev Cell Dev Biol 28:489–521

Pimental D (2004) Economic impact of insects. In: Goodman RM (ed) Encyclopedia of plant and crop science. Taylor and Francis, London, pp 407–409

Pincebourde S, Woods HA (2012) Climate uncertainty on leaf surfaces: the biophysics of leaf microclimates and their consequences for leaf-dwelling organisms. Funct Ecol 26:844–853

Piubelli GC, Hoffmann-Campo CB, Moscardi F, Miyakubo SH, De Oliveira MCN (2005) Are chemical compounds important for soybean resistance to *Anticarsia gemmatalis*? J Chem Ecol 31:1509–1525

Pohnert G, Jung V, Haukioja E, Lempa K, Boland W (1999) New fatty acid amides from regurgitant of lepidopteran (Noctuidae, Geometridae) caterpillars. Tetrahedron 55:11275–80

Pre M, Atallah M, Champion A, De Vos M, Pieterse CMJ, Memelink J (2008) The AP2/ERF domain transcription factor ORA59 integrates jasmonic acid and ethylene signals in plant defense. Plant Physiol 147:1347–1357

Prather CA, Guazzotti SA, Suess DT, Pastor SH, Coffee K (2001) New insights into the role of aerosols in affecting pollution and global climate change. Abstr Pap Am Chem Ecol Soc 221:U458–U458

Qiu JL, Fiil BK, Petersen K, Nielsen HB, Botanga CJ, Thorgrimsen S, Palma K, Suarez-Rodriguez MC, Sandbech-Clausen S, Lichota J, Brodersen P, Grasser KD, Mattsson O, Glazenbrook J, Mundy J, Petersen M (2008) Arabidopsis MAP kinase 4 regulates gene expression through transcription factor release in the nucleus. EMBO J 27:2214–2221

Rasmann S, Johnson MD, Agrawal AA (2009) Induced responses to herbivory and jasmonate in three milkweed species. J Chem Ecol 35:1326–34

Ray DK, Mueller ND, West PC, Foley JA (2013) Yield trends are insufficient to double global crop production by 2050. PLoS One 8:1–8

Ren D, Yang H, Zhang S (2002) Cell death mediated by MAPK is associated with the hydrogen peroxide production in Arabidopsis. J Biol Chem 277:559–565

Richards LA, Dyer LA, Forister ML, Smilanich AM, Dodson CD, Leonard MD, Jeffrey CS (2015) Phytochemical diversity drives plant-insect community diversity. Proc Natl Acad Sci U S A 112:10973–10978

Romeis T, Ludwig AA, Martin R, Jones JD (2001) Calcium-dependent protein kinases play an essential role in a plant defense response. EMBO J 20:5556–67

Ryan CA (1990) Protease inhibitors in plants: genes for improving defenses against insects and pathogens. Annu Rev Phytopathol 28:425–449

Schaller F, Schaller A, Stintzi A (2005) Biosynthesis and metabolism of jasmonates. J Plant Growth Regul 23:179–99

Schonhof I, Kläring HP, Krumbein A, Schreiner M (2007) Interaction between atmospheric CO_2 and glucosinolates in Broccoli. J Chem Ecol 33:105–114

Spiteller D, Boland W (2003) N-(15,16-Dihydroxylinoleoyl)-glutamine and N-(15,16-epoxylinoleoyl)-glutamine isolated from oral secretions of lepidopteran larvae. Tetrahedron 59:135–39

Spiteller D, Oldham NJ, Boland W (2004) N-(17-phosphonooxylinolenoyl) glutamine and N-(17-phosphonooxylinoleoyl) glutamine from insect gut: the first backbone-phosphorylated fatty acid derivatives in nature. J Org Chem 69:1104–9

Spoel SH, Loake GJ (2011) Redox-based protein modifications: the missing link in plant immune signaling. Curr Opin s 14:358–364

Spoel SH, Koornnef A, Claessens MC, Korzelius JP, Van Pelt JA, Mueller MJ, Buchala AJ, Metraux JP, Brown R, Kazan K, Van Loon LC, Dong X, Pieterse CMJ (2003) NPR1 modulates cross-talk between salicylate- and jasmonate-dependent defense pathways through a novel function in the cytosol. Plant Cell 15:760–770

Spoel SH, Johnson JS, Dong X (2007) Regulation of tradeoffs between plant defenses against pathogens with different lifestyles. Proc Natl Acad Sci U S A 104:18842–18847

Stiling P, Cornelissen T (2007) How does elevated carbon dioxide (CO_2) affect plant–herbivore interactions? A field experiment and meta-analysis of CO_2-mediated changes on plant chemistry and herbivore performance. Glob Chang Biol 13:1823–1842

Sun Y, Cao H, Yin J, Kang L, Ge F (2010) Elevated CO_2 changes the interactions between nematode and tomato genotypes differing in the JA pathway. Plant Cell Environ 33:729–739

Sun Y, Yin J, Cao H, Li C, Kang L, Ge F (2011) Elevated CO_2 influences nematode-induced defense responses of tomato genotypes differing in the JA pathway. PLoS One 6, e19751. doi:10.1371/journal.pone.0019751

Sun Y, Guo H, Zhu-Salzman K, Ge F (2013) Elevated CO_2 increases the abundance of the peach aphid on Arabidopsis by reducing jasmonic acid defenses. Plant Sci 210:128–140

Taub DR, Miller B, Allen H (2008) Effects of elevated CO_2 on the protein concentration of food crops: a meta-analysis. Glob Chang Biol 14:565–75

Taylor G, Tallis MJ, Giardina CP, Percy KE, Miglietta F, Gupta PS, Giolis B, Calfapietra C, Gielen B, Kubiske ME, Scarascia-Mugnozza GE, Kets K, Long SP, Karnosky DF (2008) Future atmospheric CO_2 leads to delayed autumnal senescence. Glob Chang Biol 14:264–275

Thines B, Katsir L, Melotto M et al (2007) JAZ repressor proteins are targets of the SCFCOI1 complex during jasmonate signalling. Nature 448:661–665

Uppalapati SR, Ishiga Y, Wangdi T, Kunkel BN, Anand A, Mysore KS, Bender CL (2007) The phytotoxin coronatine contributes to pathogen fitness and is required for suppression of salicylic acid accumulation in tomato inoculated with *Pseudomonas syringae* pv. tomato DC3000. Mol Plant Microbe Interact 20:955–965

Vanette RL, Hunter MD (2011) Genetic variation in the expression of defense phenotype may mediate evolutionary adaptation of *Asclepias syriaca* to elevated CO_2. Glob Chang Biol 17:1277–88

Vuorinen T, Reddy GVP, Nerg AM, Holopainen JK (2004a) Monoterpene and herbivore-induced emissions from cabbage plants grown at elevated atmospheric CO_2 concentration. Atmos Environ 38:675–682

Vuorinen T, Nerg AM, Ibrahim MA, Reddy GVP, Holopainen JK (2004b) Emission of *Plutella xylostella*-induced compounds from cabbages grown at elevated CO_2 and orientation behavior of the natural enemies. Plant Physiol 135:1984–1992

Walling LL (2000) The myriad plant responses to herbivores. J Plant Growth Regul 19:195–216

Wasternack C (2007) Jasmonates: an update on biosynthesis, signal transduction and action in plant stress response, growth and development. Ann Bot 100:681–697

Wasternack C, Hause B (2013) Jasmonates: biosynthesis, perception, signal transduction and action in plant stress response, growth and development. An update to the 2007 review in Annals of Botany. Ann Bot 111:1021–1058

Weisser WW, Siemann E (eds) (2008) The various effects of insects on ecosystem functioning. In: Insects and ecosystem function. Springer, Berlin/Heidelberg, pp 3–24

Wink M (2003) Evolution of secondary metabolites from an ecological and molecular phylogenetic perspective. Phytochemistry 64:3–19

Wu J, Baldwin IT (2010) New insights into plant responses to the attack from insect herbivores. Annu Rev Genet 44:1–24

Wu J, Hettenhausen C, Meldau S, Baldwin IT (2007) Herbivory rapidly activates MAPK signaling in attacked and un attacked leaf regions but not between leaves of *Nicotiana attenuata*. Plant Cell 19:1096–122

Yoshinaga N, Aboshi T, Ishikawa C, Fukui M, Shimoda M et al (2007) Fatty acid amides, previously identified in caterpillars, found in the cricket *Teleogryllus taiwanemma* and fruit fly *Drosophila melanogaster* larvae. J Chem Ecol 33:1376–81

Zarate SI, Kempema LA, Walling LL (2007) Silver leaf whitefly induces salicylic acid defenses and suppresses effectual jasmonic acid defences. Plant Physiol 143:866–875

Zavala JA, Baldwin IT (2004) Fitness benefits of trypsin proteinase inhibitor expression in *Nicotiana attenuata* are greater than their costs when plants are attacked. BMC Ecol 4:11. doi:10.1186/1472-6785-4-11

Zavala JA, Patankar AG, Gase K, Baldwin IT (2004) Constitutive and inducible trypsin proteinase inhibitor production incurs large fitness costs in *Nicotiana attenuata*. Proc Natl Acad Sci U S A 101:1607–1612

Zavala JA, Casteel CL, DeLucia EH, Berenbaum MR (2008) Anthropogenic increase in carbon dioxide compromises plant defense against invasive insects. Proc Natl Acad Sci U S A 105:5129–5133

Zavala JA, Casteel CL, Nabity PD, Berenbaum MR, DeLucia EH (2009) Role of cysteine proteinase inhibitors in preference of Japanese beetles (*Popillia japonica*) for soybean (*Glycine max*) leaves of different ages and grown under elevated CO_2. Oecologia 161:35–41

Zavala JA, Nabity PD, DeLucia EH (2013) An emerging understanding of mechanisms governing insect herbivory under elevated CO_2. Annu Rev Entomol 58:79–97

Zhang S, Klessig DF (2001) MAPK cascades in plant defense signaling. Trends Plant Sci 6:520–527

Zhu XG, Long SP, Ort DR (2008) What is the maximum efficiency with which photosynthesis can convert solar energy into biomass? Curr Opin Biotechnol 19:153–159

Zhu Z, An F, Feng Y et al (2011) Derepression of ethylene-stabilized transcription factors (EIN3/EIL1) mediates jasmonate and ethylene signaling synergy in Arabidopsis. Proc Natl Acad Sci U S A 108:12539–12544

GM Crops for Developing World in the Era of Climate Change: For Increase of Farmer's Income, Poverty Alleviation, Nutrition and Health

11

Manish Sainger, Poonam Ahlawat Sainger, Darshna Chaudhary, Ranjana Jaiwal, Rana Pratap Singh, Om Parkash Dhankher, and Pawan Kumar Jaiwal

Abstract

Overexpansion of population and poor availability of resources in developing countries have resulted in serious problems of poverty, malnutrition, and poor health. Rapidly changing climate as a result of global warming is making the situation worse. Farmers are deviating from agriculture due to increased input costs and low benefits. Tools of plant biotechnology like GM crops have improved the present situation and undoubtedly have contributed to the increase of farmer's income, nutrition and health, and poverty reduction leading to a step toward food and nutritional security in the developing world. As a result, GM crops have emerged as the fastest adopted crop technology in the history of modern agriculture in spite of a strong opposition initially, which is nullifying gradually.

Keywords

Climate change • GM crops • Developing countries • Farmer's income • Nutrition and Health • Poverty alleviation

M. Sainger • D. Chaudhary • P.K. Jaiwal (✉)
Centre for Biotechnology, Maharshi Dayanand University, Rohtak 124001, Haryana, India
e-mail: jaiwalpawan@rediffmail.com

P.A. Sainger
Department of Environmental Sciences, Maharshi Dayanand University, Rohtak 124001, India

R. Jaiwal
Department of Zoology, Maharshi Dayanand University, Rohtak 124001, India

R.P. Singh
Department of Environmental Science, Babasaheb Bhimrao Ambedkar University (A Central University), Lucknow 226025, Uttar Pradesh, India

O.P. Dhankher
Stockbridge School of Agriculture, University of Massachusetts Amherst, Amherst, MA 01002, USA

P.K. Jaiwal et al. (eds.), *Genetic Manipulation in Plants for Mitigation of Climate Change*,
DOI 10.1007/978-81-322-2662-8_11

11.1 Introduction

Agriculture worldwide needs a significant increase in productivity to ensure sufficient availability of food and other raw materials for ongoing population expansion which is estimated to exceed 9 billion by the year 2050. More than 3 billion people suffer from diseases caused by inadequate nutrition in one or the other way (Gomez-Galera et al. 2010; Farre et al. 2010), and around 900 million people are undersupplied with calorie worldwide (FAO 2012). The conditions are worse in developing countries which have the largest proportion of undernourished people especially in sub-Saharan Africa and South Asia (FAO 2012). Poverty in this part of the world is usually linked to low agricultural productivity as a consequence of poor availability of resources, less agricultural land, and overgrowing population. Plant biotechnology is a potent tool that can improve the present situation and contribute to the poverty reduction and food security in the developing world (Qaim 2010). Conventional breeding and development of genetically engineered plants or GM crops are two widely used biotechnological strategies to resolve the widespread problems of malnutrition and health and for increasing farmers' income and elimination of poverty in developing countries. Conventional breeding involves identifying parents with traits that complement each other and are sexually compatible. However, plant breeding has been successful but limited. It requires selection of offspring with desired traits over a long period of time, and production of improved crop lines can take as long as 10 years from the first parental crossing to generation and distribution of selected improved crops. Besides, strategies to overcome yield-limiting factors and hence enhancement in crop production by conventional breeding have been slow due to the lack of desirable level of genetic variability in germplasm (Sahoo and Jaiwal 2008). This leaves limited options of genetic improvement through selection since selection operates on existing genetic variability. Furthermore, the reproductive barriers limit the transfer of favorable alleles from interspecific and intergeneric sources. Moreover, the approach is time-consuming and labor intensive. Besides, this can also lead to crop vulnerability due to pests and disease epidemics and unpredictable climatic factors. Therefore, the development of genetically engineered plants by the introduction and/or overexpression of selected genes seems to be a viable option to enhance the tolerance to various stresses and hence stabilize yield (Kaur and Murphy 2012; Atif et al. 2013). In addition to widening the gene pool of useful genes, it also allows introgression of novel genes and traits from any living organism into elite agronomic background. Genetic engineering also avoids the complexities of linkage drag. Even for traits that can be improved by traditional breeding, genetic engineering may facilitate and speed up the process (Potrykus 2010). The first GM crop became commercially available in the mid-1990s (Qaim 2009) and was grown in 1.7 million hectares of land. Since then, farmers around the world have adopted genetically modified (GM) crops at a very rapid rate, and by the year 2013, 175.2 million hectares of GM crops were grown in 28 countries (Fig. 11.2). This GM crop revolution started in the USA where the adoption rate for soybean and maize is 95 % and 75 %, respectively (USDA 2013); however, now more than 50 % of this area is in developing countries (James 2013). These figures in themselves are indicative that GM crops have brought benefits to farmers and society by increasing agricultural productivity and reducing food costs while providing numerous economic, environmental, and nutritional benefits even in the era of rapidly changing climate.

11.2 GM Crops: An Overview of Plant Transformation

Genetic engineering is the process in which a desired gene is isolated, cloned, and inserted into a host organism. Initially, a plasmid vector is designed to transfer the candidate gene into the crop plant genome. The transformation vector often contains a cassette with a selectable marker along with the transgene expression cassette that allows for the selection of plant cells that contain

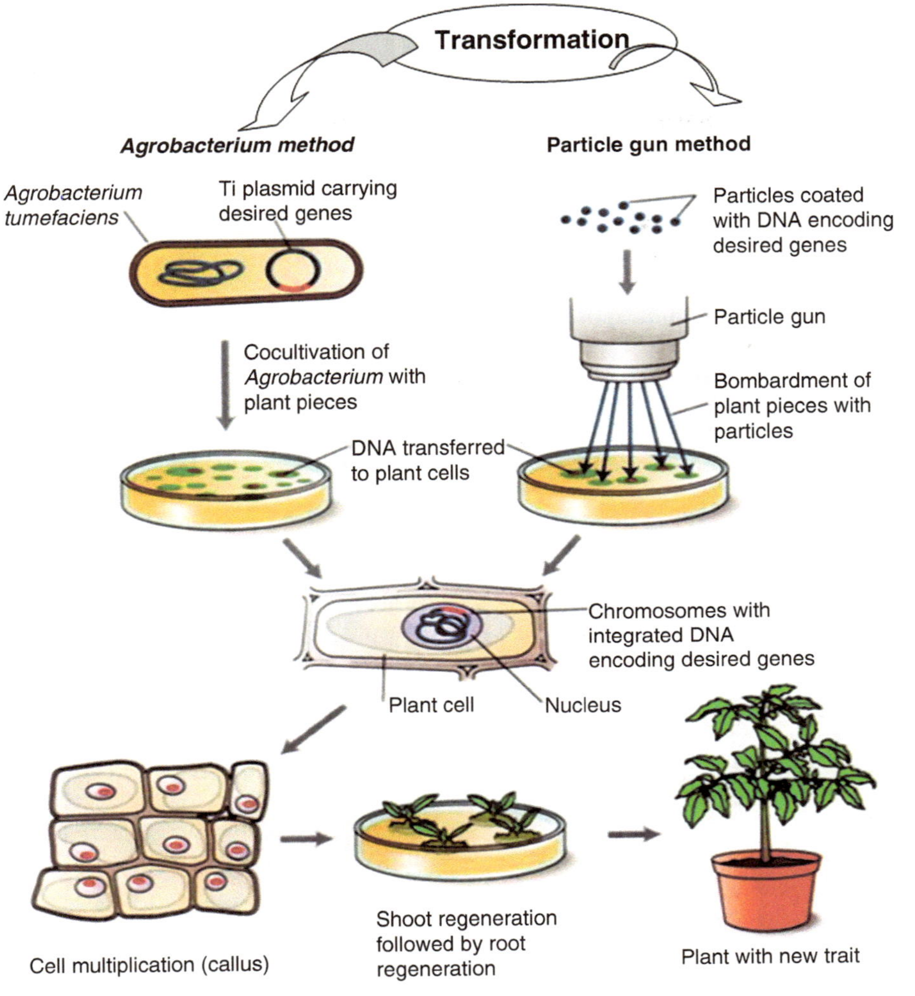

Fig. 11.1 Overview of primary methods used for plant transformation. During the transformation process, either *Agrobacterium tumefaciens* or particle bombardment is used to transfer the desired gene(s) into individual plant cells. These transferred genes then become integrated into the genome of some recipient cells. Whole new transgenic plants are regenerated from transformed cells, giving rise to a transgenic line (Source: modified after Mirkov 2003)

the transgene. Transformed plant cells are finally regenerated into plants (Fig. 11.1). Two commonly used methods that can be employed to insert the transgene into the plant genome are *Agrobacterium*-mediated transformation and transformation via bombardment with DNA-coated particles (Altpeter et al. 2005; Tzfira and Citovsky 2006; Prado et al. 2014). The first GM plants were produced using either *Agrobacterium-based* or direct gene transfer techniques, such as particle bombardment. *Agrobacterium tumefaciens* infects wounded plant tissue(s) and inserts a short section of DNA, called the transfer DNA or T-DNA, into the host plant genome (Chilton et al. 1977). This work by Mary-Dell Chilton provided evidence that plant genomes could be manipulated more precisely, and hence very recently, she has been awarded the 2013 World Food Prize (World Food Prize 2013). This technology no doubt leads to overcome the bottlenecks faced during green revolution while using conventional breeding techniques.

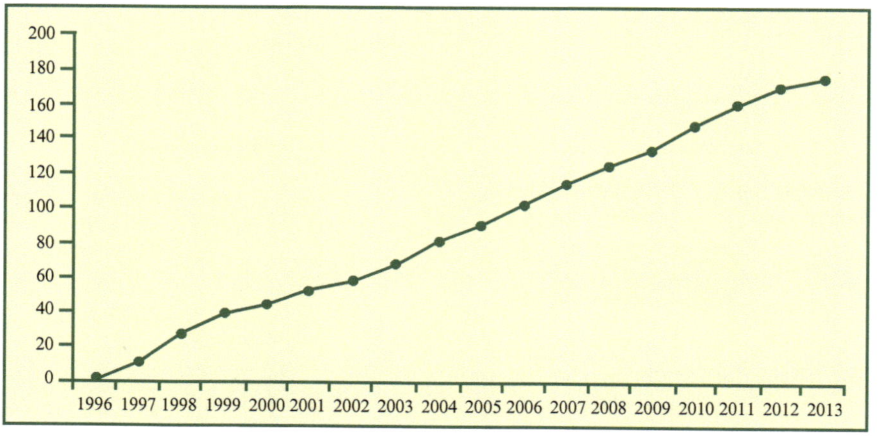

Fig. 11.2 Global area of transgenic crops in millions of hectares, since 1996 to 2013 (Source: James 2013)

11.3 GM Crops in Developing Countries

GM crops are the fastest adopted crop technology in the history of modern agriculture with 100-fold increase from 1.7 million hectares in 1996 to over 175 million hectares in 2013. Developing countries have established the use of GM crops, with Brazil, Argentina, India, China, and South Africa being the prominent players among the 19 developing countries adopting the technology in 2013. These five countries together constitute 41 % of world population and are growing 47 % of global GM crops. Over 18 million farmers across the globe planted GM crops in 2013, among which16.5 million were small, poor farmers from developing countries. More than 7.5 million small farmers from China were growing 4.2 million hectares of Bt cotton; however, the trendsetters were 7.3 million farmers from India cultivating a record of 11.0 million hectares of Bt cotton (James 2013). Bt cotton resistance to bollworms and budworms is quite relevant in developing countries. Bt crops are based on the *cry* genes of a soil bacterium, *Bacillus thuringiensis* (Bt), which produces proteins that are specifically toxic to larvae of some lepidopteran and coleopteran insect species, whereas other insect pests, especially sucking pests, remain unaffected. Different strains of the bacterium have different *cry* genes classified into groups *cryI*–*cryIV* and subgroups A, B, C, etc., and each encodes a protein that is effective against a different type of insect (Maagd et al. 1999; Halford 2012). More advanced transgenic cotton varieties such as Bollgard II, which contains two Bt genes and expresses two Cry proteins (Cry1Ac and Cry2Ab2), are now available and are becoming widely used. Bollgard I technology involved the *Cry1Ac* Bt gene. The USA which accounts for 40 % of the global GM crops is the leading adopter with 70.1 million hectares among the developed world, with Canada at the second spot; however, a developing country like Brazil is emerging as a strong global leader and is only trailing by the USA with 40.3 million hectares (James 2013). Figure 11.3 shows the relative area of biotech crops in developed and developing countries in millions of hectares, since 1996 to 2013.

11.4 Transgenic Crops for Increase of Farmer's Income

Agriculture in developing countries is mostly dependent on nature as there is a lack of appropriate agricultural technologies, the reason for which may be economic or the unavailability of apt techniques and their inefficient implementa-

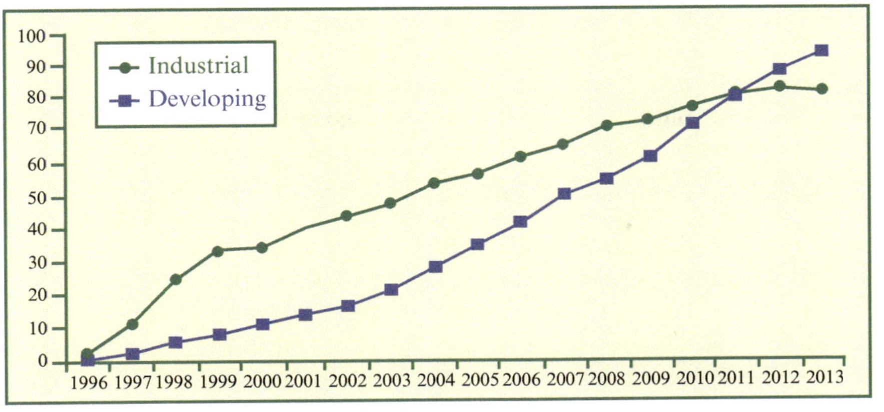

Fig. 11.3 Adoption of GM crops in developed and developing countries in millions of hectares (Source: James 2013)

tion. Farmers are threatened by extreme weather, crop pests, and hence unpredictable and low productivity which culminates into food scarcity and insecurity. The extreme weather conditions like frequent occurrences of drought and floods invariably result in low crop production and acute food shortages. As a result, poor people in developing countries suffer from different types of malnutrition. Tools of plant biotechnology provide opportunities for improving the economy of developing countries and the well-being of the people and offer a means for increasing agricultural production, improving human health, and minimizing environmental degradation by developing high-yield varieties, which requires less use of chemical pesticides and do not require mechanical tilling (Carpenter 2010).

11.4.1 The First Generation of GM Crops: Increasing Farmer's Income

The first generations of GM crops were focused on traits which directly benefit the farmers. These were called as input traits and affect the husbandry and management of a crop. Input traits include herbicide tolerance; resistance to insects or pathogens like fungus, bacteria, and virus; and the ability to survive stress conditions, such as drought (Halford 2006). Insect resistance and herbicide tolerance are two primarily and most widely targeted traits, which directly benefit the growers although consumers may benefit indirectly through lower food prices (Chen and Lin 2013; Rommens 2010; Halford 2012). These first-generation GM crops have proved to be quite promising in enhancing agricultural productivity and reducing poverty in developing countries (Christou and Twyman 2004; Farre' et al. 2010, 2011).

Development of the GM IR traits in crops has resulted in less expenditure on insecticides and lower costs of production and hence higher incomes through improved yields in all countries. Thus, gains from technology are direct, and farmers mostly in developing countries have been able to improve both their productivity and economic returns. The gains from GM HT traits on the other hand have come from a combination of effects including reduced costs of production, e.g., using low-cost, broad-spectrum herbicide (glyphosate), which is directly beneficial for farmers. Indirect gains have come by the facilitation of changes in farming systems, e.g., in both North and South America, it facilitated the moving away from conventional to low or no-tillage production systems and enabled many farmers to plant a second crop of soybeans after wheat in the same season (Brookes and Barfoot 2013). Adopting conservation tillage techniques reduces soil erosion and improves soil quality through a gradual accumulation of organic material in the soil (Park et al. 2011). Insecticide reduction and yield effects are closely related. Yield enhancement varies depending on the environment and the local intensity of pest and weed pressures. Pest pressure is often higher in the tropics and subtropics, and resource-poor farmers face more severe constraints in chemical pest control, so yield effects will be more pronounced in developing countries (Qaim and Zilberman 2003; Qaim 2009; Park et al. 2011). Adoption of Bt crops resulted in 65 % insecticide reduction and 24 % increase in effective yield in China. As a result, increase in gross margin in China reached US$470/ha (Pray et al. 2002). Similar results were later reported in India where adoption of Bt crops resulted in 41 % insecticide reduction and 37 % increase in effective yield leading to US$135/ha increase in gross margin (Qaim et al. 2006; Sadashivappa and Qaim 2009). Farmers in developing countries received $3.74 for each dollar invested in GM crop seeds in 2012 (the cost being equal to 21 % of total technology gains), while farmers in developed countries received $3.04 for each dollar invested in GM crop seed (the cost being equal to 25 % of the total technology gains) (Brookes and Barfoot 2014). The higher share of total technology gains realized by farmers in developing countries relative to farmers in developed countries mainly reflects weaker provision and enforcement of intellectual property rights coupled with higher average levels of benefits in developing countries.

Crop biotechnology undoubtedly helps farmers earn reasonable incomes for their work. The net economic benefit at the farm level in 2012 was $18.8 billion, equal to an average increase in income of $117/hectare. For the 17-year period (1996–2012), the global farm income gain has been $116.9 billion. The total farm income gain of $116.9 billion was divided equally between farmers in developing and developed countries. Fifty-eight percent (58 %) economic gains were due to reduced production costs (less plowing, fewer pesticide sprays, and less labor) and 42 % due to substantial yield gains of 377 million tons. The highest yield gains were obtained by farmers in developing countries, many of which are resource poor and farm small plots of land. The cost farmers paid for accessing crop biotechnology in 2012 ($5.6 billion payable to the seed supply chain) was equal to 23 % of the total gains (a total of $24.4 billion inclusive of the $18.8 billion income gains). Globally, farmers received an average of $3.33 for each dollar invested in GM crop seeds (Brookes and Barfoot 2014). So, crop biotechnology continues to be a good investment for farmers around the world.

11.4.2 Concern of Farmers about Planting Genetically Modified Crops

Interest of farmers lies in increased income and productivity; better allocation of labor, time, and resources: and safer practices and products for themselves and the environment. In view of these facts, major concerns of farmers while planting genetically modified crops are as follows:

1. Seeds bred for particular soils, particular temperature, and rainfall zones do not perform to the optimum in other soils and zones; therefore, farmers must have access to seeds that are suitable for the agroecological conditions of their particular fields. Farmers must be able to coexist with their neighbors in neighborly ways so that each farmer can choose what is appropriate for his/her field (Kershen 2010).
2. Laws should not be too stringent when it comes to using genetically modified seeds. They must not face discriminatory rules and regulations that limit their choices and inappropriately impose liability upon them simply because they desire to grow genetically modified crops. When laws allow farmers the choice, farmers have chosen quickly and broadly to grow genetically modified crops in their fields. In countries like India, Pakistan, and Brazil, farmers are defying the law to improve their lives and their farms (Rehman 2007; Roy et al. 2007). Illegal cultivation of Bt cotton in Pakistan forced the regulators to approve its cultivation in 2010 (Nazli et al. 2010).

   Interestingly, a small country like Bangladesh approved Bt eggplant (brinjal) for planting for the first time in 2013 and released four varieties of Bt eggplant in January 2014 (BARI 2014). In the future, it may serve as an exemplary model for other small poor countries (James 2013).

11.5 Transgenic Crops for Poverty Alleviation

For developing world, poverty is the main cause of chronic food insecurity (Wijk 2002; Christou and Twyman 2004; Yuan et al. 2011). GM crops can contribute to the alleviation of poverty in developing countries through increased income for producers as well as addressing persistent problems of hunger (Juma 2011; Brookes and Barfoot 2009, 2013). The reasons for the increase in producer's income may vary from region to region as well as from farm to farm (Finger et al. 2011). According to the UN reports of 2011, half the population in sub-Saharan Africa and in the least developed countries subsists on less than US$ 1 per day which comes to more than a quarter of the population in developing countries as a whole (UN 2011). On a larger scale, this translates into productivity losses that can account for 2–4 % of gross domestic product (GDP) as demonstrated for several countries in South Asia (FAO 2012). Developments in crop will have a direct impact on poverty alleviation as many of the poorest people and countries in the world are highly reliant on agriculture (DFID 2005). Hence, a remarkable gain in overall economic welfare can be achieved through adoption of technologies such as genetic engineering of plants. GM crops can contribute significantly to poverty reduction and rural development, when they are suited to the small farm sector and embedded in a contributory institutional environment. The effects of transgenic crops are considered in relation to crop yield, inputs such as pesticides, and their effects on overall profitability. Yield improvement, higher revenue, and lower pesticide costs are widely reported for Bt cotton, producing in most cases significant net benefit after accounting for higher seed prices.

So, GM technology has had a significant positive impact on farm income derived from a combination of enhanced productivity and efficiency gains. It was assumed that the productivity of unskilled labor would rise by 2 % following adoption of second-/next-generation GM crops. Even golden rice on its own could add $3.2 billion per year to developing countries economic welfare (Anderson 2010). Global value of GM rice has recently been estimated to be US$64 billion per year by aggregating the expected annual benefits (Demont and Stein 2013).

In 2012, the direct global farm income benefit from GM crops was $18.8 billion. This is equivalent to having added 5.6 % to the value of global production of the four main crops of soybean, maize, canola, and cotton. Positive yield impacts from the use of this technology have occurred in all user countries when compared to average yields derived from crops using conventional technology (such as application of insecticides and seed treatments). Because most of the farmers in developing countries are small-scale farmers, so any increase in their income can have a direct impact on poverty alleviation. The average yield impact across the total area planted to insect-resistant (IR) traits during the 1996–2012 period has been +10.4 % for insect-resistant corn and +16.1 % for insect-resistant cotton. In 2012, 46.2 % of the farm income benefits have been earned by developing country farmers, and the vast majority of these income gains have been from GM IR cotton and GM HT soybeans (Tables 11.1 and 11.2). In the absence of crop biotechnology, 17.3 million farmers using this technology in 2012 would not have maintained global production levels equivalent to 2012 levels and would have required additional plantings of 4.9 million ha of soybeans, 6.9 million ha of corn, 3.1 million ha of cotton, and 0.2 million ha of canola. This total area requirement is equivalent to 9 % of the arable land in the USA or 24 % of the arable land in Brazil (Brookes and Barfoot 2014). GM crops are, hence, allowing farmers to grow more without using additional land. It is, therefore, a land-saving technology (James 2010).

11.5.1 Role of Bt Cotton in Alleviating Poverty in India

India has the third largest area in the world under GM crops. There are more than 60 transgenic crops under research and more than 20 under field trials in India (ISAAA 2013). But only one crop is under cultivation, i.e., Bt cotton, since its approval in 2002 by the Genetic Engineering Approval Committee (GEAC) of the Government of India. Premature discontinuation of GM mustard in 2001 and the moratorium of Bt brinjal by the Ministry of Environment and Forests (MOEF) in 2010 raised a question on the regulatory system of GM crops in India. However, on a brighter side, Bt cotton is doing exceptionally well. In India, a case study was done, which compared the performance of over 9000 Bt and non-Bt cotton farm plots in Maharashtra, and it was found that Bt cotton varieties had a significant positive impact on average yields and on the economic performance of cotton growers. Between 2003–2004 and 2006–2007, cotton yields in India indicate a significant yield advantage of more than 30 % with Bt cotton compared with conventional varieties with corresponding increase in farm income (Karihaloo and Kumar 2009). Similar study was done on resource-poor small-scale cotton farmers of South Africa, and similar results were obtained (Bennett et al. 2006). Bt cotton produces 82 % higher aggregate incomes per hectare in India, and as a result annual consumption expenditures of Bt-adopting households increased by 18 %, during 2006–2008, in comparison with non-adopters (Qaim 2009; Klumper and Qaim 2014). The yield increases in the range of +30 % to +40 % have been confirmed later in India (Qaim 2009, 2010; Gruere and Sengupta 2011; Herring and Rao 2012). As a result, Bt cotton area increased from 0.05 million hectares in 2002 to 9.3 million hectares in 2011–2012, accounting for 88 % of total area (IGMORIS 2013), and the Indian cotton sector switched from a net import to a significant export situation. Despite this significant increase in cotton area, the use of insecticides on cotton decreased from 46 % of total insecticides used in agriculture during 2001–2002 to 20 % in 2011–2012 (Kranthi 2012). Bt technology further contributed to 24 % increase in cotton yield per acre through reduced pest damage and a 50 % increase in profit among cotton smallholders (Kathage and Qaim 2012).

Table 11.1 GM crop farm income benefits of developing countries in million US$ in the year 2012

GM crops	Farm income benefits in developing countries in 2012
GM IR cotton	4800.7
GM HT soybeans	1842.5
GM IR maize	1400.3
GM HT maize	543.9
GM HT cotton	75.8
Total	**8663.2**

Source: modified after Brookes and Barfoot (2014)

Table 11.2 GM crop farm income benefits for developing countries in million US$, 1996–2012

Countries	GM IR cotton	GM HT soybeans	GM HT maize	GM HT cotton	GM IR maize	Total
China	**15,270.4**	N/a	N/a	N/a	N/a	**15,270.4**
India	**14,557.1**	N/a	N/a	N/a	N/a	**14,557.1**
South Africa	**34.2**	**9.1**	**4.1**	**3.2**	**1100.6**	**1151.2**
Paraguay	N/a	**828**	N/a	N/a	N/a	**828.0**
Pakistan	**725.1**	N/a	N/a	N/a	N/a	**725.1**
Bolivia	N/a	**432.2**	N/a	N/a	N/a	**432.2**
Philippines	N/a	N/a	**104.7**	N/a	**273.6**	**378.3**
Burkina Faso	**186.9**	N/a	N/a	N/a	N/a	**186.9**
Honduras	N/a	N/a	N/a	N/a	**6.9**	**6.9**

Source: modified after Brookes and Barfoot (2014)

Bt cotton has significantly raised living standards of small farm households in India.

11.6 Transgenic Crops for Nutrition and Health

On the one hand, the world is facing newer challenges, such as the expansion of cultivated areas to less fertile fields and the adaptation of crops to a globally changing climate (Avni and Bla'zquez 2011), on the other hand the nutritional quality of food or balanced diet is emerging as a major problem in developing countries. In response to escalating food prices, poor households had to limit their food consumption, and poor people are unable to procure a balanced diet. Food prices throughout the world have increased sharply in the last decade. For example, prices of wheat and maize were three times higher in 2008 than at the beginning of 2003, and the price of rice was five times higher (Braun 2008, 2010). Majorities of the people living in developing countries are extremely poor and cannot afford combination of expensive foods like meat, fish, milk, pulses, etc., which usually forms essential balanced diets. According to an estimate, food production would have to be doubled by 2050 to overcome existing hunger, feed an additional 2 billion people, and accommodate rising demand from income growth (Braun 2010; Adenle et al. 2012).

Lack of balanced diet leads to micronutrient deficiencies and, hence, negative consequences on people's nutrition and health. Micronutrients are involved in all aspects of development, growth, and physiology of the human body, and their deficiencies can cause birth defects, permanent physical and mental impairment, as well as an increased risk of death by infectious and chronic diseases. The long-term consequences of insufficient amounts of essential micronutrients in the human diet can be more devastating than low-energy intake (Murgia et al. 2013). The leading micronutrient deficiencies are iron deficiency, iodine deficiency, zinc deficiency, folic acid deficiency, and vitamin A deficiency. One or more of these affect almost half of the world's population. Since children's nutrition is crucial for their physical and cognitive development and for their productivity and earnings as adults, the health and economic consequences of insufficient food and poor diets are lifelong – for the individuals as well as for society. Besides, GM crops with insect-resistant genes may reduce the need for pesticides which improves the health of farmers, especially in developing countries where pesticides are still applied with handheld sprayers (Chrispeels 2014).

11.6.1 Next Generation GM Crops: Improving Nutrition and Health

Scientific knowledge has achieved breakthrough in the field of genomics, proteomics, and metabolomics in the recent times. It has broadened our understanding of the sources and nutritional values of the products of many of food crops (Arber 2010). Next-generation GM crops target traits which affect the composition of the crop product for quality improvements for nutrition and industrial purposes and are called as output traits. These traits include improved nutritional value like staple foods with enhanced contents of essential amino acids (especially lysine and methionine) and micronutrients (vitamins A and E, iron, folate, and ascorbate); oilseeds with improved fatty acid composition (oleic acid, omega-3 fatty acid); changes in starch quality, i.e., resistant starch and antioxidants (anthocyanins); etc. (Jefferson-Moore and Traxler 2005; Pew Initiative on Food and Biotechnology 2007).

In general, one or several key genes in metabolic pathways are introduced or knocked down by genetic modification to promote the accumulation of healthy metabolites, and nutritional requirements can be addressed directly by contributing to multipoint intervention strategies (Yuan et al. 2011). The main beneficiaries of these so-called next-generation GM crops are consumers and/or food processors. One famous example is golden rice, which can prevent vitamin A deficiency that prevails in poor populations solely dependent on rice as a staple food crop. Approximately 500,000 children in developing countries become blind each year owing to vitamin A deficiency. Africa accounts for almost 50 % of the children who are clinically or subclinically deficient in vitamin A, particularly under 5 years of age (FAO/WHO 1998; WHO 2010). Golden rice contains a high content of β-carotene by introduction of a previously absent biosynthetic pathway into rice endosperm. A GM line containing β-carotene was developed by Ingo Potrykus and coworkers in 2000 at the Swiss Federal Institute of Technology, Zurich (Potrykus 2003). Rice endosperm contains geranylgeranyl diphosphate, which is converted into β-carotene by three enzymes produced from different transgenes: phytoene synthase (*psy*), lycopene β-cyclase gene from daffodil (*Narcissus pseudonarcissus*), and a phytoene desaturase (*crtI*) gene from the bacterium *Erwinia uredovora*. The GM rice producing β-carotene was crossed with another line engineered with multiple genes to improve iron availability, including a phytase-encoding gene from *Aspergillus fumigatus* (Lucca et al. 2001). The high-β-carotene/high-availability iron hybrid was called golden rice (Halford 2012). Higher β-carotene intakes will improve the vitamin A status of individuals, thus reducing the incidence of adverse health outcomes (Qaim 2010). The replacement of the daffodil genes *Zmppsy1* and *EucrtI* with its maize ortholog is the basis of Golden Rice 2, which produces up to 37 μg of carotenoids per gram dry weight (DW) of grain, of which 31 μg/g is β-carotene (Paine et al. 2005). The putative impact of golden rice was calculated as up to 40,000 lives saved per year for India alone (Khush 2012). β-Carotene in golden rice is as good as pure β-carotene in oil at providing vitamin A to children (Tang et al. 2012). It was the use of genetic engineering together with conventional breeding, i.e., combinatorial transformation method, which has enabled the production of provitamin A (PVA) in corn and rice plants as an alternative source of vitamin A to save millions of children who go blind every year (Avni and Bla'zquez 2011).

Bananas having levels of PVA greater than 15-fold higher than wild type have been developed through the overexpression of a single gene, phytoene synthase, using either constitutive promoters or fruit-preferred promoters. Two different phytoene synthase genes, one from a naturally high-PVA banana and other from maize gene used in Golden Rice 2, were differently expressed in bananas, and lines with elevated PVA have been identified (Dale et al. 2013). Anemia caused by iron deficiency is the world's most common nutritional deficiency. It affects pregnant and nursing women and young children most commonly (Earl and Woteki 1998; Swaminathan 2002). Genetic enrichment of iron in Indian rice Pusa Basmati (*Oryza sativa* L.) has also been accomplished through recombinant DNA technology (Shivprakash et al. 2006). Co-expression of endosperm-specific recombinant soybean ferritin and *Aspergillus* phytase in maize resulted in significant increases in the levels of bioavailable iron (Drakakaki et al. 2005). A similar end was achieved earlier with lettuce (Goto et al. 2000). Transgenic rice plants expressing the NAS (nicotianamine synthase) genes *Osnas1*, *Osnas2*, or *Osnas3* accumulated up to 19 lg/g of iron in the endosperm (Johnson et al. 2011). Recently, phosphate bioavailabilities of barley grains have been improved from 30 to 60 % using cisgenesis with an endogenous phytase gene (Holme et al. 2012). Barley grains are widely used for feeding monogastric animals such as chickens and pigs. A large number of rice or soybean ferritin overaccumulators in rice mega-variety IR64, including marker-free events, were generated and evaluated by introducing soybean or rice ferritin genes into the endosperm for product development. As much as a 37- and 19-fold increase in the expression of ferritin gene in single and co-transformed plants, respectively, and a 3.4-fold increase in Fe content in the grain over the IR64 wild type were achieved (Oliva et al. 2014).

Multivitamin maize expressing the rice *dhar* gene from the ascorbate recycling pathway accumulated six times the normal level of ascorbate (Naqvi et al. 2009). Similarly, the constitutive expression of two *Arabidopsis* cDNA clones encoding q-hydroxyphenylpyruvate dioxygenase (HPPD) and 2-methyl-6-phytylplastoquinol methyltransferase (MPBQ MT) increased the tocopherol content by threefold in transgenic maize (Naqvi et al. 2011). The essential fatty acids are abundant in fish, shellfish, nuts, and leafy vegetables, but they are not present in cereals (Farre et al. 2011). Genetic engineering can be used to produce oilseeds such as soybean and canola that have nutritional properties similar to fish oils (Damude and Kinney 2008). These include fatty acids associated with lowering risks of coronary heart disease (Haslam et al. 2013). For example, omega-3 fatty acid which reduces coronary heart disease and maintains heart health has been increased from 12 to 50 % in canola (Ursin 2003). The fatty acid biosynthesis pathway in plants was modulated to produce ω-3 and ω-6 PUFAs by introducing the microbial enzymes responsible for a sequence of fatty acid desaturation and elongation reactions (Domergue et al. 2005). Studies have also shown that the use of oil from transgenic soya in which the fatty acid metabolic pathways have been modified can increase the n-3 VLC-PUFAs of chicken meat (Rymer and Givens 2009).

Folate prevents neural tube defects and causes widespread megaloblastic anemia during pregnancy and often exacerbates already existing iron deficiency anemia (Rush 2000; Barber et al. 2000; Laurence et al. 1981; Rosenquist et al. 1991). Enhancing folate content in staple crops by metabolic engineering is a promising, cost-effective strategy to eradicate folate malnutrition worldwide (Blancquaert et al. 2014). The transformation of two pathway genes from *Arabidopsis thaliana* increased folate production from <1 mg/g to 17 mg/g in rice which is enough to meet the requirements necessary to combat its deficiency (Storozhenko et al. 2007). Several other traits are also under development, such as rice enriched with lactoferrin to reduce diarrhea in high-risk patients.

Similarly, biofortification is another important and widely used technique to nutritionally enhance the food crops at source (Zhu et al. 2007). Biofortification of staple food crops might be used as one of the possible strategies against micronutrient malnutrition in developing countries. Biofortification allows the poor to receive the necessary amounts of vitamin A, zinc, and iron via their regular staple food diets and, hence, delivers naturally fortified foods to people with limited access to commercially marketed fortified foods or supplements (Braun 2010). However, the desired traits for biofortification may not be present at all in a food crop; the best-known example is golden rice, in which the carotenoid biosynthetic pathway has been reconstituted in non-carotenogenic endosperm tissue, as a means to deliver provitamin A (Mayer et al. 2008). So, biofortification of staple crop plant tissues can be achieved through breeding where this is possible, while recombinant DNA technology must be applied in all other cases (Bayer 2010). Hence, the science of biotechnology, either through conventional breeding (often in conjunction with marker-assisted selection) or genetic modification approaches, has great potential to achieve biofortification for nutritional benefits (Table 11.3).

11.7 Transgenic Crops in the Era of Climate Change

A question that often comes in one's mind is whether genetic engineering can contribute to food security, as well as enhancing human nutrition and farming under a changing climate. Global climate change is increasing temperatures worldwide resulting in global warming besides rapid climate variability and unscheduled expansion or shrinkage of the extreme climates (Keer 2007; IPCC 2001, 2007; Webb et al. 2012). In other words, as a result of climatic change, there is an increase in the frequency of extreme events that are likely to decrease crop yield affecting all dimensions of crop production (Singh et al. 2015). Human activities are hugely accelerating this change in global climate. Continuously increasing human population is hence making the situation even worse. It is expected to peak before the end of the century, with 10 billion people before 2100 (Lutz et al. 2001; Duhamel and Vandenkoornhuyse 2013). Food crises are exacerbated by global warming as agricultural productivity has declined worldwide as a consequence of the hot summers experienced in

Table 11.3 Some of the important nutritionally enhanced GM crops

Crop	Nutrient/trait	Base level	Maximum level in GM crop	Main benefits	Reference(s)
Rice	Iron	–	37- and 19-fold increase in the expression of ferritin gene in single and co-transformed plants, respectively	Malnutrition causes anaemia or impaired mental development	Oliva et al. (2014))
Banana	Provitamin A (PVA)	–	15-fold higher	Vitamin A deficiency causes blindness and increased child mortality	Dale et al. (2013)
Barley	Phosphate bioavailability	30 %	60 %	Barley grains used for feeding monogastric animals such as chickens and pigs	Holme et al. (2012)
Wheat	Amylose	28 %	75 %	Benefit for some health issues associated with some chronic diseases	Regina et al. (2006); Zhu et al. (2012)
Maize	Tocopherol	–	Threefold	Powerful antioxidants that protect fatty acids, LDLs, and other components of cell membranes from oxidative stress.	Naqvi et al. (2011)
Soybean	Oleic acid	20 %	80 %	To hinder the progression of adrenoleukodystrophy, and reduce blood pressure	Mroczka et al. (2010); Wagner et al. (2011)
Maize	Ascorbate	18 mg/g	107 mg/g	Ascorbate deficiency causes scurvy in humans	Naqvi et al. (2009)
Rice	Iron	–	Increase > sixfold	Malnutrition causes anaemia or impaired mental development	Wirth et al. (2009)
Tomato	Anthocyanin	0 mg/g	2.83 mg/g	To extend the life of cancer-prone mice	Butelli et al. (2008)
Rice	Folate	<1 mg/g	17 mg/g	To prevent neural tube defects	Storozhenko et al. (2007)
Rice	β-Carotene	0 μg/g	37 μg/g	Vitamin A deficiency causes blindness and increased child mortality	Paine et al. (2005)
Canola	Omega-3	12 %	50 %	Reduces coronary heart disease and maintains heart health	Ursin (2003)

the recent past (Mittler 2006; Mittler and Blumwald 2010). It is estimated that global warming will reduce about 6 % and 5 % average yield per 1 °C rise when it comes to C_3 and C_4 crops, respectively (Yamori et al. 2013) Besides, current trends in yield increase are insufficient to double food production by 2050 (Ray et al. 2013). However, the impact of climate change will cut across all boundaries, and the most susceptible victims of this climate change are going to be the most food-insecure developing countries with a challenge posed to them to attain millennium development goals and achieve sustainable development by climate change (Singh et al. 2015).

The climate change which is generally related to the abiotic stresses due to extreme environmental changes may result in melting of portions of the Himalayan glaciers, disturb the monsoon pattern, and increase flooding/drought in Asia.

Increased uncertainty over the availability of water for irrigation and more frequent floods will affect 25 % of the world's cereal production. Most of the abiotic stress tolerance mechanisms in plants are complex due to the involvement of multiple metabolic pathways. Hence, manipulating these characters through conventional breeding remains a big challenge. The genetic transformation of plants is an efficient alternative to this problem. The genetic transformation of plants with regulatory genes, e.g., transcription factors, is a promising method for genetic engineering because many of the ways in which plants can adapt to cold, drought, oxidative stress, and extreme temperatures is through transcriptional control (Mittler and Blumwald 2010). The genetic transformation studies to overcome various types of stresses have already been discussed in detail in this book in an article by Bauddh and coworkers (Table 11.4). Genetic engineering (GE) has already contributed to the reduction of greenhouse gas (GHG) emissions as a result of less fuel use and additional soil carbon storage from reduced tillage with GM crops. When global impact of biotech crops on environmental effects was estimated between 1996 and 2010, it was found that farming with transgenic crops since 1996 has led to additional soil carbon sequestered, equivalent to 133,639 million tons of CO_2 (Brookes and Barfoot 2012). In 2012, this was equivalent to removing 27 billion kg of carbon dioxide from the atmosphere or equal to removing 11.9 million cars from the road for 1 year (Brookes and Barfoot 2014). Increased productivity from GM crops has decreased pressure for land conversion of non-cropland to cropland which is a major contributor to the greenhouse gas increase in the atmosphere. This indicates that it can play a large role in both the mitigation of and adaptation to climate change.

11.8 Conclusions and Future Prospects

In spite of the strong opposition, farmers around the world have adopted GM crops at an unprecedented rate (Herring 2008; James 2013) especially in some major developing countries like India and China. The principal beneficiaries of agricultural development from GM crops are likely to be poor farmers and poor consumers (Weale 2010). GM crops have the potential to improve food security in developing countries by improving incomes of farmers and availability of lower-priced and better-quality food for consumers (Qaim and Kouser 2013). There are several constraints to the research and application of biotechnology in developing countries like financial resources, lack of policies and absence of systems for the delivery of technologies to potential users, and finally lack of awareness, leading to misconceptions about the potential of and risks posed by biotechnology. Hence, public controversies about the risks and benefits of GM crops continue (Gilbert 2013; Fernandez-Cornejo et al. 2014). However, there is no scientific evidence that the process of transferring genes from one kind of organism to another possesses intrinsic problems. Further, there are no such reports that anyone has become ill as a result of eating GM foods. Hundreds of millions of people are regularly consuming foods produced by GM crops (Raven 2010). There is an increasing scientific consensus, even in Europe, that the GM foods and crops currently on the market have brought no documented new risks either to human health or to the environment (Paarlberg 2010; European Commission 2010; European Academies Science Advisory Council 2013; DeFrancesco 2013). However, the existing negative public attitude toward GM crops, especially in Europe, has contributed to a stringent complex regulatory framework and has limited public and private investments into GM crop research, increasing the cost of technologies making it difficult for developing countries to continue the research on large scale and, hence, reap its benefits (Qaim 2010). Besides, investments should be more in the areas of R&D, rural infrastructure, rural institutions, and information monitoring and sharing to enhance agricultural productivity (Braun 2010). The public sector needs to resource these developing country-targeted projects as they do not represent commercially valuable targets and therefore cannot be a commercial priority for the

Table 11.4 Some of the important genes transferred against major abiotic stresses

Stress	Gene	Crop/plant	Reference(s)
Cold	*CBF* (*CpCBF2*)	Papaya	Zhu et al. (2013)
	OsDREB1B	Mouse-ear cress (Arabidopsis)	Qin et al. (2007)
	ApGSMT and *ApDMT*	Mouse-ear cress (Arabidopsis)	Waditee et al. (2005)
	BNCBF5/BNCBF17	Brassica	Savitch et al. (2005)
Heat	*RcaB*	Wheat	Wang et al. (2014)
	CBF (*CpCBF2*)	Papaya	Zhu et al. (2013)
	AChE	Tobacco	Yamamoto et al. (2011)
	AtDREB1A	Chrysanthemum	Hong et al. (2009)
	GASA4 T-DNA (SALK_042431)	Maize	Ko et al. (2007)
	AtDREB2A	Mouse-ear cress (Arabidopsis)	Sakuma et al. (2006)
Drought	*codA*	Rice.	Kathuria et al. (2009)
	betA	Cotton	Lv et al. (2007)
	TPS1	Tobacco	Almeida et al. 2007
	TPS1-TPS2	Mousear Cress (Arabidopsis)	Miranda et al. (2007)
	TPS1-TPS2	Tobacco and mouse-ear cress (Arabidopsis)	Karim et al. (2007)
	GmDREB2	Tobacco and mouse-ear Cress (Arabidopsis)	Chen et al. (2007)
	AtDREB1A	Fescue	Zhao et al. (2007)
	P5CSF129A	Wheat	Vendruscolo et al. (2007)
	P5CSF129A	Sugarcane	Molinari et al. (2007)
	OsDREB1	Rice	Ito et al. (2006)
	P5CSF129A	Petunia	Yamada et al. (2005)

private sector (Bayer 2010). However, to develop novel traits and to distribute it systematically at a wider scale would require the expertise and resources of both public and private sector institutions. In the immediate future, the research should focus on development of methods avoiding antibiotic- or herbicide-resistant genes as selectable marker or use of positive selectable markers such as phosphomannose isomerase (*pmi*), xylose isomerase (*xyl* A), etc., to widen the acceptability of GM crops and selection of genes for the desirable traits for the transfer and strategies for the seed distribution system, where the end user in the developing countries is benefited and not only industries in developed countries. There is a wide scope to produce nutritionally enhanced crops such as finger millet, cassava, etc., which are widely grown in sub-Saharan countries like Africa and are nutritionally poorer crops. Once nutritionally enhanced, these crops can be easily available and, hence, beneficial to poorer local populations. The new tools of genomics, proteomics, and metabolomics would allow better understanding of vital processes and metabolic pathways for their improvement. Generation of more number of next-generation GM crops in the near future and development of new biotechnologies and non-targeted safety assessment approaches may improve public perception about the potential risk of GM crops (Chen and Lin 2013). The focus should now be on the use of new techniques like RNA interference, agro-infiltration, cisgenesis, oligonucleotide-directed mutagenesis, and zinc finger nuclease technology that may or may not come under the strict GM regulations (Halford 2012). Crops created through genome engineering might prove to be more acceptable to the public than plants that carry foreign DNA in their genomes. Crops with enhanced nutritional value can be created by altering only a few nucleotides. The use of gene knockouts to disrupt biochemical pathways should make it possible to create plants that accumulate a variety of valuable biosynthetic intermediates (Voytas and Gao 2014).

GM crops hold a significant potential to contribute to poverty reduction, better nutrition and health, and sustainable development in developing countries especially in the present scenario of climate change. So, risk–benefit analysis approach should be considered over risk assessment. Saving lives by curbing malnutrition and food security should be the first priority. Hence, the "if's and but's" should be in waiting, when it comes to saving lives. Commercialization of GM crops will have substantial implications for the alleviation of poverty, hunger, and malnutrition. They have much more to offer the developing world than the developed because when it comes to food developing world needs more and has fewer alternatives.

Acknowledgements PKJ is grateful to HSCST, Chandigarh, and UGC, Department of Biotechnology and Department of Science and Technology, New Delhi, for financial support to his laboratory for improvement of grain legumes and other crops. MS is thankful to HSCST, Chandigarh, and Department of Biotechnology, New Delhi, for JRF; Council of Scientific and Industrial Research, New Delhi, for SRF; and Department of Science and Technology, New Delhi, for the Young Scientist Award.

References

Adenle AA, Aworh OC, Akromah R, Govindan P (2012) Developing GM super cassava for improved health and food security: future challenges in Africa. Agric Food Secur 1:11. http://www.agricultureandfoodsecurity.com/content/1/1/11

Almeida AM, Silva AB, Araujo SS, Cardoso LA, Santos DMJ, Torne JM, Silva JM, Paul MJ, Fevereiro PS (2007) Responses to water withdrawal of tobacco plants genetically engineered with the *AtTPS1* gene: a special reference to photosynthetic parameters. Euphytica 154:113–126

Altpeter F, Niranjan B, Beachy R, Bock R, Capell T et al (2005) Particle bombardment and the genetic enhancement of crops: myths and realities. Mol Breed 15:305–327

Anderson K (2010) Economic impacts of policies affecting crop biotechnology and trade. New Biotechnol 27:558–564

Arber W (2010) Genetic engineering compared to natural genetic variations. New Biotechnol 27:518–521

Atif RM, Patat-Ochatt EM, Svabova L, Ondrej V, Klenoticova H, Jacas L, Griga M, Ochatt SJ (2013) Gene transfer in legumes. Progress Bot 74:37–100

Avni A, Bla' zquez MA (2011) Can plant biotechnology help in solving our food and energy shortage in the future? Editorial overview. Curr Opin Biotechnol 22:220–223

Bangladesh Agricultural Research Institute (BARI) (2014) Bt brinjal set for release to farmers in Bangladesh. www.bari.gov.bd/userfiles/BARI_Bt%20Brinjal%20in%20Bangladesh.doc

Barber R et al (2000) Investigation of folate pathway gene polymorphisms and the incidence of neural tube defects in a Texas Hispanic population. Mol Genet Metab 70:45–52

Bennett R, Kambhampati U, Morse S, Ismael Y (2006) Farm-level economic performance of genetically modified cotton in Maharashtra. India Rev Agric Econ 28:59–71

Beyer P (2010) Golden rice and 'Golden' crops for human nutrition. New Biotechnol 27:478–481

Blancquaert D, Steur HD, Gellynck X, Straeten DVD (2014) Present and future of folate biofortification of crop plants. J Exp Bot 65:895–906

Braun VW (2008) Food and financial crises: implications for agriculture and the poor. In: Food policy report. IFPRI, Washington, DC

Braun JV (2010) Food insecurity, hunger and malnutrition: necessary policy and technology changes. New Biotechnol 27:449–452

Brookes G, Barfoot P (2009) GM crops: global socio-economic and environmental impacts 1996–2007. PG Economics Ltd, Dorchester

Brookes G, Barfoot P (2012) Global impact of biotech crops- environmental effects, 1996–2010. GM Crops Food Biotechnol Agric Food Chain 3:129–137

Brookes G, Barfoot P (2013) The global income and production effects of genetically modified (GM) crops 1996–2011. GM Crops Food 4:74–83

Brookes G, Barfoot P (2014) GM crops: global socio-economic and environmental impacts 1996–2012. PG Economics Ltd, Dorchester

Butelli E, Titta L, Giorgio M, Mock HP, Matros A, Peterek S, Schijlen EGWM, Hall RD, Bovy AG, Luo J et al (2008) Enrichment of tomato fruit with health-promoting anthocyanins by expression of select transcription factors. Nat Biotechnol 26:1301–1308

Carpenter JE (2010) Peer-reviewed surveys indicate positive impact of commercialized GM crops. Nat Biotechnol 28:219–221

Chen H, Lin Y (2013) Promise and issues of genetically modified crops. Curr Opin Plant Biol 16(2):255–260

Chen M, Wang QY, Cheng XG, Xu ZS, Li LC, Ye XG, Xia LQ, Ma YZ (2007) *GmDREB2*, a soybean DRE-binding transcription factor, conferred drought and high-salt tolerance in transgenic plants. Biochem Biophy Res Co 353:299–305

Chilton MD, Drummond MH, Merlo DJ, Sciaky D, Montoya AL, Gordon MP et al (1977) Stable incorporation of plasmid DNA into higher plant cells: the molecular basis of crown gall tumorigenesis. Cell 11:263–271

Chrispeels MJ (2014) Yes indeed, most Americans do eat GMOs every day! J Integr Plant Biol 56:4–6

Christou P, Twyman R (2004) The potential of genetically enhanced plants to address food insecurity. Nutr Res Rev 17:23–42

Dale J, Harding R, Tushemeirwe W, Paul J, Namanya P, Kleidon J (2013) Use of transgene expression to elevate micronutrients in banana. In: Abstracts Plant and Animal Genome. PAG Asia 2013, Singapore, March 17–19, 2013. https://pag.confex.com/pag/asia2013/webprogram/paper8905.html

Damude HG, Kinney AJ (2008) Engineering oilseed plants for a sustainable, land-based source of long chain polyunsaturated fatty acids. Lipids 42:179–185

DeFrancesco L (2013) How safe does transgenic food need to be? Nat Biotechnol 31:794–802

Demont M, Stein AJ (2013) Global value of GM rice: a review of expected agronomic and consumer benefits. New Biotechnol 30:426–436

DFID (2005) Growth and poverty reduction: the role of agriculture. http://www.dfid.gov.uk/Documents/publications/growthpoverty-agriculture.pdf

Domergue F, Abbadi A, Heinz E (2005) Relief for fish stocks: oceanic fatty acids in transgenic oilseeds. Trends Plant Sci 10:112–116

Drakakaki G, Marcel M, Glahn R, Lund L, Periagh S, Christou P, Stoger E (2005) Endosperm-specific co-expression of recombinant soybean ferritin and *Aspergillus* phytase in maize results in significant increases in the levels of bioavailable iron. Plant Mol Biol 59:869–880

Duhamel M, Vandenkoornhuyse P (2013) Sustainable agriculture: possible trajectories from mutualistic symbiosis and plant neodomestication. Trends Plant Sci 18:597–600

Earl R, Woteki CE (eds) (1998) Recommended guidelines for the prevention, detection and management of iron deficiency anemia among U.S. children and women of childbearing age, Institute of Medicine, National Academy Press, Washington, DC

European Academies Science Advisory Council (2013) Planting the future: opportunities and challenges for using crop genetic improvement technologies for sustainable agriculture. EASAC, Halle

European Commission (2010) A decade of EU-funded GMO research 2001–2010. European Commission, Brussels

FAO (2012) The state of food insecurity in the world. Food and Agriculture Organization of the United Nations, Rome

FAO/WHO (1998) Vitamin and mineral requirements in human nutrition. Report of a joint FAO/WHO expert consultation, Bangkok, Thailand

Farre G, Ramessar K, Twyman RM, Capell T, Christou P (2010) The humanitarian impact of plant biotechnology: recent breakthroughs vs bottlenecks for adoption. Curr Opin Plant Biol 13:219–225

Farre G, Twyman RM, Zhu C, Capell T, Christou P (2011) Nutritionally enhanced crops and food security: scientific achievements versus political expediency. Curr Opin Biotechnol 22:245–251

Fernandez-Cornejo J, Wechsler JJ, Livingston M, Mitchell L (2014) Genetically engineered crops in the United States. Economic research report ERR-162, United States Department of Agriculture, Washington, DC

Finger R, Benni E, Kaphengst T, Evans C, Herbert S, Lehmann B, Morse S, Stupak N (2011) A meta-analysis on farm-level costs and benefits of GM crops. Sustainability 3:743–762

Gilbert N (2013) A hard look at GM crops. Nature 497:24–26

Go'mez-Galera S, Rojas E, Durajalagraja S, Zhu C, Pelacho AM, Capell T, Christou P (2010) Critical evaluation of strategies for mineral fortification of staple food crops. Transgenic Res 19:165–180

Goto F, Yoshihara T, Saiki H (2000) Iron accumulation and enhanced growth in transgenic lettuce plants expressing the iron-binding protein ferritin. Theor Appl Genet 100:658–664

Gruere G, Sengupta D (2011) Bt cotton and farmer suicides in India: an evidence-based assessment. J Dev Stud 47:316–337

Halford NG (ed) (2006) Plant biotechnology: current and future uses of genetically modified crops. Wiley, Chichester

Halford NG (2012) Toward two decades of plant biotechnology: successes, failures, and prospects. Food Energy Secur 1:9–28

Haslam RP, Ruiz-Lopez N, Eastmond P, Moloney M, Sayanova O, Napier JA (2013) The modification of plant oil composition via metabolic engineering-better nutrition by design. Plant Biotechnol J 11:157–168

Herring RJ (2008) Whose numbers count? Probing discrepant evidence on transgenic cotton in Warangal district of India. Int J Mult Res Approach 2:145–159

Herring R, Rao C (2012) On the 'failure of Bt cotton': analysing a decade of experience. Econ Polit Wkly XLVII(18):45–53

Holme IB, Dionisio G, Brinch-Pedersen H, Wendt T, Madsen CK, Vincze E, BachHolm P (2012) A cisgenic approach for improving the bioavailability of phosphate in the barley grain. ISB News report

Hong B, Ma C, Yang Y, Wang T, Yamaguchi-Shinozaki K, Gao J (2009) Over-expression of *AtDREB1A* in chrysanthemum enhances tolerance to heat stress. Plant Mol Biol 70:231–240

IGMORIS (2013) Indian GMO Research Information System. http://igmoris.nic.in/

Intergovernmental Panel on Climate Change (IPCC) (2007) Summary for policymakers. In: Parry ML, Canziani OF, Palutiko JP, van der Linden PJ, Hanson CE (eds) Climate change: impacts, adaptation and vulnerability. Contribution of working group II to the fourth assessment report of the Intergovernmental Panel on Climate Change. Cambridge University Press, Cambridge

Intergovernmental Panel on Climate Change (IPCC): synthesis report (2001) Contribution of working group I, II and III to the third assessment report of the Intergovernmental Panel on Climate Change, Cambridge University Press, Cambridge

International Service for the Acquisition of Agri-biotech Applications (ISAAA) (2013) Global status of commercialized biotech/GM crops 2013. http://isaaa.org/resources/publications/briefs/46/executivesummary/default.asp

Ito Y, Katsura K, Maruyama K, Taji T, Kobayashi M, Seki M, Shinozaki K, Yamaguchi-Shinozaki K (2006) Functional analysis of rice DREB1/CBF-type transcription factors involved in cold-responsive gene expression in transgenic rice. Plant Cell Physiol 47:141–153

James C (2010) Global status of commercialized biotech/GM crops: 2010 – executive summary – ISAAA Brief No. 42, ISAAA, Ithaca

James C (2013) Global Status of Commercialized Biotech/GM Crops: 2013. ISAAA Brief No. 46. ISAAA, Ithaca

Jefferson-Moore KY, Traxler G (2005) Second-generation GMOs: where to from here? AgBioForum 8:143–150

Johnson AAT, Kyriacou B, Callahan DL, Carruthers L, Stangoulis J, Lombi E, Tester M (2011) Constitutive overexpression of the OsNAS gene family reveals single-gene strategies for effective iron- and zinc-biofortification of rice endosperm. PLOSone 6:e24476

Juma C (2011) Preventing hunger: biotechnology is key. Nature 479:471–472

Karihaloo JL, Kumar PA (2009) Bt cotton in India – a status report, 2nd edn. Asia-Pacific Consortium on Agricultural Biotechnology (APCoBA), New Delhi

Karim S, Aronsson H, Ericson H, Pirhonen M, Leyman B, Welin B, Mantyla E, Palva ET, van Dijck P, Holstrom KO (2007) Improved drought tolerance without undesired side effects in transgenic plants producing trehalose. Plant Mol Biol 64:371–386

Kathage J, Qaim M (2012) Economic impacts and impact dynamics of Bt (Bacillus thuringiensis) cotton in India. Proc Natl Acad Sci U S A 11652–11656. www.pnas.org/cgi/doi/10.1073/pnas.1203647109

Kathuria H, Giri J, Tyagi H, Tyagi AK (2009) Advances in transgenic rice biotechnology. Crit Rev Plant Sci 26:65–103

Kaur N, Murphy JB (2012) Enhanced isoflavone biosynthesis in transgenic cowpea (*Vigna unguiculata* l.) Callus. Plant Mol Biol Biotechnol 3:1–8

Keer RA (2007) Global warming is changing the world. Science 316:188–190

Kershen DL (2010) Trade and commerce in improved crops and food. An essay in food security. New Biotechnol 27:623–627

Khush GS (2012) Genetically modified crops: the fastest adopted crop technology in the history of modern agriculture. Agric Food Secur 1:14

Klumper W, Qaim M (2014) A meta-analysis of the impacts of genetically modified crops. PLoS One 9:e111629. doi:10.1371/journal.pone.0111629

Ko C, Woo Y, Lee DJ, Lee M, Kim CS (2007) Enhanced tolerance to heat stress in transgenic plants expressing the *GASA4* gene. Plant Physiol Biochem 45:722–728

Kranthi KR (2012) Bt cotton-questions and answers. Indian Society for Cotton Improvement, Mumbai

Laurence KM et al (1981) Double-blind randomised controlled trial of folate treatment before conception to prevent recurrence of neural-tube defects. Br Med J (Clin Res Ed) 282:1509–1511

Lucca P, Hurrell R, Potrykus I (2001) Genetic engineering approach to improve bioavailability and the level of iron in rice grains. Theor Appl Genet 102:392–397

Lutz W et al (2001) The end of world population growth. Nature 412:543–545

Lv S, Yang A, Zhang K, Wang L, Zhang J (2007) Increase of glycine betaine synthesis improves drought tolerance in cotton. Mol Breed 20:233–248

Maagd DRA, Bosch D, Stiekema W (1999) *Bacillus thuringiensis* toxin-mediated insect resistance in plants. Trends Plant Sci 4:9–13

Mayer JE, Pfeiffer WF, Beyer P (2008) Biofortified crops to alleviate micronutrient malnutrition. Curr Opin Plant Biol 11:166–170

Miranda JA, Avonce N, Suarez R, Thevelein JM, Dijck PV, Iturriaga G (2007) A bifunctional TPS–TPP enzyme from yeast confers tolerance to multiple and extreme abiotic conditions in transgenic Arabidopsis. Planta 226:1411–1421

Mirkov TE (2003) The molecular basis of genetic modification and improvement of crops. In: Chrispeels MJ, Sadava DE (eds) Plants, genes, and crop biotechnology, 2nd edn. Jones and Bartlett, Boston, pp 124–151

Mittler R (2006) Abiotic stress, the field environment and stress combination. Trends Plant Sci 11:15–19

Mittler R, Blumwald E (2010) Genetic engineering for modern agriculture: challenges and perspectives. Annu Rev Plant Biol 61:443–462

Molinari HBC, Marur CJ, Filho JCB, Kobayashi AK, Pileggi M, Junior RPL, Pereira LFP, Vieira LGE (2007) Osmotic adjustment in transgenic citrus rootstock Carrizo citrange (*Citrus sinensis* Osb. *Poncirus trifoliata* L. Raf.) overproducing proline. Plant Sci 167:1375–1381

Mroczka A, Roberts PD, Fillatti JJ, Wiggins BE, Ulmasov T, Voelker T (2010) An intron sense suppression construct targeting soybean FAD2-1 requires a double-stranded RNA-producing inverted repeat T-DNA insert. Plant Physiol 153:882–891

Murgia I, De Gara L, Grusak MA (2013) Biofortification: how can we exploit plant science and biotechnology to reduce micronutrient deficiencies? Front Physiol 4:429. doi:10.3389/fpls.00429

Naqvi S, Zhu C, Farre' G, Ramessar K, Bassie L, Breitenbach J, Perez Conesa D, Ros G, Sandmann G, Capell T, Christou P (2009) Transgenic multivitamin corn through biofortification of endosperm with three vitamins representing three distinct metabolic pathways. Proc Natl Acad Sci U S A 106:7762–7767

Naqvi S, Zhu C, Farre G, Sandmann G, Capell T, Christou P (2011) Synergistic metabolism in hybrid corn reveals bottlenecks in the carotenoid pathway and leads to the accumulation of extraordinary levels of the nutritionally important carotenoid zeaxanthin. Plant Biotechnol J 9:384–393

Nazli H, Sarela R, Meilke KD, Orden D (2010) Economic performance of Bt cotton varieties in Pakistan. Contributed paper at the Agricultural and Applied Economics Association (AAEA) Annual meetings, Denver, CO, July 25–27

Oliva N, Mohanty PC, Poletti S, Abrigo E, Atienza G, Torrizo L, Garcia R et al (2014) Large-scale production and evaluation of marker-free indica rice IR64 expressing phytoferritin genes. Mol Breed 33:23–37

Paarlberg R (2010) GMO foods and crops: Africa's choice. New Biotechnol 27:609–613

Paine J, Shipton C, Chaggar S, Howells R, Kennedy M, Vernon G, Wright S, Hinchliffe E, Adams J, Silverstone A, Drake R (2005) Improving the nutritional value of golden rice through increased provitamin A content. Nat Biotechnol 23:482–487

Park JR, McFarlane I, Phipps RH, Ceddia G (2011) The role of transgenic crops in sustainable development. Plant Biotechnol J 9:2–21

Pew Initiative on Food and Biotechnology (2007) Applications of biotechnology for functional foods-a report. Pew initiative on food biotechnology Washington, DC

Potrykus I (2003) Nutritionally enhanced rice to combat malnutrition disorders of the poor. Nutr Rev 61:101–104

Potrykus I (2010) Lessons from the 'Humanitarian Golden Rice' project: regulation prevents development of public good genetically engineered crop products. New Biotechnol 27:466–472

Prado JR, Segers G, Voelker T, Carson D, Dobert R, Phillips J, Cook K, Cornejo C, Monken J, Grapes L, Reynolds T, Martino-Catt S (2014) Biotech crop development: from idea to product. Annu Rev Plant Biol 65:21.1–21.22. doi:10.1146/annurev-arplant-050213-040039

Pray CE, Huang J, Hu R, Rozelle S (2002) Five years of Bt cotton in China-the benefits continue. Plant J 31:423–430

Qaim M (2009) The economics of genetically modified crops. Annu Rev Resour Econ 1(665):693

Qaim M (2010) Benefits of genetically modified crops for the poor: household income, nutrition, and health. New Biotechnol 27:552–557. doi:10.1016/j.nbt.2010.07.009

Qaim M, Kouser S (2013) Genetically modified crops and food security. PLOSOne 8:e64879. doi:10.1371/journal.pone.0064879

Qaim M, Zilberman D (2003) Yield effects of genetically modified crops in developing countries. Science 299:900–902

Qaim M, Subramanian A, Naik G, Zilberman D (2006) Adoption of Bt cotton and impact variability: insights from India. Rev Agric Econ 28:48–58

Qin Q, Liu J, Zhang Z, Peng R, Xiong A, Yao Q, Chen J (2007) Isolation, optimization, and functional analysis of the cDNA encoding transcription factor *OsDREB1B* in *Oryza Sativa* L. Mol Breed 19:329–340

Raven PH (2010) Does the use of transgenic plants diminish or promote biodiversity? New Biotechnol 27:528–533

Ray DK, Mueller ND, West PC, Foley JA (2013) Yield trends are insufficient to double global crop production by 2050. PLoS One 8(6):e66428. doi:10.1371/journal.pone.0066428

Regina A, Bird A, Topping D, Bowden S, Freeman J, Barsby T, Kosar-Hashemi B, Li Z, Rahman S, Morell M (2006) High amylose wheat generated by RNA interference improves indices of large-bowel health in rats. Proc Natl Acad Sci U S A 103:3546–3551

Rehman UMS (2007) Pakistan agricultural situation cotton update. USDA GAIN report no. PK 7026, U.S. Embassy, Washington, DC

Rommens CM (2010) Barriers and paths to market for genetically engineered crops. Plant Biotechnol J 8:101–111

Rosenquist TH, Anne Ratashak S, Selhub J et al (1991) Homocysteine induces congenital defects of the heart and neural tube: effect of folic acid. Proc Natl Acad Sci U S A 93:15227–15232

Roy D, Herring RJ, Geisler C (2007) Naturalizing transgenics: loose seeds, official seeds, and risk in the decision matrix of Gujarati cotton farmers. J Dev Stud 43:158–176

Rush D (2000) Nutrition and maternal mortality in the developing world. Am J Clin Nutr 72:212–240

Rymer C, Givens DI (2009) The effect of feeding stearidonic acid enriched soya oil to broilers on the fatty acid composition and sensory characteristics of chicken meat. Br Poult 5:44

Sadashivappa P, Qaim M (2009) Effects of Bt cotton in India during the first five years of adoption. Presented at international association of agricultural economics triennial conference, Beijing

Sahoo L, Jaiwal PK (2008) Asiatic beans. In: Kole C, Hall T (eds) A compendium of transgenic crop plants. Blackwell Publication, Oxford, 115–132

Sakuma Y, Maruyama K, Qin F, Osakabe Y, Shinozaki K, Yamaguchi-Shinozaki K (2006) Dual function of an Arabidopsis transcription factor *DREB2A* in water-stress-responsive and heat-stress-responsive gene expression. Proc Natl Acad Sci U S A 103:18822–18827

Savitch LV, Allard G, Seki M, Robert LS, Tinker NA, Huner NPA, Shinozaki K, Singh J (2005) The effect of over-expression of two *Brassica* CBF/DREB1-like transcription factors on photosynthetic capacity and frost tolerance in *Brassica napus*. Plant Cell Physiol 46:1525–1539

Singh RP, Varaprasad PV, Reddy KR (2015) Climate change: implications for stakeholders in genetic resources and seed sector. Adv Agron 129:117–180. doi:10.1016/bs.agron.2014.09.002

Sivaprakash KR et al (2006) Tissue-specific histochemical localization of iron and ferritin gene expression in transgenic indica rice Pusa Basmati (*Oryza sativa* L.). J Genet 85:157–160

Storozhenko S, Brouwer VD, Volckaert M, Navarrete O, Blancquaert D, Zhang GF, Lambert W, Van Der Straeten D (2007) Folate fortification of rice by metabolic engineering. Nat Biotechnol 25:1277–1279

Swaminathan MS (2002) Building a national nutrition security system. India- ASEAN eminent persons lecture series. FAO, Bangkok

Tang W, Hu Y, S-a Y, Wang Y, Dallal GE, Grusak MA, Russell RM (2012) β-carotene in golden rice is as good as β-carotene in oil at providing vitamin to children. Am J Clin Nutr 96:658–664

Tzfira T, Citovsky V (2006) *Agrobacterium*-mediated genetic transformation of plants: biology and biotechnology. Curr Opin Biotechnol 17:147–154

UN (2011) The millennium development goals report . United Nations, New York

Ursin VM (2003) Modification of plant lipids for human health: development of functional land-based omega-3 fatty acids. J Nutr 133:4271–4274

USDA-ERS (2013) Economic research service – Data products. United States Department of Agriculture ERS, Washington DC. http://www.ers.usda.gov/data-products.aspx#.UoqklKJ7yA

Vendruscolo ECG, Schuster I, Pileggi M, Scapim CA, Molinari HBC, Marur CJ, Vieira LGE (2007) Stress-induced synthesis of proline confers tolerance to water deficit in transgenic wheat. J Plant Physiol 164:1367–1376

Voytas DF, Gao C (2014) Precision genome engineering and agriculture: opportunities and regulatory challenges. PLoS Biol 12:e1001877. doi:10.1371/journal.pbio.1001877

Waditee R, Bhuiyan MNH, Rai V, Aoki K, Tanaka Y, Hibino T, Suzuki S, Takano J, Jagenodorf AT, Takabe T, Takabe T (2005) Genes for direct methylation of glycine provide high levels of glycinebetaine and abiotic-stress tolerance in *Synechococcus* and Arabidopsis. Proc Natl Acad Sci-U S A 102:1318–1323

Wagner N, Mroczka A, Roberts PD, Schreckengost W, Voelker T (2011) RNAi trigger fragment truncation attenuates soybean FAD2-1 transcript suppression and yields intermediate oil phenotypes. Plant Biotechnol J 9:723–728

Wang X, Cai J, Liu F, Dai T, Cao W, Wollenweber B, Jiang D (2014) Multiple heat priming enhances thermo-tolerance to a later high temperature stress via improving subcellular antioxidant activities in wheat seedlings. Plant Physiol Biochem 74:185–192

Weale A (2010) Ethical arguments relevent to the GM crops. New Biotechnol 27:583–587

Webb NP, Stokes CJ, Scanlan JC (2012) Interacting effects of vegetation, soils and management on the sensitivity of savanna rangelands to climate change. Clim Change 112:925–943

WHO (2010) Micronutrient deficiencies: iron deficiency anaemia. WHO, Geneva. http://www.who.int/nutrition/topics/ida/en/

Wijk JV (2002) Food insecurity: prevalence, causes, and the potential of transgenic 'Golden Rice'. Phytochem Rev 1:141–151

Wirth J, Poletti S, Aeschlimann B, Yakandawala N, Drosse B, Osorio S, Tohge T, Fernie AR, Gunther D, Gunther W, Sautter C (2009) Rice endosperm iron biofortification by targeted and synergistic action of nicotianamine synthase and ferritin. Plant Biotechnol J 7:631–644

World Food Prize (2013) Three biotechnology scientists awarded 2013 world food prize, press release. World Food Prize Foundation, Iowa

Yamada M, Morishita H, Urano K, Shiozaki N, Yamaguchi-Shinozaki K, Shinozaki K, Yoshiba Y (2005) Effects of free proline accumulation in petunias under drought stress. J Exp Bot 56:1975–1981

Yamamoto K, Sakamoto H, Momonoki YS (2011) Maize acetylcholinesterase is a positive regulator of heat tolerance in plants. J Plant Physiol 168:1987–1992

Yamori W, Hikosaka K, Way DA (2013) Temperature response of photosynthesis in C_3, C_4 and CAM plants. Photosynth Res 119:101–117

Yuan D, Bassie L, Sabalza M, Miralpeix B, Dashevskaya S, Farre' G, Rivera SM, Banakar R, Bai C, Sanahuja G, Arjo' G, Avilla E, Zorrilla-Lo'pez U, Ugidos-Damboriena N, Lo'pez A, Almacellas D, Zhu C, Capell T, Hahne G, Twyman RM, Christou P (2011) The potential impact of plant biotechnology on the millennium development goals. Plant Cell Rep 30:249–265

Zhao J, Ren W, Zhi D, Wang L, Xia G (2007) Arabidopsis DREB1A/CBF3 bestowed transgenic tall fescue increased tolerance to drought stress. Plant Cell Rep 26:1521–1528

Zhu C, Naqvi S, Gomez-Galera S, Pelacho AM, Capell T, Christou P (2007) Transgenic strategies for the nutritional enhancement of plants. Trends Plant Sci 12:548–555

Zhu L, Meng GM, Cheung SCK, Yu H, Huang J, Sun Y, Shi Y, Liu Q (2012) High-amylose rice improves indices of animal health in normal and diabetic rats. Plant Biotechnol J 10(353):362

Zhu XF, Jin YH, Yoo CY, Lin XL, Kim WY, Yun DJ et al (2013) *CYCLINH;1* regulates drought stress responses and blue light-induced stomatal opening by inhibiting Reactive Oxygen Species accumulation in Arabidopsis. Plant Physiol 162:1030–1041